Studies in Systems, Decision and Control

Volume 636

The series "Studies in Systems, Decision and Control" (SSDC) covers both new developments and advances, as well as the state of the art, in the various areas of broadly perceived systems, decision making and control–quickly, up to date and with a high quality. The intent is to cover the theory, applications, and perspectives on the state of the art and future developments relevant to systems, decision making, control, complex processes and related areas, as embedded in the fields of engineering, computer science, physics, economics, social and life sciences, as well as the paradigms and methodologies behind them. The series contains monographs, textbooks, lecture notes and edited volumes in systems, decision making and control spanning the areas of Cyber-Physical Systems, Autonomous Systems, Sensor Networks, Control Systems, Energy Systems, Automotive Systems, Biological Systems, Vehicular Networking and Connected Vehicles, Aerospace Systems, Automation, Manufacturing, Smart Grids, Nonlinear Systems, Power Systems, Robotics, Social Systems, Economic Systems and other. Of particular value to both the contributors and the readership are the short publication timeframe and the world-wide distribution and exposure which enable both a wide and rapid dissemination of research output.

Indexed by SCOPUS, DBLP, WTI Frankfurt eG, zbMATH, SCImago.

All books published in the series are submitted for consideration in Web of Science.

Yekai Yang · Yugang Niu · Jiarui Li

Analysis and Synthesis of Interval Type-2 Fuzzy Systems Under Communication Protocols

Springer

Yekai Yang
Donghua University
Shanghai, China

Yugang Niu
East China University of Science and Technology
Shanghai, China

Jiarui Li
Shanghai Maritime University
Shanghai, China

ISSN 2198-4182 ISSN 2198-4190 (electronic)
Studies in Systems, Decision and Control
ISBN 978-981-95-2354-2 ISBN 978-981-95-2355-9 (eBook)
https://doi.org/10.1007/978-981-95-2355-9

This Springer imprint is published by the registered company Springer Nature Singapore Pte Ltd.
The registered company address is: 152 Beach Road, #21-01/04 Gateway East, Singapore 189721, Singapore

To our families.

Preface

Physical systems in real-world applications are inherently affected by nonlinear behaviors and uncertain parameters, making the analysis and control of nonlinear systems highly challenging. These nonlinear dynamics and uncertainties often complicate system modeling and controller design, so that nonlinear control theory remains a critical and challenging area in modern control research. Takagi-Sugeno (T-S) fuzzy model has proven to be an effective method for analyzing nonlinear systems. It represents complex nonlinear dynamics using a combination of simpler linear subsystems with membership functions. While traditional type-1 T-S fuzzy systems have been widely applied, they cannot directly handle parameter uncertainties commonly found in practical systems. To address this limitation, interval type-2 (IT2) T-S fuzzy models have been developed. By using upper and lower membership functions to address parameter uncertainties, the IT2 fuzzy system has become an indispensable tool in advanced control theory and practice.

The widespread application of networked control systems (NCSs), driven by advancements in communication technologies, has revolutionized modern industrial control area by enabling cost-effective remote monitoring and control through shared wireless networks. By interconnecting sensors, controllers, and actuators via digital channels, NCSs reduce hardware complexity, lower installation costs, and enhance operational flexibility. However, the bandwidth-constrained networks introduces critical challenges, including transmission delays, data loss, and vulnerability to cyber threats, which degrade system reliability and complicate control design. These limitations are particularly pronounced in IT2 T-S fuzzy systems with inherent uncertainties, where traditional control methods often fail to ensure stability under restricted communication resources. Aiming at the limited communication resources, communication protocol is an important approach to regulate the transmission process of measurement signal or control input.

This book offers comprehensive and systematic exploration of control design and analysis for IT2 T-S fuzzy systems under constrained communication networks. Various communication protocols are developed, including a membership-function-based dynamic event-triggered scheme, self-triggered scheme, multi-node stochastic scheduling protocols, and event-triggered weighted-try-once-discard (WTOD) protocols. Special effort is placed on establishing a unified theoretical framework for the design, analysis, and optimization of IT2 fuzzy systems under communication protocols. The book provides diverse control schemes such as state-feedback control, sliding mode control (SMC), static output-feedback control, and fuzzy observer design, incorporating the characteristics and information of membership functions in both analysis and synthesis. Additionally, some membership-function-dependent analysis methods are developed under communication protocols, leveraging global and local boundary information of membership functions. This monograph is structured into two parts. Part I focuses on control design and analysis for IT2 fuzzy systems under event-triggered schemes, addressing the following topics: Chap. 2 investigates dynamic event-triggered SMC under unmeasured states and fading channels; Chap. 3 considers component-based dynamic event-triggered SMC problem; Chap. 4 proposes a quantized SMC scheme and a membership-function-based event-triggered scheme; Chap. 5 addresses security event-triggered SMC subject to randomly occurring denial-of-service attacks; Chap. 6 proposes a self-triggered fuzzy control design strategy. Part II focuses on control design and analysis for IT2 fuzzy systems under channel-scheduling protocols, including: Chap. 7 investigates the fuzzy SMC design under multi-node stochastic scheduling protocols; Chap. 8 proposed the co-design scheme of scheduling protocols and the fuzzy sliding mode controller; Chap. 9 studies the co-design problem of multi-node scheduling protocols and the output-feedback fuzzy controller; Chap. 10 develops a fuzzy SMC scheme under event-triggered WTOD protocols.

This publication is a research monograph whose intended audience include researchers, senior undergraduate, graduate students, and postgraduate who wish to explore the area of fuzzy systems and networked control systems.

Shanghai, China
June 2025

Yekai Yang
Yugang Niu
Jiarui Li

Acknowledgements There are numerous individuals without whose constructive comments, useful suggestions, and wealth of ideas, this monograph could not have been completed. Special thanks go to Prof. James Lam from The University of Hong Kong and Prof. Hak-Keung Lam from King's College London, for their valuable suggestions, constructive guidance, and support. Additionally, we extend our acknowledgments to East China University of Science and Technology, Donghua University, and Shanghai Maritime University for providing an environment conducive to the research presented in this monograph. Our appreciation also goes to the members of our research group and colleagues who have provided invaluable support and encouragement throughout this research effort. Finally, we would like to thank the editors at Springer for their professional and efficient handling of this project.

The monograph was supported in part by the National Natural Science Foundation of China (62333006, 62403304, 62503108), in part by Natural Science Foundation of Shanghai (24ZR1403000, 24ZR1454700), in part by Shanghai Pujiang Programme (24PJD041), and in part by the Fundamental Research Funds for the Central Universities of China (2232024D-36).

Competing Interests The authors have no conflicts of interest to declare that are relevant to the content of this book.

Contents

Notations and Abbreviations

Notations

$\mathbb{R}$	Field of real numbers
$\mathbb{R}^n$	n-dimensional real number vector space
$\mathbb{R}^{n\times m}$	$n \times m$ real matrix space
$\mathbb{N}(\mathbb{N}^+)$	Field of nonnegative (positive) integral numbers
$\mathbf{E}\{\cdot\}$	Mathematical expectation operator
$\mathbf{Pr}\{x \mid y\}$	Conditional probability function of x on y
$\max(\cdot)$	Maximal function, i.e., the maximum element of a set
$\min(\cdot)$	Minimal function, i.e., the minimum element of a set
$\inf(\cdot)$	Infimum function, i.e., the infimum of a set
$\lambda_{\min}(A)$	Minimum eigenvalue of a matrix A
$\lambda_{\max}(A)$	Maximum eigenvalue of a matrix A
$\text{sign}(s)$	The sign function: $\text{sign}(s) = 1$ for $s > 0$, $\text{sign}(0) = 0$, and $\text{sign}(s) = -1$ for $s < 0$
$\text{He}(A)$	The matrix operation $A + A^T$
P^T	Transpose of matrix P
P^{-1}	Inverse of matrix P
$P > (<)0$	P is real symmetric positive (negative) definite
$P \geq (\leq)0$	P is real symmetric positive (negative) semi-definite
$|x|$	Absolute value (or modulus) of a scalar x
$\|x\|$	Euclidean norm of a vector $x = (x_1, \ldots, x_n)$, i.e., $\|x\| = \sqrt{\sum_{i=1}^{n} x_i^2}$
$\|A\|$	Euclidean norm (or spectral norm) of a matrix A, i.e., $\|A\| = \sqrt{\lambda_{\max}(A^T A)}$
$*$	Symmetric terms in a symmetric matrix
$\text{diag}\{\cdots\}$	Block diagonal matrix
I	An identity matrix of compatible dimension
0	A zero matrix of compatible dimension
$\mathbf{1}_n$	An n dimensional column vector with all ones
$\otimes$	The Kronecker product of matrices

$\delta(a)$	The Kronecker delta function that equals 1 if $a = 0$ and equals 0 otherwise
$u(t) \in \mathcal{L}_\infty^l$	The essentially bounded function $u(t) : \mathbb{R}^+ \to \mathbb{R}^l$ with $\|u\|_\infty \triangleq \text{ess}\sup_{t\in[0,\infty)}\{\|\|u(t)\|\|\} < +\infty$
$\mathcal{K}$	The continuous strictly increasing function $\mu : \mathbb{R}^+ \to \mathbb{R}^+$ with $\mu(0) = 0$
$\mathcal{K}_\infty$	The function $\mu \in \mathcal{K}$ with $\mu(r) \to \infty$ as $r \to \infty$
$\mathcal{KL}$	The function $\beta : \mathbb{R}^+ \times \mathbb{R}^+ \to \mathbb{R}^+$ if the mapping $\beta(\cdot, t) \in \mathcal{K}$ for each fixed t and the function $\beta(s, t)$ is decreasing to zero on t as $t \to \infty$ for each fixed s.

Abbreviations

ETM	Event-Triggered Mechanism
ETS	Event-Triggered Scheme
FLS	Fuzzy Logic System
IT2	Interval Type-2
LMI	Linear Matrix Inequality
MFD	Membership-Function-Dependent
MFI	Membership-Function-Independent
MFs	Membership Functions
PDC	Parallel Distributed Compensation
SMC	Sliding Mode Control
T-S	Takagi–Sugeno

List of Figures

List of Tables

Chapter 1
Introduction

1.1 Background

Nonlinearities widely exist in practical physical systems, so nonlinear control and analysis become a research hotspot in the control field. The Takagi-Sugeno (T-S) fuzzy model provides an important tool for modeling nonlinear dynamics. By mean of the type-1 fuzzy set, nonlinear structures can be represented via the local linear models and their membership functions (MFs) [1]. Because of its good nonlinear modeling ability, the type-1 T-S fuzzy model has been widely utilized in practical applications, such as, DC–DC converters [2], wheeled vehicles [3], unmanned marine vehicles [4]. However, it should be pointed out that practical systems usually involve some uncertain parameters so that the research of uncertain nonlinear systems attract more attentions. Different from the type-1 T-S fuzzy model, the interval type-2 (IT2) T-S fuzzy model not only has the ability to deal with nonlinearity, but also shows the advantage of dealing with uncertainties by means of the upper and lower MFs. Recently, the IT2 T-S fuzzy system has become a research hotspot, and it has been widely used to deal with modeling and control problems of practical systems, including vehicle active suspensions [5], autonomous ground vehicles [6], inverted pendulums [7], magnetic bearing systems [8], and hypersonic vehicles [9].

With the rapid development of network communication technology and computer science, the networked control systems (NCSs) have shown important application prospects due to its low cost and easy to remote transmission of wireless network communication. Compared with the traditional point-to-point control systems, the signals of system components (such as, sensor, controller, and actuator) in NCSs are transmitted via the network communication channels, which will be inevitably limited by communication resource constraints in practice. To relieve the communication resource constraints and improve the resource utilization, a proper signal transmission scheme is quite important for determining the transmission of signal and govern the resource allocation, which has significant meaning for ensuring the desired control systems of NCSs. However, it should be pointed out that when signals

Y. Yang et al., *Analysis and Synthesis of Interval Type-2 Fuzzy Systems Under Communication Protocols*, Studies in Systems, Decision and Control 636,
https://doi.org/10.1007/978-981-95-2355-9_1

are transmitted via network communication channels, which poses more difficulties on control design and performance analysis of T-S fuzzy systems, especially, for the IT2 T-S fuzzy systems. Besides, there are some undesire phenomena to affect the signal transmission and degrade the control performance, such as channel fading, cyber-attacks, and quantization error. Thus, this book focus on control design and analysis for the IT2 T-S fuzzy systems under communication protocols, and a series of control schemes are proposed to reduce the communication resources and improve the control performance.

1.2 Fuzzy-Model-Based T-S System and Control Design

Fuzzy-model-based system includes the type-1 fuzzy model and IT2 T-S fuzzy model, whose main feature lies in representing nonlinear systems via local linear subsystems and MFs. However, the main difference is the ability of handling uncertainties, that is, the IT2 fuzzy systems show better ability to dealing with uncertainties. In this section, we shall firstly introduce the type-1 T-S fuzzy model and its control design methods. Then, the IT2 T-S fuzzy model and its control design methods are further introduced.

1.2.1 Type-1 T-S Fuzzy System

The development of fuzzy system originates from the fundamental fuzzy set, and the fuzzy set theory is established by Zadeh in 1965 [10]. By means of fuzzy rules, linguistic variables, and reasoning mechanisms, a complex nonlinear dynamics can be effectively represented and addressed [11]. Based on fuzzy theory and fuzzy reasoning, Prof. Ebrahim Mamdani developed fuzzy logic system (FLS) and proposed the fuzzy controller to deal with the control issues of nonlinear systems [12–14]. As shown in Fig. 1.1, an FLS consists of four components, i.e., fuzzifier, rule base, inference engine, and defuzzifier. Due to its simple structure and strong ability to address nonlinearities, fuzzy control has attracted considerable attention from both academic and industrial communities. In particular, the Mamdani-type fuzzy control method has been successfully applied in engineering applications such as sludge wastewater treatment processes [15] and cement kiln control systems [16]. However, it should be pointed out that the Mamdani-type fuzzy controller exhibits two notable limitations in practical applications. Firstly, its control methodology heavily relies on heuristic design principles rather than formal mathematical frameworks, which poses challenges for rigorous theoretical stability analysis. This inherent characteristic makes it difficult to guarantee the stability of the closed-loop system through conventional control theory approaches. Secondly, the system performance demonstrates significant inconsistency due to its knowledge-based architecture, which is mainly because

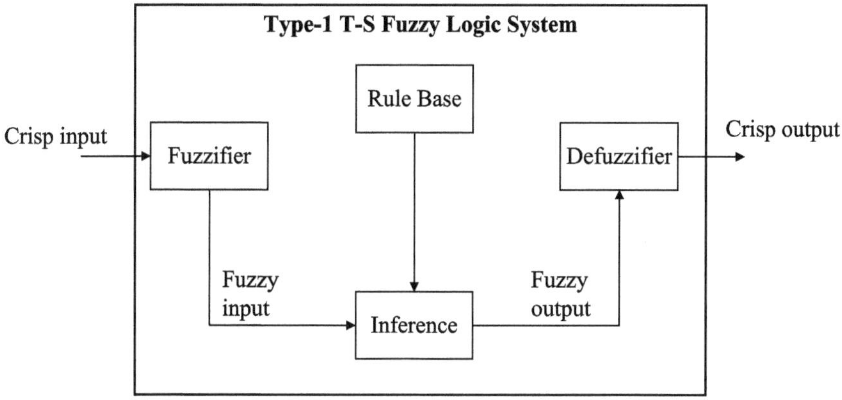

Fig. 1.1 Structure of type-1 T-S FLS

that different control strategies and rule formulations proposed by various domain experts for identical processes inevitably lead to divergent rule bases.

Compared to the Mamdani-type fuzzy control approach, the fuzzy-model-based (FMB) control methods via T-S fuzzy model have shown significant disadvantages [17], which provides an important theoretical analysis frame for the modeling, control, and synthesis of nonlinear systems. The T-S fuzzy model was formally established in 1985 [1, 18], which is composed of a family of fuzzy IF-THEN rules representing local linear input/output relations of local subsystem. The overall fuzzy model is obtained by smoothly blending these local linear models together through membership functions. The type-1 T-S fuzzy model provides a frame of fuzzy-model-based control design and analysis method, which has significant meanings for modeling and control of nonlinear systems.

In the following subsections, we will introduce the type-1 T-S fuzzy model, and discuss its control design methods.

1.2.1.1 Type-1 T-S Fuzzy Model

Generally, there are two principal methodologies for constructing type-1 T-S fuzzy models. The first methodology is to employ the system identification techniques based on input-output data, which is advantageous for the case that the mathematical formulation of a nonlinear system is unknown but input-output data are available [1, 18]. The second methodology is to utilize model transformation techniques, which is applicable when an explicit mathematical description of the nonlinear system is available. This approach systematically converts the original nonlinear dynamics into a fuzzy-model-based structure by means of the sector nonlinearity or local approximation methods, which can generally ensure mathematical consistency between the fuzzy representation and the original system [19, 20].

In the following, a continuous-time type-1 T-S fuzzy model is shown:
Rule i: IF $f_1(\theta(t))$ is $\mathscr{N}_1^i$,$\cdots$, $f_\hbar(\theta(t))$ is $\mathscr{N}_\hbar^i$, THEN

$$\dot{x}(t) = A_i x(t) + B_i u(t), \tag{1.1}$$

where $x(t) \in \mathbb{R}^n$ and $u(t) \in \mathbb{R}^m$ are the system state and control input, respectively. The matrices A_i and B_i $(i = 1, \ldots, r)$ are known with appropriate dimensions, where r is the number of fuzzy rules. $f_\iota(\theta(t))$ $(\iota = 1, \ldots, \hbar)$ is the premise variable, which is assumed to be independent of the control input $u(t)$, and $\mathscr{N}_\iota^i$ is a type-1 fuzzy set, where $\hbar$ is the number of premise variables.

Through single point fuzzification, product reasoning, and weighted defuzzification, the type-1 T-S fuzzy model is inferred as

$$\dot{x}(t) = \sum_{i=1}^{r} m_i(\theta(t))[A_i x(t) + B_i u(t)], \tag{1.2}$$

where $m_i(\theta(t))$ is the membership function with

$$m_i(\theta(t)) = \frac{\prod_{\iota=1}^{\hbar} \mu_{\mathscr{N}_\iota^i}(f_\iota(\theta(t)))}{\sum_{i=1}^{r} \prod_{\iota=1}^{\hbar} \mu_{\mathscr{N}_\iota^i}(f_\iota(\theta(t)))},$$

and

$$\sum_{i=1}^{r} m_i(\theta(t)) = 1, \quad m_i(\theta(t)) \in [0, 1],$$

where $\mu_{\mathscr{N}_\iota^i}(f(\theta(t))) \geq 0$ is the grade of membership with respect to fuzzy set $\mathscr{N}_\iota^i$.

Similarly, the discrete-time type-1 T-S fuzzy model can be represented as

$$x(k+1) = \sum_{i=1}^{r} m_i(\theta(k))[A_i x(k) + B_i u(k)], \tag{1.3}$$

where $m_i(\theta(k))$ is membership function with

$$m_i(\theta(k)) = \frac{\prod_{\iota=1}^{\hbar} \mu_{\mathscr{N}_\iota^i}(f_\iota(\theta(k)))}{\sum_{i=1}^{r} \prod_{\iota=1}^{\hbar} \mu_{\mathscr{N}_\iota^i}(f_\iota(\theta(k)))} \in [0, 1], \quad \sum_{i=1}^{r} m_i(\theta(k)) = 1$$

and the membership grade $\mu_{\mathscr{N}_\iota^i}(f(\theta(k))) \geq 0$ with respect to fuzzy set $\mathscr{N}_\iota^i$.

The continuous and discrete-time type-1 T-S fuzzy models (1.2) and (1.3) are obtained via the integration of local linear models with nonlinear MFs $m_i(\theta(t))$ and $m_i(\theta(k))$, which forms a semi-linear structure. This structural characteristic enables some existing design and analysis methods for linear systems to be applied to the FMB analysis framework through rigorous mathematical processes.

1.2.1.2 Control Design and Stability Analysis for Type-1 T-S Fuzzy System

For the FMB systems, the design of controller is essential for ensuring system performance. Control design problems for type-1 T-S fuzzy systems involve the local linear controllers and the MFs. The design methodology depends on the relationship between the system's MFs and those of the controller, which can be divided into two principal categories, i.e., parallel distributed compensation (PDC) and non-parallel distributed compensation (non-PDC) methods.

The classical PDC approach is to construct a series of local linear controllers under a similar frame of fuzzy model, which is an effective control design method [21–23]. Under the PDC design methodology, a local linear controller is designed to ensure the stability of the corresponding local linear subsystem, and the fuzzy controller is obtained by weighting all the local linear controllers with the same MFs as the T–S fuzzy system. However, its implementation becomes infeasible when system's premise variables or MFs are unavailable for the fuzzy controller. To overcome these limitations, the non-PDC method is proposed by independently designing MFs different from the fuzzy systems. According to the relation of the number of fuzzy rules between the controller and fuzzy system, the non-PDC method can be further divided into *partially matched* MFs and *completely mismatched* MFs. To be more specific, the design approach of partially matched MFs is to design new controller's MFs and keep the same number of fuzzy rules as the one of fuzzy systems. The method of completely mismatched MFs can allow that the controller has different MFs and fuzzy rules, which can improve the design flexibility. In [24, 25], the controller and the system share an identical number of fuzzy rules, and by reconstructing the controller's MFs under each rule, a partially matched design methodology is proposed. However, when the fuzzy system contains a large number of rules, the practical implementation of such partially matched controllers becomes highly complex. To overcome this shortcoming, Lam et al. [26, 27] further relaxes the requirement on the matched number of fuzzy rules, enabling the design of controllers with a different number of fuzzy rules from the fuzzy system, i.e., completely mismatched fuzzy controllers. Compared to the PDC method and the partially matched non-PDC approach, the above completely mismatched control design eliminates the requirement for identical rule counts between the controller and the system, which significantly enhances the control design flexibility.

Under the aforementioned control design methods, it is necessary to analyze the control performance of closed-loop T-S fuzzy systems. Due to the semi-linear structural characteristics of the type-1 T-S fuzzy system, the stability analysis typically employs Lyapunov stability theory and linear matrix inequality (LMI) technique. It is worth pointing out that the information of MFs may also be involved in the stability analysis process, including the construction of Lyapunov functions and the analysis of LMI conditions. For the choice of Lyapunov functions, there are generally three typical approaches. The first approach is to construct the *common quadratic Lyapunov function* $V(x(t)) = x^T(t)Px(t)$ with a positive definite matrix P, whose feature is that all information of MFs is ignored and a common Lyapunov matrix P is needed to satisfy all subsystems under each fuzzy rule. The second Lyapunov function is *fuzzy Lyapunov function* $V(x(t)) = \sum_{i=1}^{r} m_i(x(t))x^T(t)P_i x(t)$, which contains system's MFs and can reduce the conservatism in the stability analysis [28, 29]. The third method is to construct the *piecewise Lyapunov function*. This method strategically partitions the operating domain into multiple subregions based on the geometric properties and dynamic characteristics of MFs, that is, a Lyapunov function is constructed within each sub-region, and the piecewise Lyapunov function is analyzed to guarantee the stability of the global system. In [30, 31], the fuzzy Lyapunov function was proposed by embedding the system's MFs, which can obtain the stability conditions for reducing conservatism. Besides, according to the shape and characteristics of MFs, the fuzzy modeling domain in [32–35] was divided into multiple sub-regions, and the piecewise Lyapunov functions on the sub-regions were designed to ensure the stability of the closed-loop systems. Compared with the previous two methods, the piecewise Lyapunov functions can use more information on MFs, which is conducive to reduce the conservatism of stability conditions. However, the effectiveness of the piecewise Lyapunov function approach fundamentally depends on both the configuration of MFs and the partition of the fuzzy operating domain. Notably, the computational complexity and challenges of Lyapunov function construction and stability analysis grows substantially for the intricate membership function structures or the extensive sub-region divisions

The stability analysis of the T-S fuzzy systems can be categorized into two distinct approaches based on the utilization of membership function information during the analysis process, i.e., membership-function-independent (MFI) method and membership-function-dependent (MFD) method. The MFI stability analysis method does not contain any information of MFs, which implies that stability conditions need to ensure the desired performance for any MFs. The characteristic of this method is simple analysis, but the obtained results are conservative. To reduce the conservatism of stability conditions, by means of the membership function property under the PDC control design frame, the Pòlya's lemma was proposed to introduce the relation of fuzzy rule between the fuzzy model and the controller [36–39].

Compared with the MFI analysis method, the MFD analysis approach can reduce the conservatism by introducing the information of MFs into stability conditions [40]. According to the utilized information of MFs, the MFD analysis approach can be mainly divided into three handling techniques, i.e., global-boundary-information-dependent, local-boundary-information-dependent, and membership-function-approximation approaches. In [41], the upper bounds of MFs in fuzzy systems and controller were utilized to integrate slack matrices so that more relax conditions were obtained. On the basic of the result in [41], the polynomial function was employed to describe the nonlinear MFs and obtain more relax stability conditions. Besides, in [24], a linear relation was established between MFs in fuzzy model and controller, the partially matched MFs were converted to the product of the matched MFs and the time-varying functions, and the upper and lower bounds of MFs were utilized to obtained stability conditions. In [42], the affine functions were proposed to describe the relation between MFs in the fuzzy model and the controller, and it further extended the result in [24]. By constructing the fuzzy Lyapunov function, the upper bounds of MFs and their derivatives were utilized in [29] to analyze the stability of the closed-loop system. The aforementioned analysis methods primarily rely on the global information or the general shape of MFs in fuzzy systems and controller. In contrast, the approach based on local boundary information can leverage more precise membership function characteristics to enhance analytical accuracy [43–45]. As demonstrated in [43], the modeling domain was partitioned into multiple subregions and local upper and lower boundaries of MFs was obtained within each subregion. By incorporating these subregions and local boundary information into stability analysis, the relax stability conditions exhibit less conservatism compared to the results via the global approaches.

Unlike the global and local boundary information approaches, the membership-function-approximation approach is to utilize the constant, linear functions, or nonlinear functions to approximate the MFs of fuzzy systems and controller. Based on the membership-function-approximation methods, the trapezoidal functions were utilized in [46–48] to approximate MFs of both the system and the controller. By incorporating the approximated MFs and their approximation errors into stability analysis, this method can achieve more precise characterization of membership function properties to some extent. As the number of approximation steps increases, the method enhances the approximation accuracy of the MFs, while simultaneously increasing the computational complexity of the derived stability conditions. Besides, piecewise polynomial functions were proposed in [49–51] to approximate membership functions and integrates the properties of polynomial functions into stability analysis. Compared to the MFI analysis methods, the MFD approaches utilize richer information of MFs, which can obtain stability conditions with less conservative.

1.2.2 Interval Type-2 T-S Fuzzy System

In practical physical systems, uncertainties are inherent and inevitably complicate the analysis of nonlinear systems. It should be noted that although type-1 T-S fuzzy system, which are based on the type-1 fuzzy sets, demonstrates strong capabilities in handling nonlinearities, it lacks effective ability to dealing with uncertainties.

Type-2 fuzzy sets, proposed by Prof. Zadeh in 1975, fundamentally extend type-1 fuzzy sets through a three-dimensional structure comprising primary and secondary MFs [52]. The main difference between the type-2 fuzzy set and the type-1 fuzzy set is that the values of the primary MFs in the type-2 fuzzy set are type-1 fuzzy sets rather than crisp numbers, which enables it to deal with uncertainty [53, 54]. The special structure also makes the analysis more complicated in the research of type-2 fuzzy set and type-2 FLSs. Therefore, more researches have focused on the IT2 fuzzy sets and IT2 FLSs, where the secondary MFs are simplified to intervals instead of type-1 fuzzy sets. This simplification not only improves the computational efficiency, but also ensures the ability to deal with uncertainty. As shown in Fig. 1.2, the IT2 FLS is similar to the type-1 FLSs, and the major difference is that its rule base incorporates at least one IT2 fuzzy set. This structural feature results in the inference engine generating type-2 fuzzy outputs, and a type-reducer is necessary to convert them into a type-1 fuzzy set before defuzzification process [55, 56]. Besides, similar to the FMB type-1 T-S fuzzy systems, the IT2 T-S fuzzy system, as a computationally efficient variant of general type-2 fuzzy systems, constrains its secondary MFs to unity while maintaining interval-valued primary MFs [57–59]. This special structure achieves a balance between the capability of dealing with uncertainties and the computational complexity. Due to the broad applicability, the IT2 T-S fuzzy system has garnered considerable research interest, thereby making the control design and stability analysis of IT2 T-S fuzzy models as a key research focus within the control community [60–62].

1.2.2.1 Interval Type-2 T-S Fuzzy Model

An IT2 T-S fuzzy system employs IT2 fuzzy sets in their rule antecedents while maintaining local linear subsystems as consequents, which can effectively deal with the system uncertainties. For a continuous-time IT2 T-S fuzzy system with p rules, the local dynamics of each fuzzy rule can be described as

Rule i**:** IF $f_1(\theta(t))$ is $\mathscr{F}_1^i, \cdots,$ and $f_\hbar(\theta(t))$ is $\mathscr{F}_\hbar^i$, THEN

$$\dot{x}(t) = A_i x(t) + B_i u(t), \tag{1.4}$$

where $f_\iota(\theta(t))$ $(\iota = 1, \ldots, \hbar)$ is the premise variable, $\mathscr{F}_\iota^i$ $(i = 1, \ldots, r)$ is the IT2 fuzzy set, and other parameters are the same as the system (1.1). For the i-th fuzzy rule, the emission interval is represented as

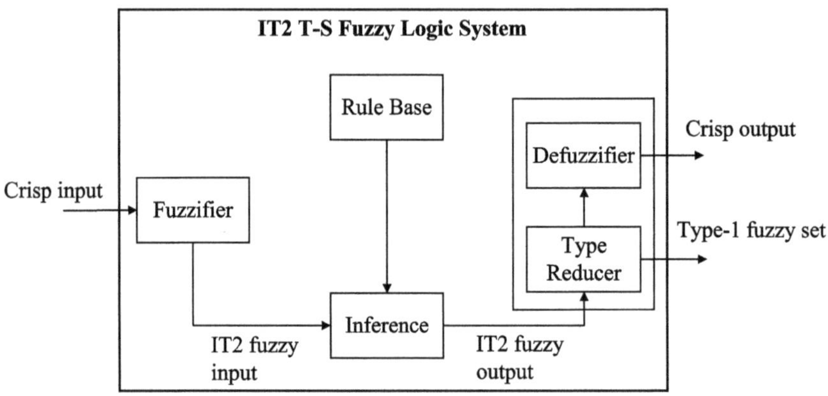

Fig. 1.2 Structure of IT2 T-S FLS

$$\vec{m}_i(\theta(t)) = \left[\underline{m}_i(\theta(t)), \overline{m}_i(\theta(t))\right]$$

and

$$\underline{m}_i(\theta(t)) = \prod_{\iota=1}^{\hbar} \underline{\mu}_{\mathscr{F}_\iota^i}(f_\iota(\theta(t))) > 0,$$
$$\overline{m}_i(\theta(t)) = \prod_{\iota=1}^{\hbar} \overline{\mu}_{\mathscr{F}_\iota^i}(f_\iota(\theta(t))) > 0,$$

where $\underline{\mu}_{\mathscr{F}_\iota^i}(f_\iota(\theta(t)))$ and $\overline{\mu}_{\mathscr{F}_\iota^i}(f_\iota(\theta(t))) \in [0, 1]$ denote the upper and lower MFs, respectively, and satisfy $\underline{\mu}_{\mathscr{F}_\iota^i}(f_\iota(\theta(t))) \le \overline{\mu}_{\mathscr{F}_\iota^i}(f_\iota(\theta(t)))$.

Through single point fuzzification, product reasoning and weighted defuzzification, the global model is inferred as

$$\dot{x}(t) = \sum_{i=1}^{r} m_i(\theta(t))[A_i x(t) + B_i u(t)], \tag{1.5}$$

where membership function, $m_i(\theta(t))$, can be represented as

$$m_i(\theta(t)) = \overline{m}_i(\theta(t))\overline{\alpha}_i(\theta(t)) + \underline{m}_i(\theta(t))\underline{\alpha}_i(\theta(t)), \quad \sum_{i=1}^{r} m_i(\theta(t)) = 1,$$

with the weighting coefficients $\overline{\alpha}_i(\theta(t))$ and $\underline{\alpha}_i(\theta(t))$ satisfying

$$\overline{\alpha}_i(\theta(t)), \underline{\alpha}_i(\theta(t)) \in [0, 1], \quad \overline{\alpha}_i(\theta(t)) + \underline{\alpha}_i(\theta(t)) = 1.$$

Similarly, the discrete-time IT2 T-S fuzzy systems can be represented as

$$x(k+1) = \sum_{i=1}^{r} m_i(\theta(k))[A_i x(k) + B_i u(k)], \tag{1.6}$$

where membership function $m_i(\theta(k))$ satisfies

$$m_i(\theta(k)) = \overline{m}_i(\theta(k))\overline{\alpha}_i(\theta(k)) + \underline{m}_i(\theta(k))\underline{\alpha}_i(\theta(k)), \quad \sum_{i=1}^{r} m_i(\theta(k)) = 1$$
$$\overline{\alpha}_i(\theta(k)), \underline{\alpha}_i(\theta(k)) \in [0, 1], \quad \overline{\alpha}_i(\theta(k)) + \underline{\alpha}_i(\theta(k)) = 1,$$

in which weighting coefficients $\overline{\alpha}_i(\theta(k))$ and $\underline{\alpha}_i(\theta(k))$ are nonlinear functions dependent on parameter uncertainties, which are not necessarily known but exist. By means of weighting coefficients $\overline{\alpha}_i(\theta(k))$ and $\underline{\alpha}_i(\theta(k))$, the upper and lower MFs $\underline{m}_i(\theta(k))$ and $\overline{m}_i(\theta(k))$ can effectively capture the uncertainties, which also render

MFs $m_i(\theta(k))$ uncertain. Consequently, control design methods based on matched MFs for type-1 T-S fuzzy systems cannot be directly extended to IT2 T-S fuzzy systems owing to the inherent uncertainties in the MFs.

1.2.2.2 Control Design and Stability Analysis for Interval Type-2 T-S Fuzzy System

Due to the uncertain MFs in IT2 T-S fuzzy systems, the fuzzy controller cannot utilize the same MFs in fuzzy system so that the famous PDC approach is invalid. Consequently, only non-PDC approaches are applicable for IT2 fuzzy controller design, which contain two kinds of membership function design approaches, i.e., partially matched MFs and completely mismatched MFs. The former keeps the same number of fuzzy rules as the one of fuzzy systems, and only redesign/reconstruct the controller's MFs. It should be pointed out that upper and lower MFs of the IT2 T-S fuzzy systems are known and well-defined, and they can be utilized in the controller's MFs. Based on the partially matched membership function approaches, an IT2 fuzzy controller was designed in [60], where the controller's MFs were constructed by normalizing the upper and lower MFs of fuzzy systems, and MFD stability conditions were derived by leveraging the inequality relation between upper and lower MFs. Furthermore, based on the completely mismatched MFs, an IT2 fuzzy controller was design in [61] without the requirement for identical rule numbers and MFs between the controller and the system, where the operation domain was partitioned into multiple subregions so that their local information was utilized in stability analysis for reducing the conservatism. In [63], an IT2 fuzzy tracking controller based on the completely mismatched MFs approach was proposed, and the piecewise linearization approximation method was utilized to integrate the approximated MFs and their approximation errors into stability analysis for reducing the conservatism. It should be noted that in the aforementioned control design methods, the controller's MFs inherently rely on the availability of full-state information.

For IT2 T-S fuzzy systems subject to unmeasurable system state, a dynamic output feedback fuzzy controller was developed in [64], where the controller's MFs were designed by using measurable premise variables with the same number of fuzzy rules as the one in the fuzzy system. Besides, in [65], a dynamic output feedback controller was designed, and trapezoidal MFs were utilized to approximate the controller's and system's MFs, where the approximation precision will increase with the increase of steps number and decrease of the width of each step. An observer-based IT2 fuzzy output-feedback control strategies were proposed in [66, 67], where the MFs in the observer and controller were constructed by replacing the unmeasurable states in the system's MFs with the estimated states. Based on the similar method, the observer-based fault detection was considered in [68] for the IT2 T-S fuzzy systems, and a fuzzy Lyapunov function was chosen to ensure a desired system performance.

Due to its intrinsic benefits of inherent robustness and structural simplicity, sliding mode control (SMC) has attracted sustained attention in both theoretical research and engineering applications, and numerous remarkable research advancements have been yielded [69–78]. However, the structural complexity of the IT2 T-S fuzzy systems poses substantial challenges for the SMC design and analysis [79–81]. By introducing the assumption that all local subsystems of fuzzy rules have the same input matrix, an adaptive sliding mode fault-tolerant control was investigated in [82] for the IT2 T-S fuzzy time-delay systems. Based on the reduced-order SMC method, an adaptive SMC law was developed to ensure uniform ultimate boundedness of the closed-loop systems and the stability of the sliding mode dynamics. However, the prerequisite of identical input matrices severely limits practical applicability of the IT2 fuzzy SMC method. In [83], an integral sliding surface was constructed by utilizing the lower MFs in the IT2 fuzzy system, which removes the assumption on identical input matrices, and the proposed SMC law can guarantee the boundedness of the closed-loop systems and the stability of the sliding mode dynamics. Subsequently, the fuzzy SMC approach in [83] was extended to the stochastic IT2 T-S fuzzy systems subject to actuator faults in [84], where an adaptive SMC law was designed for guaranteeing the reachability of the sliding surface and the stochastic stability of the sliding mode dynamics. Along the similar approach, the SMC design problem was addressed in [85] for IT2 T-S fuzzy time-delay systems with unmeasurable states, where an output-dependent integral sliding surface was constructed by using auxiliary variables and the system's upper MFs for attaining the H_2 and H_∞ performances. Meanwhile, an adaptive SMC law based on the system's known upper MFs ensured uniform ultimate boundedness. The finite-time observer-based SMC for the IT2 T-S fuzzy systems was considered in [86], where a fuzzy observer was designed under completely mismatched MFs, and the integral sliding surface and the fuzzy sliding mode controller were constructed by using the estimated states.

1.3 Fuzzy Control and Stability Analysis Under Communication Protocols

Rapid developments in computer science and network communication technologies have enabled the interconnection of control system components via communication network to realize the remote data transmission and exchange, which makes networked control systems (NCSs) widely used in practical applications [87–89]. Compared with conventional point-to-point control configurations, NCSs exhibit several advantages, such as superior flexibility, high scalability, enhanced reliability, and improved maintainability. Recently, NCSs have been implemented in various applications including automotive engineering [90], power grid management [91], and aerospace flight control [92].

Although NCSs offer significant advantages, the signal transmission over communication network is inherently constrained by finite communication resources. Under bandwidth-limited network, the inevitable network congestion may result in adverse network-induced phenomena including transmission delays and packet dropouts, which may have the effects on control performance [93–97]. Therefore, it is important and critical to properly regulate the signal transmission process and improve the utilization of constrained network resources.

Communication protocols have emerged as important solutions for addressing resource-constrained issue in NCSs. Currently, the most widely utilized protocols are event-triggered protocols (ETPs) and channel-scheduling protocols (CSPs). Under these protocols, research on control design and analysis for T-S fuzzy systems has garnered increasing attention. However, due to the special structure of T-S fuzzy systems, especially, for IT2 T-S fuzzy systems, introducing communication protocols to regulate the signal transmission process will inevitably affect control design and stability analysis, which will pose new challenges for analysis and synthesis of fuzzy systems.

1.3.1 Event-Triggered Fuzzy Control

ETPs, also termed as event-triggered schemes (ETSs), are effective methods to regulate the signal transmission process and improve the utilization of communication resources. Its main idea is that the signal is transmitted only when the pre-defined triggering condition is satisfied [98]. Compared with the conventional period sampling, ETSs can dynamically regulate the transmission instants and reduce the transmission frequency. By designing the triggering conditions and thresholds, some typical ETSs have been developed, such as, static ETSs [99–101], dynamic ETSs [102, 103], adaptive ETSs [104, 105], and self-triggered schemes (STSs) [106–109]. Note that when an event-triggering mechanism (ETM) is utilized to determine the transmission of measurement signal or control signal, an important issue is to ensure that there exists a positive minimum inter-event time [110–112], i.e., avoiding Zeno behavior, which is actually difficult. To address this difficulty, the sampled-based ETSs were developed and received increasing attention [113, 114]. Under a sample-based ETS, the triggering conditions are judged only at the sampling instant, which can absolutely avoid Zeno phenomenon.

Recently, some advancements have been gained in event-triggered fuzzy control for T-S fuzzy systems [115–129]. For continuous-time T-S fuzzy systems, a dynamic ETM was introduced in [118] into sensor-controller (S/C) channels to regulate the transmission of system state, and a fuzzy event-triggered controller and its MFs were designed using the triggered signals. In [128, 130], a memory-based ETM was developed by using historical data for designing dynamic and adaptive parameter updating rules, and state/output-feedback event-triggered fuzzy control schemes were proposed. Specifically, in [128], a minimum positive time-interval was introduced into the triggering condition to prevent Zeno phenomenon. Furthermore, by employing SMC technique, event-triggered control schemes were investigated for T-S fuzzy systems in [116, 117]. These schemes utilized a static ETM with constant thresholds and a hybrid ETM integrating both state errors and sliding variable errors, and the proposed control schemes can guarantee the bounded stability of the sliding motions. Meanwhile, an event-triggered fuzzy SMC law was formulated based on triggered states, which can rigorously eliminate Zeno phenomena. It should be noted that existing continuous-time event-triggered control frameworks face two critical challenges: (1) the requirement to guarantee a positive minimum triggering interval for avoiding Zeno phenomena; (2) the necessity of dedicated hardware for continuous monitoring of event-triggering conditions. To address these limitations, sampled-data-based ETSs have also been consequently explored for T-S fuzzy systems.

A periodic event-triggered fuzzy control scheme was proposed in [122], where system state was sampled at a constant period, and the triggering condition was judged only at sampling instants. Notably, this control design scheme contained an implicit assumption that the premise variables between fuzzy controllers and system models are the same. This constraint was subsequently overcome in [121, 125] by using received information to design controller's MFs, and finite-time event-triggered fuzzy control and security fuzzy control schemes were developed, respectively. Subsequently, an observer-based output-feedback control scheme was established in [123] under periodic ETSs, and premise variables in fuzzy observer and controller were constructed by using estimated state. Adaptive parameter updating rules were used to adjust triggering thresholds in [131, 132], where fuzzy controllers and MFs were constructed by using system state and premise variables at triggering instants. By employing SMC technique, a periodic event-triggered fuzzy control scheme was introduced in [127] by designing integral sliding surface and fuzzy SMC law to guarantee the ultimate boundedness stability of the sliding mode dynamics. Contrasting with these periodic sampled-data-based approaches, recent advancements in [126, 133] have developed aperiodic sampled-data ETSs employing time-varying sampling period with the upper and lower bounds.

Discrete-time T-S fuzzy systems inherently circumvent Zeno phenomena due to their temporal discretization characteristics, which has motivated the development of event-triggered fuzzy control frameworks. For example, the research in [124] considered a finite-time event-triggered fuzzy control scheme for fuzzy Markov jump singularly perturbed systems. Besides, a reachable-set-based event-triggered control framework was established in [129] to evaluate control performance degradation when introducing an ETM into S/C channels. For addressing hybrid communication constraints, a quantized event-triggered static output-feedback tracking control strategy was investigated in [119], where reference signals and system states were transmission via network channels with quantization and ETMs, and fuzzy controller was designed based on the quantized triggered signals. Furthermore, by constructing linear sliding functions, event-triggered SMC issues were considered in [115, 120], where fuzzy controller and its MFs were designed by integrating triggered state and system's MFs.

The inherent uncertainty in MFs of IT2 T-S fuzzy systems fundamentally challenges conventional event-triggered control design methods, because system's MFs over triggering interval becomes unavailable for controller, which significantly complicates both controller synthesis and stability analysis. To address this challenge, a fuzzy control scheme was established in [134] under periodic ETSs, leveraging triggered state signals together with known upper and lower MFs in fuzzy model to ensure an implementable fuzzy controller while resolving mismatched MFs problems through establishing inequality relation between controller's and system's MFs. Besides, an aperiodic adaptive event-triggered fuzzy control strategy was proposed in [135], where new MFs in fuzzy controller were constructed independent of the system model for attaining higher design freedom. In [136], dynamic periodic event-triggered control problem was considered for fuzzy singularly perturbed systems, where a fuzzy disturbance observer was designed to estimate external disturbance, and a fuzzy SMC scheme was developed. It should be noted that the proposed fuzzy disturbance observer utilized the same premise variables and MFs as the ones in fuzzy model, which may lead to some difficulties in practical implementation. In [137], a periodic event-triggered fuzzy control scheme was developed, and a fuzzy-rule-dependent adaptive ETM was designed under each fuzzy rule. Furthermore, security event-triggered fuzzy control scheme was developed in [138], where the triggered signal may be subject to aperiodic Denial-of-Service (DoS) attacks and random deception attacks. In [139], a dynamic event-triggered fuzzy control scheme was considered for discrete-time IT2 T-S fuzzy systems under fading channels, and MFD analysis method was employed to deal with mismatched MFs and obtain sufficient conditions.

1.3.2 Channel-Scheduling-Based Fuzzy Control

CSPs, also termed as communication scheduling protocols, can regulate the data transmission for bandwidth-constrained multi-nodes NCSs through activating one transmission node at each instant. Its basic principle is to determine which transmission nodes access the network according to different selection mechanisms, thereby reducing transmission loads and mitigating network congestion in limited-bandwidth communication environments [140–144]. There are three typical channel scheduling protocols investigated widely in the literatures, i.e., Round-Robin (RR) protocol [145–147], stochastic scheduling (SS) protocol [148–152], and weighted-try-once-discard (WTOD) protocol [153–156].

The design and analysis of control systems under channel scheduling protocols fundamentally address two core challenges: node selection strategies and signal compensation mechanisms. Node selection strategies determine transmission node access to network channels through three principal methodologies: the RR protocol implements deterministic cyclic channel allocation via fixed temporal sequencing [146], the SS protocol employs probabilistic node selection governed by independent and identically distributed (i.i.d.) sequence [141] or discrete-time Markov chain [143], while the WTOD protocol performs dynamic node prioritization through error-dependent weighting coefficients obeying maximum-error-first (MEF) principle [140]. For those unselected nodes, general signal compensation techniques are either the Zero-Order-Hold (ZOH) strategies, which retain the last transmitted value, or Zero-Input (ZI) strategies that nullify untransmitted data through signal discarding.

In recent years, the issues of channel-scheduling-based control have been extended to T-S fuzzy systems. When CSPs are implemented in the communication channels of T-S fuzzy systems, which may affect the design of fuzzy controller and its MFs, leading to more difficulties in design and analysis. The RR protocol was adopted for type-1 T-S fuzzy systems in [157] to periodically select a sensor node for transmitting measured output signals, and a token-dependent output-feedback fuzzy controller was designed to guarantee the H_∞ performance of closed-loop systems. Besides, by employing a Markov chain-based SS protocol to regulate the transmission of measurement output signal, a dynamic output-feedback fuzzy controller was developed in [158, 159], and the scheduling signal of protocol was estimated in [159] for the control design. Additionally, by means of WTOD protocol for transmitting the measured output signals, a token-dependent output-feedback fuzzy controller was designed in [160]. However, it should be pointed out that the proposed fuzzy controllers in [157–160] directly utilized the system's MFs, which poses significant challenges for practical implementation. The similar limitation can also be found in [161], where observer-based fuzzy control scheme was proposed under WTOD protocol, and fuzzy observer shared the same premise variables as the ones in fuzzy model. Recently, channel-scheduling-based SMC problems have been investigated in [162–166]. In [162], the WTOD protocol was utilized to arrange the transmission of control signal, under which only one node was permitted to obtain access to the communication network, and an MFD sliding surface and a token-dependent sliding mode controller were designed. It should be pointed out that system's MFs can be utilized to design fuzzy controller, since WTOD protocol was only implemented as the controller-actuator (C/A) channels. To ensure the fuzzy controller to be implementable when CSPs are utilized in S/C channels, a feasible control design method is to utilize the received signal by controller to construct premise variables and MFs in fuzzy controller as in [163–166].

The introduction of CSPs into the IT2 T-S fuzzy systems presents compounded challenges for control design and stability analysis, primarily due to the inherent uncertainty of MFs in IT2 fuzzy systems and the intermittent transmission of signals induced by scheduling mechanisms. When the SS protocol was implemented in the C/A channels to regulate the transmission of control signal, a fuzzy controller was designed in [167] with a different number of fuzzy rules from the one in fuzzy system. In [168], the RR protocol was employed to select sensor nodes, and a token-dependent output-feedback fuzzy controller was developed to ensure the H_∞ performance. Beside, the measurement output signal was transmitted via SS protocols in [169, 170] and WTOD protocol in [171], and observer-based fuzzy control schemes were proposed by using system's MFs. An effective fuzzy control design scheme was developed in [172] for the case that system state is subject to a RR protocol, where the proposed IT2 fuzzy controller and its MFs depended on the scheduled state information. A RR protocol was used to select sensor nodes, a sliding surface and a token-dependent SMC law were proposed in [173] to guarantee the reachability of sliding domain and the input-to-state stability of closed-loop systems. Furthermore, the sliding mode security control problem subject to SS protocols and potential DoS attacks was also considered in [174], where a secure sliding mode controller was designed to maintain a desired control performance.

1.4 Outline of This Book

The content of this monograph contains two parts. The first part focuses on event-triggered control design and stability analysis for IT2 T-S fuzzy systems. The second part focuses on the design of channel scheduling protocols and fuzzy controller for IT2 fuzzy systems, and proposes novel event-triggered WTOD protocols. The organization structure of this monograph is shown in Fig. 1.3.
This chapter presents the research background and motivations, which involves basic control design and stability analysis for type-1 and IT2 T-S fuzzy systems, various event-triggered and channel scheduling protocols.

Part I focuses on the event-triggered fuzzy control issue for IT2 T-S fuzzy systems, including developing various event-triggered protocols, addressing mismatched membership functions issue due to event-triggering communication, proposing control design and analysis methods. This part contains five chapters from Chaps. 2–6.

Chapter 2 investigates the dynamic event-triggered SMC problem for continuous-time IT2 T-S fuzzy systems subject to unmeasure state and fading channels. The transmission of measurement output is determined via a dynamic ETM in the S/C channels. A fuzzy state observer is designed to estimate the unmeasure state, based on which a integral fuzzy sliding function is constructed, and observer-based fuzzy sliding mode controller is designed for ensuring the reachability of sliding surface and input-to-state practical stability of closed-loop systems. Besides, a positive minimum inter-event time is proved to avoid Zeno behavior.

Chapter 3 considers the issue of the component-based dynamic event-triggered SMC and performance optimization for discrete-time IT2 T-S fuzzy systems. For each sensor of state component, an individual dynamic ETM is designed to determine the signal transmission instant. Based on the triggered state, a fuzzy sliding mode controller and its MFs are designed. A relationship between MFs in fuzzy model and controller is established to deal with the mismatched membership function phenomenon, and sufficient conditions are derived for ensuring stability of closed-loop systems and reachability of specified sliding surface. Besides, an optimization algorithm is given to minimize the sliding domain for improving control performance.

Chapter 4 proposes the quantized SMC scheme for IT2 T-S fuzzy systems under ETS. A novel membership-function-based dynamic ETS is proposed to judge the variation of system sates and MFs, and determine the transmission of system state. Based on the triggered states, an event-triggered fuzzy controller and its MFs are designed. Meanwhile, the MFs' local information of both controller and fuzzy sys-

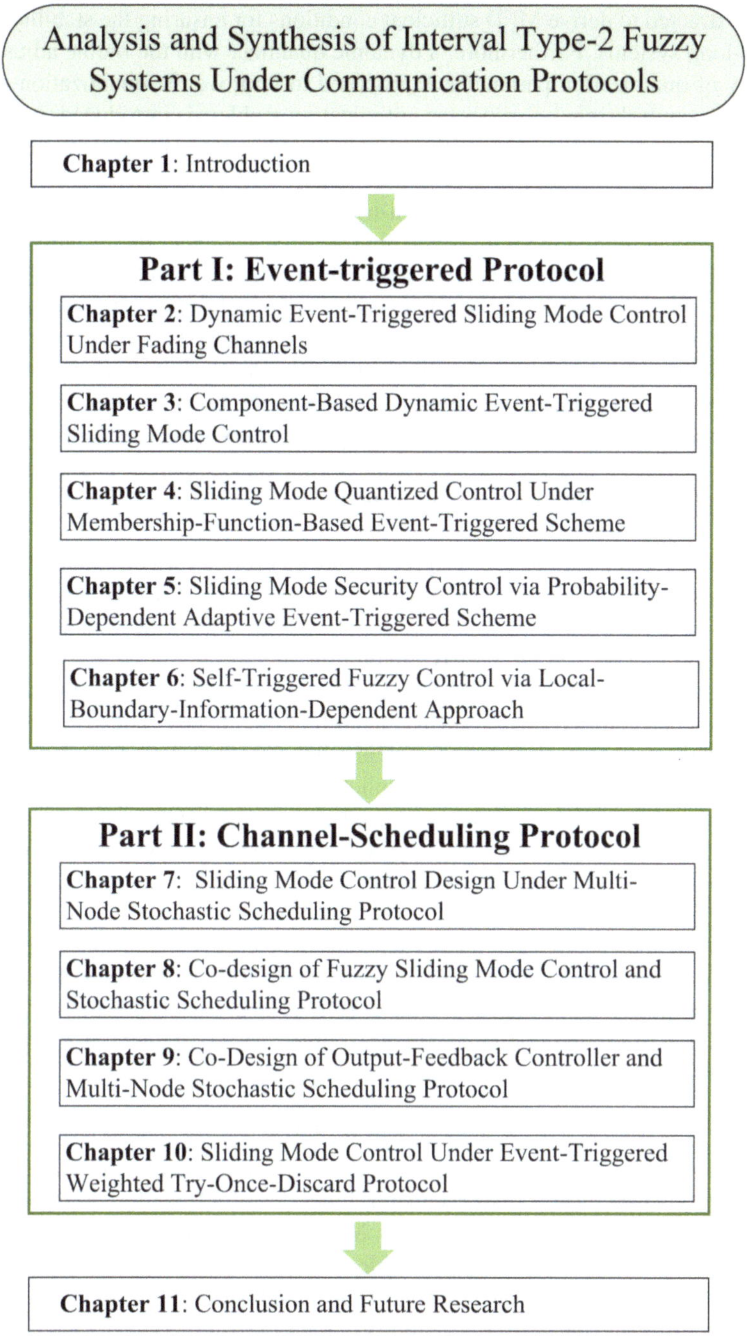

Fig. 1.3 The organization structure of this book

tem is extracted to derive MFD sufficient conditions for ensuring the stability of the closed-loop systems. Furthermore, a dynamic quantizer with the online adjustment strategy of quantization parameter is proposed to improve the quantization procedure of the control signal. Besides, an optimization problem is introduced to improve the comprehensive performance of the convergence domain and triggering frequency.

Chapter 5 focuses on the security event-triggered SMC problem for IT2 T-S fuzzy systems subject to the randomly occurring DoS attacks. Under the resource-limited and unreliable network environments, a probability-dependent adaptive ETM is proposed, in which the triggering parameter can be adaptively adjusted according to the attack probability and the convergence trend of the system state. By using the compromised triggered states, a security fuzzy sliding mode controller is designed, which does not require the same fuzzy rules as the ones in the fuzzy system. Furthermore, by employing the global boundaries of MFs in the fuzzy system and controller, the co-design scheme of the adaptive ETS and the fuzzy controller are developed so that the desired SMC performance can be ensured.

Chapter 6 investigates the self-triggered control problem for continuous-time IT2 T-S fuzzy systems under limited communication resources. A self-triggered scheme is proposed to determine the transmission of system states, whose main feature is that there is no need to continuously monitor the pre-specified triggering condition, and the next triggering instant can be calculated in advance according to the latest triggered states. Furthermore, a self-triggered fuzzy controller is designed to ensure the boundedness of the closed-loop systems. Besides, the relationship between the triggered state and the system state is explicitly established, and the local boundary information of MFs is analyzed and utilized to obtain the stability conditions with less conservatism. Meanwhile, the Zeno-free phenomenon is analyzed to ensure the feasibility of self-triggered control scheme.

Part II focuses on the design channel scheduling protocols and fuzzy controller for IT2 T-S fuzzy systems, where the co-design scheme of SS protocols and fuzzy controller is proposed, and event-triggered WTOD protocol is established to improve the utilization of limited communication resources. This part contains five chapters from Chaps. 7–10.

Chapter 7 investigates the SMC problem for IT2 fuzzy systems under multi-node SS protocol. A Markov chain-driven multi-node scheduling protocol is developed to selectively activate partial sensor/actuator nodes. Two independent Markov chains are utilized to describe the access status of sensor and actuator nodes, which are unified into a composite Markov variable via the mapping techniques. A token-dependent fuzzy sliding mode controller is designed, and MFD stability conditions are derived to ensure the stochastic stability of the closed-loop systems and the reachability of the sliding surface. An optimization algorithm is further proposed to search the sliding matrix for reducing the control energy consumption.

Chapter 8 addresses the co-design issue of scheduling protocols and fuzzy sliding mode controllers for IT2 T-S fuzzy systems under bandwidth-limited networks. To alleviate communication congestion, a Markov chain-based SS protocol is proposed to regulate the transmission of system state by activating only one sensor node at each instant. Distinct from conventional approaches of given Markov chain's transition probability, this proposed protocol is co-designed with the control gains. Based on the available system states scheduled by the protocol, a token-dependent sliding mode controller is designed. By establishing the relation between MFs in fuzzy system and controller, MFD sufficient conditions are obtained to attain the stochastic stability of the closed-loop systems and the reachability of the sliding surface. An iterative optimization algorithm is further developed to optimize the token-dependent control parameters and sliding matrices with the objective of minimizing sliding domain.

Chapter 9 investigates the co-design problem of multi-node scheduling protocol and the SMC design for IT2 T-S fuzzy systems subject to unmeasure states. The transmission of measurement output is determined via a multi-node SS protocol govern by a discrete-time Markov chain, under which only a part of sensor nodes are allowed to access communication channels. A token-dependent static output-feedback fuzzy sliding mode controller is designed so that the stochastic stability and the desired SMC performance are ensured. Besides, the transition probability of Markov chain is co-designed with the fuzzy control gains, and an optimization algorithm is proposed to obtain optimal design and scheduling parameters.

Chapter 10 investigates the SMC problem for IT2 T-S fuzzy systems under resource-constrained communication networks. An event-triggered WTOD (ET-WTOD) protocol is formulated by introducing two threshold parameters, which can determine the transmission of the state signal for reducing the communication burden. The proposed ET-WTOD protocol can dynamically adjust the number of transmitted nodes, and permits only a part of sensor nodes obtain access to network channels. By analyzing different transmission cases, a new scheduling signal set is established, and token-dependent fuzzy sliding controller is designed. With the aid of the MFD analysis approach, the mismatched MFs in the fuzzy systems and the controller are effectively handled, and stability conditions are derived for attaining a desired control performance. Furthermore, an optimization algorithm is provided to minimize the sliding domain for enhancing control performance.

References

1. Takagi, T., Sugeno, M.: Fuzzy identification of systems and its applications to modeling and control. IEEE Trans. Syst. Man Cybern. **15**(1), 116–132 (1985)
2. Wang, Y., Jiang, B., Wu, Z.G., Xie, S., Peng, Y.: Adaptive sliding mode fault-tolerant fuzzy tracking control with application to unmanned marine vehicles. IEEE Trans. Syst. Man Cybern. Syst. **51**(11), 6691–6700 (2021)

3. Zhang, C., Lam, H.K., Qiu, J., Qi, P., Chen, Q.: Fuzzy-model-based output feedback steering control in autonomous driving subject to actuator constraints. IEEE Trans. Fuzzy Syst. **29**(3), 457–470 (2019)
4. Mehran, K., Giaouris, D., Zahawi, B.: Stability analysis and control of nonlinear phenomena in boost converters using model-based Takagi-Sugeno fuzzy approach. IEEE Trans. Circuits Syst. I Regul. Pap. **57**(1), 200–212 (2010)
5. Cao, J., Li, P., Liu, H.: An interval fuzzy controller for vehicle active suspension systems. IEEE Trans. Intell. Transp. Syst. **11**(4), 885–895 (2010)
6. Li, W., Xie, Z., Wong, P.K., Zhang, X., Zhao, J., Zhao, J.: Interval type-2 fuzzy path tracking control for autonomous ground vehicles under switched triggered and sensor attacks. IEEE Trans. Intell. Transp. Syst. **25**(11), 16024–16035 (2024)
7. Huang, J., Ri, M., Wu, D., Ri, S.: Interval type-2 fuzzy logic modeling and control of a mobile two-wheeled inverted pendulum. IEEE Trans. Fuzzy Syst. **26**(4), 2030–2038 (2018)
8. Ren, G.P., Chen, Z., Zhang, H.T., Wu, Y., Meng, H., Wu, D., Ding, H.: Design of interval type-2 fuzzy controllers for active magnetic bearing systems. IEEE/ASME Trans. Mechatron. **25**(5), 2449–2459 (2020)
9. Tao, X., Yi, J., Pu, Z., Xiong, T.: Robust adaptive tracking control for hypersonic vehicle based on interval type-2 fuzzy logic system and small-gain approach. IEEE Trans. Cybern. **51**(5), 2504–2517 (2021)
10. Zadeh, L.: Fuzzy sets. Inf. Control **8**(3), 338–353 (1965)
11. Zadeh, L.A.: Outline of a new approach to the analysis of complex systems and decision processes. IEEE Trans. Syst. Man Cybern. **SMC-3**(1), 28–44 (1973)
12. Mamdani, E.H.: An experiment in linguistic synthesis with a fuzzy logic controller. Int. J. Man Mach. Stud. **7**(1), 1–13 (1975)
13. Mamdani, E.H.: Advances in the linguistic synthesis of fuzzy controllers. Int. J. Man Mach. Stud. **8**(6), 669–678 (1976)
14. Kickert, W.J.M., Mamdani, E.H.: Analysis of a fuzzy logic controller. Fuzzy Sets Syst. **1**(1), 29–44 (1978)
15. Tong, R.M., Beck, M.B., Latten, A.: Fuzzy control of the activated sludge wastewater treatment process. Automatica **16**(6), 695–701 (1980)
16. Holmblad, L., Ostergaard, J.J.: Control of a cement kiln by fuzzy logic techniques. IFAC Proc. **14**, 809–814 (1981)
17. Feng, G.: A survey on analysis and design of model-based fuzzy control systems. IEEE Trans. Fuzzy Syst. **14**(5), 676–697 (2006)
18. Sugeno, M., Kang, G.: Structure identification of fuzzy model. Fuzzy Sets Syst. **28**(1), 15–33 (1988)
19. Kim, E., Park, M., Ji, S., Park, M.: A new approach to fuzzy modeling. IEEE Trans. Fuzzy Syst. **5**(3), 328–337 (1997)
20. Tsai, S.H.: An improved fuzzy modeling method for a class of multi-input non-affine nonlinear systems. J. Optim. Theory Appl. **157**, 287–296 (2013)
21. Wang, H.O., Tanaka, K., Griffin, M.F.: An approach to fuzzy control of nonlinear systems: stability and design issues. IEEE Trans. Fuzzy Syst. **4**(1), 14–23 (1996)
22. Tanaka, K., Ikeda, T., Wang, H.: Fuzzy regulators and fuzzy observers: relaxed stability conditions and LMI-based designs. IEEE Trans. Fuzzy Syst. **6**(2), 250–265 (1998)
23. Teixeira, M., Assuncao, E., Avellar, R.: On relaxed LMI-based designs for fuzzy regulators and fuzzy observers. IEEE Trans. Fuzzy Syst. **11**(5), 613–623 (2003)
24. Lam, H.K., Leung, F.H.: Stability analysis of fuzzy control systems subject to uncertain grades of membership. IEEE Trans. Syst. Man Cybern. Part B **35**(6), 1322–1325 (2005)
25. Lam, H.K., Narimani, M.: Stability analysis and performance design for fuzzy-model-based control system under imperfect premise matching. IEEE Trans. Fuzzy Syst. **17**(4), 949–961 (2008)
26. Lam, H.K., Leung, F.H.F., Tam, P.K.S.: Stable and robust fuzzy control for uncertain nonlinear systems. IEEE Trans. Syst. Man Cybern. Part A Syst. Hum. **30**(6), 825–840 (2000)

27. Lam, H.K., Leung, F.H.F., Tam, P.K.S.: Nonlinear state feedback controller for nonlinear systems: stability analysis and design based on fuzzy plant model. IEEE Trans. Fuzzy Syst. **9**(4), 657–661 (2001)
28. Lam, H.K., Lauber, J.: Membership-function-dependent stability analysis of fuzzy-model-based control systems using fuzzy Lyapunov functions. Inf. Sci. **232**, 253–266 (2013)
29. Lam, H.K., Leung, F.H.: LMI-based stability and performance conditions for continuous-time nonlinear systems in Takagi-Sugeno's form. IEEE Trans. Syst. Man Cybern. Part B **37**(5), 1396–1406 (2007)
30. Tanaka, K., Hori, T., Wang, H.: A multiple Lyapunov function approach to stabilization of fuzzy control systems. IEEE Trans. Fuzzy Syst. **11**(4), 582–589 (2003)
31. Zhou, S., Lam, J., Zheng, W.X.: Control design for fuzzy systems based on relaxed non-quadratic stability and $\mathcal{H}_\infty$ performance conditions. IEEE Trans. Fuzzy Syst. **15**(2), 188–199 (2007)
32. Johansson, M., Rantzer, A., Arzen, K.E.: Piecewise quadratic stability of fuzzy systems. IEEE Trans. Fuzzy Syst. **7**(6), 713–722 (1999)
33. Feng, M., Harris, C.J.: Piecewise Lyapunov stability conditions of fuzzy systems. IEEE Trans. Syst. Man Cybern. Part B **31**(2), 259–262 (2001)
34. Feng, G.: Controller synthesis of fuzzy dynamic systems based on piecewise Lyapunov functions. IEEE Trans. Fuzzy Syst. **11**(5), 605–612 (2003)
35. Lam, H.K., Narimani, M., Li, H., Liu, H.: Stability analysis of polynomial-fuzzy-model-based control systems using switching polynomial Lyapunov function. IEEE Trans. Fuzzy Syst. **21**(5), 800–813 (2013)
36. Powers, V., Reznick, B.: A new bound for Polya's theorem with applications to polynomials positive on polyhedra. J. Pure Appl. Algebra **164**(1), 221–229 (2001)
37. Montagner, V.F., Oliveira, R.C.L.F., Peres, P.L.D.: Convergent LMI relaxations for quadratic stabilizability and $\mathcal{H}_\infty$ control of takagi-sugeno fuzzy systems. IEEE Trans. Fuzzy Syst. **17**(4), 863–873 (2009)
38. Fang, C.H., Liu, Y.S., Kau, S.W., Hong, L., Lee, C.H.: A new LMI-based approach to relaxed quadratic stabilization of T-S fuzzy control systems. IEEE Trans. Fuzzy Syst. **14**(3), 386–397 (2006)
39. Sala, A., Ariño, C.: Asymptotically necessary and sufficient conditions for stability and performance in fuzzy control: applications of Polya's theorem. Fuzzy Sets Syst. **158**(24), 2671–2686 (2007)
40. Lam, H.: A review on stability analysis of continuous-time fuzzy-model-based control systems: from membership-function-independent to membership-function-dependent analysis. Eng. Appl. Artif. Intell. **67**, 390–408 (2018)
41. Sala, A., Ariño, C.: Relaxed stability and performance conditions for Takagi-Sugeno fuzzy systems with knowledge on membership function overlap. IEEE Trans. Syst. Man Cybern. Part B **37**(3), 727–732 (2007)
42. Ariño, C., Sala, A.: Extensions to stability analysis of fuzzy control systems subject to uncertain grades of membership. IEEE Trans. Syst. Man Cybern. Part B **38**(2), 558–563 (2008)
43. Narimani, M., Lam, H.K.: Relaxed LMI-based stability conditions for Takagi-Sugeno fuzzy control systems using regional-membership-function-shape-dependent analysis approach. IEEE Trans. Fuzzy Syst. **17**(5), 1221–1228 (2009)
44. Lam, H.: Design of stable fuzzy controller for non-linear systems subject to imperfect premise matching based on grid-point approach. IET Control Theory Appl. **4**, 2770–2780 (2010)
45. Kruszewski, A., Sala, A., Guerra, T.M., Ariño, C.: A triangulation approach to asymptotically exact conditions for fuzzy summations. IEEE Trans. Fuzzy Syst. **17**(5), 985–994 (2009)
46. Lam, H.K., Narimani, M.: Quadratic-stability analysis of fuzzy-model-based control systems using staircase membership functions. IEEE Trans. Fuzzy Syst. **18**(1), 125–137 (2009)
47. Lam, H.K.: LMI-based stability analysis for fuzzy-model-based control systems using artificial T-S fuzzy model. IEEE Trans. Fuzzy Syst. **19**(3), 505–513 (2011)
48. Cao, K., Gao, X., Lam, H.K., Vasilakos, A., Pedrycz, W.: A new relaxed stability condition for Takagi-Sugeno fuzzy control systems using quadratic fuzzy Lyapunov functions and staircase membership functions. Int. J. Fuzzy Syst. **16**, 327–337 (2014)

49. Narimani, M., Lam, H.K., Dilmaghani, R., Wolfe, C.: LMI-based stability analysis of fuzzy-model-based control systems using approximated polynomial membership functions. IEEE Trans. Syst. Man Cybern. Part B (Cybern.) **41**(3), 713–724 (2010)
50. Lam, H.K., Wu, L., Zhao, Y.: Linear matrix inequalities-based membership-function-dependent stability analysis for non-parallel distributed compensation fuzzy-model-based control systems. IET Control Theory Appl. **8**(8), 614–625 (2014)
51. Yang, Y., Li, J., Niu, Y., Xu, K.: Disturbance-observer-based sliding mode control for polynomial fuzzy systems. Int. J. Syst. Sci. (2025). https://doi.org/10.1080/00207721.2025.2491780
52. Zadeh, L.: The concept of a linguistic variable and its application to approxiamte reasoning. Inf. Sci. **8**(3), 199–249 (1975)
53. Karnik, N.N., Mendel, J.M., Liang, Q.: Type-2 fuzzy logic systems. IEEE Trans. Fuzzy Syst. **7**(6), 643–658 (1999)
54. Liang, Q., Mendel, J.M.: Interval type-2 fuzzy logic systems: theory and design. IEEE Trans. Fuzzy Syst. **8**(5), 535–550 (2000)
55. Wu, D., Mendel, J.M.: On the continuity of type-1 and interval type-2 fuzzy logic systems. IEEE Trans. Fuzzy Syst. **19**(1), 179–192 (2011)
56. Wu, D.: On the fundamental differences between interval type-2 and type-1 fuzzy logic controllers. IEEE Trans. Fuzzy Syst. **20**(5), 832–848 (2012)
57. Mendel, J.M.: Advances in type-2 fuzzy sets and systems. Inf. Sci. **177**(1), 84–110 (2007)
58. Chen, C., Wu, D., Garibaldi, J.M., John, R.I., Twycross, J., Mendel, J.M.: A comprehensive study of the efficiency of type-reduction algorithms. IEEE Trans. Fuzzy Syst. **29**(6), 1556–1566 (2021)
59. Wu, D.: Approaches for reducing the computational cost of interval type-2 fuzzy logic systems: overview and comparisons. IEEE Trans. Fuzzy Syst. **21**(1), 80–99 (2013)
60. Lam, H.K., Seneviratne, L.D.: Stability analysis of interval type-2 fuzzy-model-based control systems. IEEE Trans. Syst. Man Cybern. Part B **38**(3), 617–628 (2008)
61. Lam, H.K., Li, H., Deters, C., Secco, E.L., Wurdemann, H.A.: Control design for interval type-2 fuzzy systems under imperfect premise matching. IEEE Trans. Ind. Electron. **61**(2), 956–968 (2014)
62. Li, H., Wu, C., Wu, L., Lam, H.K., Gao, Y.: Filtering of interval type-2 fuzzy systems with intermittent measurements. IEEE Trans. Cybern. **46**(3), 668–678 (2015)
63. Zeng, Y., Lam, H.K., Xiao, B., Wu, L.: Tracking control for nonlinear systems with actuator saturation via interval type-2 T-S fuzzy framework. IEEE Trans. Cybern. **53**(11), 7085–7094 (2023)
64. Li, H., Sun, X., Wu, L., Lam, H.K.: State and output feedback control of interval type-2 fuzzy systems with mismatched membership functions. IEEE Trans. Fuzzy Syst. **23**(6), 1943–1957 (2015)
65. Zeng, Y., Lam, H.K., Xiao, B., Wu, L.: ℓ_2-ℓ_∞ control of discrete-time state-delay interval type-2 fuzzy systems via dynamic output feedback. IEEE Trans. Cybern. **52**(6), 4198–4208 (2022)
66. Li, H., Wu, C., Yin, S., Lam, H.K.: Observer-based fuzzy control for nonlinear networked systems under unmeasurable premise variables. IEEE Trans. Fuzzy Syst. **24**(5), 1233–1245 (2016)
67. Zhang, Z., Dong, J.: Observer-based interval type-2 $\mathcal{L}_2$–$\mathcal{L}_\infty/\mathcal{H}_\infty$ mixed fuzzy control for uncertain nonlinear systems under measurement outliers. IEEE Trans. Syst. Man Cybern. Syst. **51**(12), 7652–7662 (2020)
68. Hassani, H., Zarei, J., Razavi-Far, R., Saif, M.: Robust interval type-2 fuzzy observer for fault detection of networked control systems subject to immeasurable premise variables. IEEE Syst. J. **13**(3), 2954–2965 (2019)
69. Utkin, V.I.: Sliding mode control design principles and applications to electric drives. IEEE Trans. Ind. Electron. **40**(1), 23–36 (1993)
70. Bartolini, G., Fridman, L., Pisano, A., Usai, E.: Modern sliding mode control theory: new perspectives and applications. Springer-Verlag, Germany Berlin (2008)

71. Utkin, V.I.: Sliding modes in control and optimization. Springer Science & Business Media (2013)
72. Drakunov, S.V., Utkin, V.I.: Sliding mode control in dynamic systems. Int. J. Control **55**(4), 1029–1037 (1992)
73. Edwards, C., Spurgeon, S.K.: Sliding mode control: theory and applications. CRC Press (1998)
74. Shi, P., Xia, Y., Liu, G., Rees, D.: On designing of sliding-mode control for stochastic jump systems. IEEE Trans. Autom. Control **51**(1), 97–103 (2006)
75. Song, J., Niu, Y., Zou, Y.: Finite-time stabilization via sliding mode control. IEEE Trans. Autom. Control **62**(3), 1478–1483 (2016)
76. Niu, Y., Ho, D.W., Lam, J.: Robust integral sliding mode control for uncertain stochastic systems with time-varying delay. Automatica **41**(5), 873–880 (2005)
77. Niu, Y., Ho, D.W., Wang, X.: Sliding mode control for Itô stochastic systems with Markovian switching. Automatica **43**(10), 1784–1790 (2007)
78. Niu, Y., Ho, D.W., Wang, X.: Robust $\mathcal{H}_\infty$ control for nonlinear stochastic systems: a sliding-mode approach. IEEE Trans. Automat. Control **53**(7), 1695–1701 (2008)
79. Li, R., Qi, W., Park, J.H., Cheng, J., Shi, K.: Sliding mode control for discrete interval type-2 fuzzy semi-Markov jump models with delay in controller mode switching. Fuzzy Sets Syst. **483**, 108915 (2024)
80. Sun, X., Zhang, Q.: Admissibility analysis for interval type-2 fuzzy descriptor systems based on sliding mode control. IEEE Trans. Cybern. **49**(8), 3032–3040 (2019)
81. Kim, H.S., Hwang, S., Joo, Y.H.: Interval type-2 fuzzy-model-based fault-tolerant sliding mode tracking control of a quadrotor UAV under actuator saturation. IET Control Theory Appl. **14**(20), 3663–3675 (2020)
82. Zhao, Y., Wang, J., Yan, F., Shen, Y.: Adaptive sliding mode fault-tolerant control for type-2 fuzzy systems with distributed delays. Inf. Sci. **473**, 227–238 (2019)
83. Li, H., Wang, J., Lam, H.K., Zhou, Q., Du, H.: Adaptive sliding mode control for interval type-2 fuzzy systems. IEEE Trans. Syst. Man Cybern. Syst. **46**(12), 1654–1663 (2016)
84. Zhang, Z., Niu, Y.: Adaptive sliding mode control for interval type-2 stochastic fuzzy systems subject to actuator failures. Int. J. Syst. Sci. **49**(15), 3169–3181 (2018)
85. Li, H., Wang, J., Wu, L., Lam, H.K., Gao, Y.: Optimal guaranteed cost sliding-mode control of interval type-2 fuzzy time-delay systems. IEEE Trans. Fuzzy Syst. **26**(1), 246–257 (2017)
86. Liu, Y., Xia, J., Wang, J., Shen, H.: Finite-time stability for interval type-2 fuzzy nonlinear systems via an observer-based sliding mode control. J. Syst. Sci. Compl. **35**(6), 2223–2247 (2022)
87. Zhang, L., Gao, H., Kaynak, O.: Network-induced constraints in networked control systems-a survey. IEEE Trans. Ind. Inf. **9**(1), 403–416 (2012)
88. Qiu, J., Gao, H., Ding, S.X.: Recent advances on fuzzy-model-based nonlinear networked control systems: a survey. IEEE Trans. Ind. Electron. **63**(2), 1207–1217 (2016)
89. Zhang, X.M., Han, Q.L., Ge, X., Ding, D., Ding, L., Yue, D., Peng, C.: Networked control systems: a survey of trends and techniques. IEEE/CAA J. Autom. Sinica. **7**(1), 1–17 (2020)
90. Zhang, D., Han, Q.L., Zhang, X.M.: Network-based modeling and proportional-integral control for direct-drive-wheel systems in wireless network environments. IEEE Trans. Cybern. **50**(6), 2462–2474 (2019)
91. Wang, S., Meng, X., Chen, T.: Wide-area control of power systems through delayed network communication. IEEE Trans. Control Syst. Technol. **20**(2), 495–503 (2011)
92. Xue, R., Cai, G.: Formation flight control of multi-UAV system with communication constraints. J. Aerosp. Technol. Manag. **8**, 203–210 (2016)
93. Huang, D., Nguang, S.K.: State feedback control of uncertain networked control systems with random time delays. IEEE Trans. Autom. Control **53**(3), 829–834 (2008)
94. Zhao, Y.B., Liu, G.P., Rees, D.: Actively compensating for data packet disorder in networked control systems. IEEE Trans. Circuits Syst. II Exp. Briefs **57**(11), 913–917 (2010)
95. Zhang, W.A., Yu, L.: Modelling and control of networked control systems with both network-induced delay and packet-dropout. Automatica **44**(12), 3206–3210 (2008)

96. Tan, C., Li, L., Zhang, H.: Stabilization of networked control systems with both network-induced delay and packet dropout. Automatica **59**, 194–199 (2015)
97. Pang, Z.H., Liu, G.P., Zhou, D., Sun, D.: Data-based predictive control for networked nonlinear systems with network-induced delay and packet dropout. IEEE Trans. Ind. Electron. **63**(2), 1249–1257 (2015)
98. Peng, C., Li, F.: A survey on recent advances in event-triggered communication and control. Inf. Sci. **457**, 113–125 (2018)
99. Tabuada, P.: Event-triggered real-time scheduling of stabilizing control tasks. IEEE Trans. Autom. Control **52**(9), 1680–1685 (2007)
100. Heemels, W., Sandee, J., Van Den Bosch, P.: Analysis of event-driven controllers for linear systems. Int. J. Control Autom. **81**(4), 571–590 (2008)
101. Zhang, K., Braverman, E., Gharesifard, B.: Event-triggered control for discrete-time delay systems. Automatica **147**, 110688 (2023)
102. Girard, A.: Dynamic triggering mechanisms for event-triggered control. IEEE Trans. Autom. Control **60**(7), 1992–1997 (2015)
103. Zhao, D., Wang, Z., Wei, G., Han, Q.L.: A dynamic event-triggered approach to observer-based PID security control subject to deception attacks. Automatica **120**, 109128 (2020)
104. Zhang, H., Wang, Z., Yan, H., Yang, F., Zhou, X.: Adaptive event-triggered transmission scheme and $\mathcal{H}_\infty$ filtering co-design over a filtering network with switching topology. IEEE Trans. Cybern. **49**(12), 4296–4307 (2019)
105. Wang, Y.L., Lim, C.C., Shi, P.: Adaptively adjusted event-triggering mechanism on fault detection for networked control systems. IEEE Trans. Cybern. **47**(8), 2299–2311 (2017)
106. Wang, X., Lemmon, M.D.: Self-triggered feedback control systems with finite-gain $\mathcal{L}_2$ stability. IEEE Trans. Autom. Control **54**(3), 452–467 (2009)
107. Wan, H., Luan, X., Karimi, H.R., Liu, F.: Dynamic self-triggered controller codesign for Markov jump systems. IEEE Trans. Autom. Control **66**(3), 1353–1360 (2021)
108. Peng, C., Han, Q.L.: On designing a novel self-triggered sampling scheme for networked control systems with data losses and communication delays. IEEE Trans. Ind. Electron. **63**(2), 1239–1248 (2016)
109. Qi, Y., Zeng, P., Bao, W.: Event-triggered and self-triggered $\mathcal{H}_\infty$ control of uncertain switched linear systems. IEEE Trans. Syst.Man Cybern. Part A Syst. Hum. **50**(4), 1442–1454 (2020)
110. Borgers, D.P., Heemels, W.M.H.: Event-separation properties of event-triggered control systems. IEEE Trans. Autom. Control **59**(10), 2644–2656 (2014)
111. Donkers, M.C.F., Heemels, W.P.M.H.: Output-based event-triggered control with guaranteed $\mathcal{L}_\infty$-gain and improved and decentralized event-triggering. IEEE Trans. Autom. Control **57**(6), 1362–1376 (2012)
112. Dolk, V.S., Borgers, D.P., Heemels, W.P.M.H.: Output-based and decentralized dynamic event-triggered control with guaranteed $\mathcal{L}_p$- gain performance and zeno-freeness. IEEE Trans. Autom. Control **62**(1), 34–49 (2017)
113. Heemels, W.P.M.H., Donkers, M.C.F., Teel, A.R.: Periodic event-triggered control for linear systems. IEEE Trans. Autom. Control **58**(4), 847–861 (2013)
114. Yue, D., Tian, E., Han, Q.L.: A delay system method for designing event-triggered controllers of networked control systems. IEEE Trans. Autom. Control **58**(2), 475–481 (2012)
115. Cao, Z., Niu, Y., Lam, H.K., Zhao, J.: Sliding mode control of Markovian jump fuzzy systems: a dynamic event-triggered method. IEEE Trans. Fuzzy Syst. **29**(10), 2902–2915 (2021)
116. Fan, X., Li, T.: Fuzzy switching sliding mode control of T-S fuzzy systems via an event-triggered strategy. IEEE Trans. Fuzzy Syst. **32**(11), 6172–6184 (2024)
117. Fan, X., Wang, Z.: Event-triggered sliding-mode control for a class of T-S fuzzy systems. IEEE Trans. Fuzzy Syst. **28**(10), 2656–2664 (2020)
118. Fei, Z., Shi, S., Ahn, C.K., Basin, M.V.: Finite-time control for switched T-S fuzzy systems via a dynamic event-triggered mechanism. IEEE Trans. Fuzzy Syst. **29**(12), 3899–3909 (2020)
119. Li, Z.M., Chang, X.H., Park, J.H.: Quantized static output feedback fuzzy tracking control for discrete-time nonlinear networked systems with asynchronous event-triggered constraints. IEEE Trans. Syst. Man Cybern. Syst. **51**(6), 3820–3831 (2021)

120. Liu, H., Hu, J., Zhang, H., Zuo, Z.: Event-triggered sliding mode control for networked T-S fuzzy delayed systems against random injection attacks. J. Frankl. Inst. **361**(18), 107333 (2024)
121. Liu, J., Yin, T., Cao, J., Yue, D., Karimi, H.R.: Security control for T-S fuzzy systems with adaptive event-triggered mechanism and multiple cyber-attacks. IEEE Trans. Syst. Man Cybern. Syst. **51**(10), 6544–6554 (2021)
122. Peng, C., Han, Q.L., Yue, D.: To transmit or not to transmit: a discrete event-triggered communication scheme for networked Takagi-Sugeno fuzzy systems. IEEE Trans. Fuzzy Syst. **21**(1), 164–170 (2012)
123. Peng, C., Ma, S., Xie, X.: Observer-based non-PDC control for networked T-S fuzzy systems with an event-triggered communication. IEEE Trans. Cybern. **47**(8), 2279–2287 (2017)
124. Qi, W., Zhang, C., Zong, G., Su, S.F., Chadli, M.: Finite-time event-triggered stabilization for discrete-time fuzzy Markov jump singularly perturbed systems. IEEE Trans. Cybern. **53**(7), 4511–4520 (2023)
125. Shen, H., Li, F., Yan, H., Karimi, H.R., Lam, H.K.: Finite-time event-triggered $\mathcal{H}_\infty$ control for T-S fuzzy Markov jump systems. IEEE Trans. Fuzzy Syst. **26**(5), 3122–3135 (2018)
126. Tan, Y., Yuan, Y., Xie, X., Niu, B.: Dynamic event-triggered security control for networked T-S fuzzy system with non-uniform sampling. Fuzzy Sets Syst. **452**, 91–109 (2023)
127. Wang, Z., Fan, X., Shi, Z.: Periodic event-triggered integral sliding-mode control for T-S fuzzy systems. IEEE Trans. Cybern. **52**(8), 7669–7681 (2022)
128. Yan, S., Gu, Z., Park, J.H., Xie, X.: Adaptive memory-event-triggered static output control of T-S fuzzy wind turbine systems. IEEE Trans. Cybern. **30**(9), 3894–3904 (2022)
129. Yang, Y., Lam, J., Niu, Y., Fan, C.: Balancing fuzzy control performance and reachable-set-dependent event-triggered communication. IEEE Trans. Cybern. **54**(9), 5152–5163 (2024)
130. Hou, Q., Dong, J.: Finite-time membership function-dependent $\mathcal{H}_\infty$ control for T-S fuzzy systems via a dynamic memory event-triggered mechanism. IEEE Trans. Fuzzy Syst. **31**(11), 4075–4084 (2023)
131. Gu, Z., Yue, D., Tian, E.: On designing of an adaptive event-triggered communication scheme for nonlinear networked interconnected control systems. Inf. Sci. **422**, 257–270 (2018)
132. Gu, Z., Tian, E., Liu, J.: Adaptive event-triggered control of a class of nonlinear networked systems. J. Frankl. Inst. **354**(9), 3854–3871 (2017)
133. Wang, Y., Xia, Y., Ahn, C.K., Zhu, Y.: Exponential stabilization of Takagi-Sugeno fuzzy systems with aperiodic sampling: an aperiodic adaptive event-triggered method. IEEE Trans. Syst., Man, Cybern. Syst. **49**(2), 444–454 (2019)
134. Pan, Y., Yang, G.H.: A novel event-based fuzzy control approach for continuous-time fuzzy systems. Neurocomputing **338**, 55–62 (2019)
135. Li, Z., Yan, H., Zhang, H., Lam, H.K., Wang, M.: Aperiodic sampled-data-based control for interval type-2 fuzzy systems via refined adaptive event-triggered communication scheme. IEEE Trans. Fuzzy Syst. **29**(2), 310–321 (2021)
136. Shen, H., Liu, Y.A., Wang, J., Yan, H., Chadli, M.: Sliding-mode control for IT2 fuzzy nonlinear singularly perturbed systems and its application to electric circuits: a dynamic event-triggered mechanism. IEEE Trans. Syst. Man Cybern. Syst. **53**(7), 4077–4090 (2023)
137. Xing, J.W., Peng, C., Xie, X.: Switched fuzzy control for nonlinear systems via a fuzzy-rule-dependent adaptive event-triggered mechanism. IEEE Trans. Fuzzy Syst. **32**(3), 870–882 (2024)
138. Xing, J.W., Peng, C., Cao, Z., Xie, W.B.: Security-based control for networked interval type-2 fuzzy systems with multiple cyber-attacks: an improved dynamic event-triggered scheme. IEEE Trans. Fuzzy Syst. **31**(8), 2747–2760 (2023)
139. Zhang, Z., Su, S.F., Niu, Y.: Dynamic event-triggered control for interval type-2 fuzzy systems under fading channel. IEEE Trans. Cybern. **51**(11), 5342–5351 (2021)
140. Walsh, G., Ye, H., Bushnell, L.: Stability analysis of networked control systems. IEEE Trans. Control Syst. Technol. **10**(3), 438–446 (2002)
141. Tabbara, M., Nešić, D.: Input–output stability of networked control systems with stochastic protocols and channels. IEEE Trans. Autom. Control **53**(5), 1160–1175 (2008)

142. Donkers, M., Heemels, W., Van de Wouw, N., Hetel, L.: Stability analysis of networked control systems using a switched linear systems approach. IEEE Trans. Autom. Control **56**(9), 2101–2115 (2011)
143. Donkers, M., Heemels, W., Bernardini, D., Bemporad, A., Shneer, V.: Stability analysis of stochastic networked control systems. Automatica **48**(5), 917–925 (2012)
144. Liu, K., Fridman, E.: Discrete-time network-based control under scheduling and actuator constraints. Int. J. Robust Nonlinear Control **25**(12), 1816–1830 (2015)
145. Ding, D., Wang, Z., Han, Q.L., Wei, G.: Neural-network-based output-feedback control under round-robin scheduling protocols. IEEE Trans. Cybern. **49**(6), 2372–2384 (2018)
146. Liu, K., Fridman, E., Hetel, L.: Stability and L_2-gain analysis of networked control systems under round-robin scheduling: a time-delay approach. Syst. Control Lett. **61**(5), 666–675 (2012)
147. Liu, K., Fridman, E., Johansson, K.H., Xia, Y.: Quantized control under round-robin communication protocol. IEEE Trans. Ind. Electron. **63**(7), 4461–4471 (2016)
148. Ding, D., Wang, Z., Han, Q.L.: Neural-network-based output-feedback control with stochastic communication protocols. Automatica **106**, 221–229 (2019)
149. Liu, K., Fridman, E., Johansson, K.H.: Networked control with stochastic scheduling. IEEE Trans. Autom. Control **60**(11), 3071–3076 (2015)
150. Zhang, J., Fridman, E.: Dynamic event-triggered control of networked stochastic systems with scheduling protocols. IEEE Trans. Autom. Control **66**(12), 6139–6147 (2021)
151. Zou, L., Wang, Z., Gao, H.: Observer-based $\mathcal{H}_\infty$ control of networked systems with stochastic communication protocol: the finite-horizon case. Automatica **63**, 366–373 (2016)
152. Zou, L., Wang, Z., Han, Q.L., Zhou, D.: Recursive filtering for time-varying systems with random access protocol. IEEE Trans. Autom. Control **64**(2), 720–727 (2018)
153. Shen, Y., Wang, Z., Shen, B., Dong, H.: Outlier-resistant recursive filtering for multisensor multirate networked systems under weighted try-once-discard protocol. IEEE Trans. Cybern. **51**(10), 4897–4908 (2020)
154. Wang, J., Song, Y.: Resilient RMPC for cyber-physical systems with polytopic uncertainties and state saturation under TOD scheduling: an ADT approach. IEEE Trans. Ind. Inform. **16**(7), 4900–4908 (2019)
155. Zou, L., Wang, Z., Gao, H.: Set-membership filtering for time-varying systems with mixed time-delays under round-robin and weighted try-once-discard protocols. Automatica **74**, 341–348 (2016)
156. Zou, L., Wang, Z., Han, Q., Zhou, D.: Ultimate boundedness control for networked systems with try-once-discard protocol and uniform quantization effects. IEEE Trans. Autom. Control **62**(12), 6582–6588 (2017)
157. Wang, Y., Wang, Z., Zou, L., Dong, H.: $\mathcal{H}_\infty$ PID control for discrete-time fuzzy systems with infinite-distributed delays under round-robin communication protocol. IEEE Trans. Fuzzy Syst. **30**(6), 1875–1888 (2021)
158. Wu, B., Chang, X.H., Zhao, X.: Fuzzy $\mathcal{H}_\infty$ output feedback control for nonlinear ncss with quantization and stochastic communication protocol. IEEE Trans. Fuzzy Syst. **29**(9), 2623–2634 (2020)
159. Cheng, J., Yan, H., Park, J.H., Zong, G.: Output-feedback control for fuzzy singularly perturbed systems: a nonhomogeneous stochastic communication protocol approach. IEEE Trans. Cybern. **53**(1), 76–87 (2023)
160. Shen, H., Liu, X., Wang, J., Shi, K.: Non-fragile output feedback control for PDT-switched fuzzy systems under weighted try-once-discard protocol and its application. Inf. Sci. **626**, 390–406 (2023)
161. Wang, J., Yang, C., Xia, J., Wu, Z.G., Shen, H.: Observer-based sliding mode control for networked fuzzy singularly perturbed systems under weighted try-once-discard protocol. IEEE Trans. Fuzzy Syst. **30**(6), 1889–1899 (2022)
162. Zhang, Z., Niu, Y., Lam, H.K.: Sliding-mode control of T-S fuzzy systems under weighted try-once-discard protocol. IEEE Trans. Cybern. **50**(12), 4972–4982 (2020)

163. Liu, S., Cheng, J., Yan, H., Zhang, D.: Adaptive neural sliding-mode control for fuzzy singularly perturbed systems: Sojourn-probability-based stochastic communication protocol. IEEE Trans. Cybern. **54**(10), 6095–6104 (2024)
164. Wu, Y., Niu, Y., Yang, Y.: Sliding mode resilient control for T-S fuzzy systems under unreliable communication environment: round-robin scheduling strategy. Proc. Inst. Mech. Eng. Part I J. Syst. Control Eng. **239**(2), 260–271 (2025)
165. Wu, Y., Niu, Y., Yang, Y.: Security sliding mode control for T-S fuzzy systems under attack probability-dependent dynamic round-robin protocol. Int. J. Syst. Sci. **55**(5), 926–940 (2024)
166. Wu, Y., Niu, Y., Yang, Y.: Sliding mode control for T-S fuzzy systems under WTOD protocol and injection attacks. Int. J. Robust Nonlinear Control **34**(13), 9204–9221 (2024)
167. Dong, Y., Song, Y., Wei, G.: Efficient model-predictive control for networked interval type-2 T-S fuzzy system with stochastic communication protocol. IEEE Trans. Fuzzy Syst. **29**(2), 286–297 (2021)
168. Zhang, S., Yang, Y.: Non-fragile control for interval type-2 T-S fuzzy systems with mixed time-delays under round-robin protocol. Int. J. Control Autom. Syst. **20**(11), 3759–3770 (2022)
169. Liu, J., Ke, J., Liu, J., Xie, X., Tian, E.: Secure event-triggered control for IT-2 fuzzy networked systems with stochastic communication protocol and FDI attacks. IEEE Trans. Fuzzy Syst. **32**(3), 1167–1180 (2024)
170. Gao, X., Deng, F., Su, C.Y., Zeng, P.: Protocol-based fuzzy control of networked systems under joint deception attacks. IEEE Trans. Fuzzy Syst. **31**(3), 1052–1063 (2023)
171. Liu, J., Gong, E., Zha, L., Tian, E., Xie, X.: Observer-based security fuzzy control for nonlinear networked systems under weighted try-once-discard protocol. IEEE Trans. Fuzzy Syst. **31**(11), 3853–3865 (2023)
172. Dong, Y., Song, Y., Wei, G.: Efficient model-predictive control for nonlinear systems in interval type-2 T-S fuzzy form under round-robin protocol. IEEE Trans. Fuzzy Syst. **30**(1), 63–74 (2022)
173. Zhang, Z., Niu, Y., Karimi, H.R.: Sliding mode control of interval type-2 fuzzy systems under round-robin scheduling protocol. IEEE Trans. Syst. Man Cybern. Syst. **51**(12), 7602–7612 (2021)
174. Zhang, Z., Niu, Y., Cao, Z., Song, J.: Security sliding mode control of interval type-2 fuzzy systems subject to cyber attacks: the stochastic communication protocol case. IEEE Trans. Fuzzy Syst. **29**(2), 240–251 (2020)

Part I
Event-Triggered Protocols

Chapter 2
Dynamic Event-Triggered Sliding Mode Control Under Fading Channels

2.1 Introduction

As the pointed out in Chap. 1 that the event-triggered scheme provides an effective approach to save finite communication resources. Its main idea is to design event-triggering conditions for reducing signal transmission frequency and alleviating communication burden, and some event-triggered control schemes have been proposed for IT2 T-S fuzzy systems [1]. However, in practical wireless network channels, multipath transmission and shadow effects will cause signal fluctuations and potential channel fading, degrading the system performance [2–6]. In [7], dynamic event-triggered control problem was considered for IT2 T-S fuzzy systems under fading channels, and a state-feedback controller was designed to guarantee the stochastic stability of closed-loop systems. However, in some special situations, system states often cannot be directly measured. How to design the state observer of the IT2 fuzzy system and overcome the effect of event-triggered communication and channel fading phenomena will be a challenging problem.

In this chapter, the event-triggered SMC problem is investigated for continuous-time IT2 T-S fuzzy systems under resource-limited communication network, in which system state cannot be measured directly, meanwhile, multipath transmission and shadow effects may cause signal fluctuations and potential channel fading phenomenon. Firstly, to relieve the communication burden of resource-limited network channels, a dynamic event-triggered protocol is proposed in the S/C channels to regulate the transmission of the output signal. In addition, considering the channel fading phenomenon over communication network, a fuzzy state observer is designed to estimate the unmeasured states by only utilizing the available triggered system output signals suffering from the influence of channel fading. Based on the estimated state, an integral sliding function is constructed and an observer-based fuzzy SMC law is designed. Under the proposed event-triggered SMC scheme, the reachability of the sliding surface and the input-to-state practical stability in mean of the closed-

Y. Yang et al., *Analysis and Synthesis of Interval Type-2 Fuzzy Systems Under Communication Protocols*, Studies in Systems, Decision and Control 636,
https://doi.org/10.1007/978-981-95-2355-9_2

loop systems can be guaranteed, simultaneously. Besides, the existence of a positive minimum triggering interval is also proven to avoid Zeno phenomena.

2.2 Problem Description

2.2.1 Interval Type-2 T-S Fuzzy System

Consider the following continuous-time IT2 T-S fuzzy systems subject to external disturbances, whose i-th rule is described as

$Rulei$: IF $f_1(x(t))$ is $\mathscr{F}_1^i$ and … and $f_\hbar(x(t))$ is $\mathscr{F}_\hbar^i$, THEN

$$\begin{cases} \dot{x}(t) = A_i x(t) + B_i u(t) + D_i d(t), \\ y(t) = Cx(t), \end{cases} \tag{2.1}$$

where $f_\iota(x(t))$ is the premise variable, F_ι^i is the IT2 fuzzy set, $\iota \in \{1, \ldots, \hbar\}$, $i \in \{1, \ldots, r\}$, and $\hbar$ and r are the numbers of premise variables and fuzzy rules, respectively. $x(t) \in \mathbb{R}^n$, $u(t) \in \mathbb{R}^m$, and $y(t) \in \mathbb{R}^p$ represent the system state, control input, and measurable output. A_i, B_i, C, and D_i are known matrices with appropriate dimensions, $d(t) \in \mathbb{R}^q$ is the external disturbance with a known upper bound $\varpi \geq 0$ satisfying $||d(t)|| \leq \varpi$.

The firing strength of the i-th fuzzy rule satisfies the following interval set:

$$\overrightarrow{m}_i(x(t)) = [\underline{m}_i(x(t)), \overline{m}_i(x(t))],$$

with

$$\underline{m}_i(x(t)) = \prod_{\iota=1}^{\hbar} \underline{\mu}_{\mathscr{F}_\iota^i}(f_\iota(x(t))) > 0, \quad \overline{m}_i(x(t)) = \prod_{\iota=1}^{\hbar} \overline{\mu}_{\mathscr{F}_\iota^i}(f_\iota(x(t))) > 0,$$

where $\underline{\mu}_{\mathscr{F}_\iota^i}(f_\iota(x(t)))$ and $\overline{\mu}_{\mathscr{F}_\iota^i}(f_\iota(x(t))) \in [0, 1]$ are the lower and upper grade of the MFs and satisfy $\overline{\mu}_{\mathscr{F}_\iota^i}(f_\iota(x(t))) \geq \underline{\mu}_{\mathscr{F}_\iota^i}(f_\iota(x(t))) \geq 0$. Then, the fuzzy system (2.1) can be blended as

$$\begin{cases} \dot{x}(t) = \sum_{i=1}^{r} m_i(x(t))[A_i x(t) + B_i u(t) + D_i d(t)], \\ y(t) = Cx(t), \end{cases} \tag{2.2}$$

with

$$m_i(x(t)) = \underline{m}_i(x(t))\underline{\alpha}_i(x(t)) + \overline{m}_i(x(t))\overline{\alpha}_i(x(t)),$$

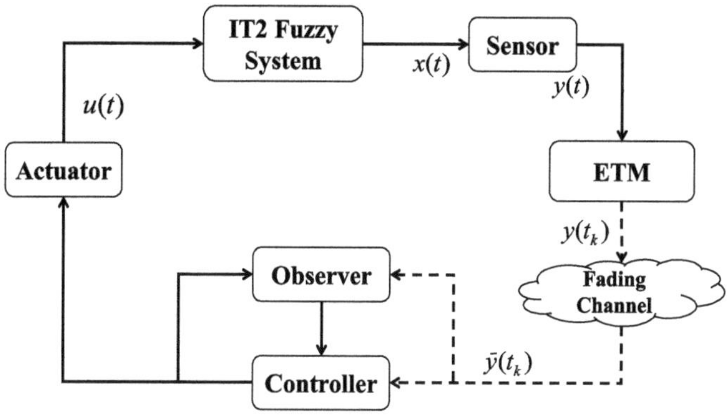

Fig. 2.1 IT2 fuzzy control systems under dynamic ETM and fading channels

$$\underline{\alpha}_i(x(t)), \overline{\alpha}_i(x(t)) \in [0, 1], \ \ \underline{\alpha}_i(x(t)) + \overline{\alpha}_i(x(t)) = 1, \ \sum_{i=1}^{r} m_i(x(t)) = 1,$$

where weighting coefficients $\underline{\alpha}_i(x(t))$ and $\overline{\alpha}_i(x(t))$ are nonlinear functions, which can capture the parameter uncertainties.

In the IT2 T-S fuzzy system (2.2), the membership function $m_i(x(t))$ $(i = 1, \ldots, r)$ is a time-varying function. Notably, only these upper and lower membership functions $\overline{m}_i(x(t))$ and $\underline{m}_i(x(t))$ are known for control design.

2.2.2 Dynamic Event-Triggered Protocol

To relieve communication burden and improve the utilization of limited communication resources, a novel dynamic ETM is introduced to determine the transmission process of the measurement output $y(t)$ (as depicted in Fig. 2.1). Define $e(t) \triangleq y(t_k) - y(t)$ as the difference between the previous triggered signal $y(t_k)$ and the current measurement signal $y(t)$. The next triggering instant t_{k+1} is decided by the following event-triggering condition:

$$t_{k+1} = \inf\left\{ t \mid t > t_k, \ e^T(t)Ne(t) \geq \frac{1}{\theta}\eta(t) + \sigma y^T(t)My(t) + \kappa \right\}, \tag{2.3}$$

with given parameters $\sigma \in (0, 1), \theta > 0, \kappa > 0$, matrices $M > 0$ and $N > 0$.

The dynamic variable $\eta(t)$ in (2.3) is updated as the following dynamical rule:

$$\dot{\eta}(t) = -\gamma\eta(t) + \sigma y^T(t)My(t) - e^T(t)Ne(t) + \kappa, \tag{2.4}$$

with a given scalar $\gamma > 0$ and initial condition $\eta(0) = \eta_0 \geq 0$.

For the proposed dynamic event-triggered protocol (2.3), the dynamic variable $\eta(t)$ can be ensured to be non-negative via the following lemma.

Lemma 2.1 *For given scalars $\theta > 0$, $\gamma > 0$, and $\eta(0) = \eta_0 > 0$, the dynamic variable $\eta(t)$ in dynamic rule (2.4) satisfies the following condition:*

$$\eta(t) \geq 0. \tag{2.5}$$

Proof For $\forall t \in [t_k, t_{k+1})$, one has

$$\frac{1}{\theta}\eta(t) + \sigma y^T(t)My(t) - e^T(t)Ne(t) + \kappa \geq 0, \tag{2.6}$$

which, together with expression (2.4), yields

$$\begin{aligned}\dot{\eta}(t) &= -\gamma\eta(t) + \sigma y^T(t)My(t) - e^T(t)Ne(t) + \kappa \\ &\geq -\gamma\eta(t) - \frac{1}{\theta}\eta(t) = -\left(\gamma + \frac{1}{\theta}\right)\eta(t).\end{aligned}$$

By means of Comparison Lemma in [10], it can be obtained

$$\eta(t) \geq \eta_0 e^{-(\gamma+\frac{1}{\theta})t}.$$

Thus, one has $\eta(t) \geq 0$ for all $t \geq 0$. □

2.2.3 Channel Fading

Under the above dynamic ETS in (2.3), the measurement signal $y(t)$ is sent only at the triggering instants, which can save the limited communication resources via reducing the transmission frequency. Considering multipath transmission and shadow effects in wireless channels that cause signal fluctuations, the released signal $y(t_k)$ may suffer from the channel fading phenomena, and the available output signal for controller and observer can be represented as

$$\bar{y}(t_k) = \mu(t_k)y(t_k), \tag{2.7}$$

where $\mu(t_k) \in (0, 1]$ is a random variable with mathematical expectation $\bar{\mu}$, reflecting the random fading phenomena.

In this chapter, the control objective is to propose a feasible fuzzy SMC scheme for ensuring the input-to-state practical stability in mean for IT2 fuzzy systems (2.2) in the presence of event-triggering communication and channel fading phenomena.

To perform the subsequent analysis, the following definitions are provided according to Refs. [8, 9].

Definition 2.1 The system $\dot{x}(t) = g(x(t), u(t))$ with $x(t) \in \mathbb{R}^n$ and $u(t) \in \mathbb{R}^l$ is said to be input-to-state stable in mean (ISSiM), if there exist functions $\gamma_1 \in \mathcal{KL}$, $\gamma_2 \in \mathcal{K}$ such that the solution $x(t)$ satisfies

$$\mathbf{E}[\|x(t)\|] \leq \gamma_1(\|x(0)\|, t) + \gamma_2(\|u\|_\infty), \quad \forall t \geq 0 \tag{2.8}$$

for any essentially bounded input $u(t) \in \mathcal{L}_\infty^l$ and any initial value $x(0) \in \mathbb{R}^n$.

Definition 2.2 The system $\dot{x}(t) = g(x(t), u(t))$ with $x(t) \in \mathbb{R}^n$ and $u(t) \in \mathbb{R}^l$ is said to be input-to-state practically stable in mean (ISpSiM), if there exist functions $\gamma_1 \in \mathcal{KL}$, $\gamma_2 \in \mathcal{K}$, and a scalar $\varrho \geq 0$ such that the solution $x(t)$ satisfies

$$\mathbf{E}[\|x(t)\|] \leq \gamma_1(\|x(0)\|, t) + \gamma_2(\|u\|_\infty) + \varrho, \quad \forall t \geq 0 \tag{2.9}$$

for any essentially bounded input $u(t) \in \mathcal{L}_\infty^l$ and any initial value $x(0) \in \mathbb{R}^n$.

Remark 2.1 Compared with Definition 2.1, the function $\gamma_2(\|u\|_\infty) + \varrho$ is more general to reflect the dynamical behavior of practical systems under event-triggering communication suffering from random fading phenomena. Notably, if $\varrho = 0$, Definition 2.2 will be reduced to Definition 2.1 as a special case. Thus, Definition 2.2 is more conducive to describing the dynamic performance of some practical control systems.

2.3 Fuzzy State Observer and Sliding Mode Controller

In order to construct an implementable fuzzy state observer, two aspects need to be considered: ① Since system state $x(t)$ of the IT2 fuzzy system (2.2) is unavailable, the premise variables in (2.1) cannot be utilized directly to construct the fuzzy observer. ② Due to the dynamic ETM (2.3) and the random channel fading phenomena (2.7), only the faded signal $\bar{y}(t_k)$ is available for the fuzzy observer. Thus, the following fuzzy observer is designed by utilizing the available faded output $\bar{y}(t_k)$ and the estimated states $\hat{x}(t)$ for $t \in [t_k, t_{k+1})$:

Rule i : IF $f_1(\hat{x}(t))$ is $\mathscr{F}_1^i$ and ... and $f_\hbar(\hat{x}(t))$ is $\mathscr{F}_\hbar^i$, THEN

$$\begin{cases} \dot{\hat{x}}(t) = A_i\hat{x}(t) + B_i u(t) + L_i(\bar{y}(t_k) - \hat{y}(t)), \\ \hat{y}(t) = C\hat{x}(t) \end{cases} \tag{2.10}$$

where $\hat{x}(t)$ is the estimation of state $x(t)$, and the observer gains L_i $(i = 1, \ldots, r)$ will be designed later.

Thus, the fuzzy state observer can be inferred as follows:

$$\begin{cases} \dot{\hat{x}}(t) = \sum_{i=1}^{r} h_i(\hat{x}(t))[A_i\hat{x}(t) + B_iu(t) + L_i(\bar{y}(t_k) - \hat{y}(t))], \\ \hat{y}(t) = C\hat{x}(t), \end{cases} \tag{2.11}$$

where $h_i(\hat{x}(t))$ is membership function and can be constructed as $h_i(\hat{x}(t)) = \underline{m}_i(\hat{x}(t))\underline{\beta}_i(\hat{x}(t)) + \overline{m}_i(\hat{x}(t))\overline{\beta}_i(\hat{x}(t))$ with the weighting coefficients $\underline{\beta}_i(\hat{x}(t)) \in [0, 1]$ and $\overline{\beta}_i(\hat{x}(t)) \in [0, 1]$ satisfying $\underline{\beta}_i(\hat{x}(t)) + \overline{\beta}_i(\hat{x}(t)) = 1$. It should be noted that, by means of estimated state $\hat{x}(t)$ and the known upper and lower MFs in fuzzy model, the membership function $h_i(\hat{x}(t))$ in fuzzy observer (2.11) is designed to replace the unavailable membership function $m_i(x(t))$.

Define the estimate error $\tilde{x}(t) \triangleq x(t) - \hat{x}(t)$, one can obtain the following error system via system (2.2) and state observer (2.11):

$$\begin{aligned} \dot{\tilde{x}}(t) = &\sum_{i=1}^{r} h_i(\hat{x}(t))[(A_i - L_iC)\tilde{x}(t) - \mu(t_k)L_ie(t) + (1 - \mu(t_k))L_iy(t)] \\ &+ \bar{w}(t) \end{aligned} \tag{2.12}$$

with

$$\bar{w}(t) = \sum_{i=1}^{r}[(m_i(x(t)) - h_i(\hat{x}(t)))(A_ix(t) + B_iu(t)) + m_i(x(t))D_id(t)].$$

It should be pointed out that $\bar{w}(t)$ can be taken as a residual error term satisfying $||\bar{w}(t)|| < +\infty$. In fact, due to the fuzzy modeling for nonlinear system over a compact set, the system state $x(t)$ is always bounded. Beside, the control signal $u(t)$ dependent on the system states is also bounded in practical situations. Thus, one has

$$||\bar{w}(t)|| \leq 2 \max_{1\leq i\leq r}\{\|A_i\|\|x(t)\| + \|B_i\|\|u(t)\| + \|D_i\|\varpi\} < +\infty.$$

In order to design sliding mode controller, a sliding function in the estimate space is constructed as follows:

$$s(t) = G\hat{x}(t) - G\int_0^t \sum_{i=1}^{r} h_i(\hat{x}(\tau))(A_i + B_iK)\hat{x}(\tau)d\tau \tag{2.13}$$

where control gain matrix $K \in \mathbb{R}^{m\times n}$ is designed later, $G \in \mathbb{R}^{m\times n}$ can be chosen to satisfy $GB_i > 0$ (or $GB_i < 0$) for $\forall i\{1, \ldots, r\}$.

By means of the estimated state $\hat{x}(t)$, an SMC law is designed as

$$u(t) = K\hat{x}(t) - \hat{G}^{-1}(h)\sum_{i=1}^{r} h_i(\hat{x}(t))GL_i(\bar{y}(t_k) - \hat{y}(t)) - \rho\text{sign}(\hat{G}^T(h)s(t)), \tag{2.14}$$

with $\hat{G}(h) \triangleq \sum_{j=1}^{r} h_j(\hat{x}(t))GB_j$ and a small scalar $\rho > 0$ to be chosen. It can be seen that the fuzzy SMC law (2.14) is only dependent on the state estimate $\hat{x}(t)$ and the faded triggered output $\bar{y}(t_k)$, which ensure the fuzzy controller to be implementable in practice.

2.4 Performance Analysis Under Event-Triggered Control Scheme

In this section, it shall be proved that the SMC law (2.14) can guarantee the reachability of the sliding surface $s(t) = 0$ and the input-to-state practical stability of the closed-loop system.

2.4.1 Reachability of Sliding Surface

Theorem 2.1 *For the IT2 T-S fuzzy systems (2.2) under the dynamic ETM (2.3) and fading channels (2.7), the proposed SMC law (2.14) with state observer (2.11) with sliding function (2.13) can ensure that the state trajectories in estimation space are driven onto the sliding surface $s(t) = 0$ in mean sense.*

Proof Choose the Lyapunov function as

$$V_s(t) = \frac{1}{2}s^T(t)s(t). \tag{2.15}$$

According to sliding function (2.13) and state observer (2.11), it can be obtained from (2.15)

$$\begin{aligned}\mathbf{E}[\dot{V}_s(t)] = \mathbf{E}\Big\{s^T(t)\sum_{i=1}^{r} h_i(\hat{x}(t))[GB_i(u(t) - K\hat{x}(t)) \\ + GL_i(\bar{y}(t_k) - C\hat{x}(t))]\Big\}. \end{aligned}\tag{2.16}$$

Substituting (2.14) into (2.16) yields

$$
\begin{aligned}
&\mathbf{E}\{\dot{V}_s(t)\} \\
&= \mathbf{E}\Big\{s^T(t)\sum_{i=1}^{r} h_i(\hat{x}(t))\big[GB_iK\hat{x}(t) \\
&\quad - GB_i\hat{G}^{-1}(h)\sum_{j=1}^{r} h_j(\hat{x}(t))GL_j(\bar{y}(t_k)-\hat{y}(t))\big] \\
&\quad + GL_i(\bar{y}(t_k)-\hat{y}(t)) - GB_iK\hat{x}(t) - \rho GB_i\mathrm{sign}(\hat{G}^T(h)s(t))\Big\} \\
&= -\rho s^T(t)\sum_{i=1}^{r} h_i(\hat{x}(t))GB_i\mathrm{sign}\Big(\big(\sum_{i=1}^{r} h_i(\hat{x}(t))GB_i\big)^T s(t)\Big) \\
&\le -\rho\|s^T(t)\sum_{i=1}^{r} h_i(\hat{x}(t))GB_i\|' \le 0.
\end{aligned}
\tag{2.17}
$$

Due to $GB_i > 0$ (or $GB_i < 0$), we have

$$
\mathbf{E}[\dot{V}_s(t)] \le -\rho\|s^T(t)\hat{G}(h)\| < 0, \text{ for } s(t) \neq 0, \tag{2.18}
$$

which implies that under the designed event-triggering SMC law (2.14), state trajectories in estimate space can arrive at the sliding surface $s(t) = 0$ in the mean sense. □

2.4.2 Input-to-State Practical Stability

In the following, the control performance of closed-loop system on the sliding surface $s(t) = 0$ shall be analyzed.

According to the state observer (2.11) and sliding function (2.13), it can be obtained that

$$
\dot{s}(t) = \sum_{i=1}^{r} h_i(\hat{x}(t))[GB_iu(t) - GB_iK\hat{x}(t) + GL_i(\bar{y}(t_k) - \hat{y}(t))]. \tag{2.19}
$$

Besides, by means of SMC theory, one can get the following equivalent control law from (2.19):

$$
u_{eq}(t) = K\hat{x}(t) - \left(\sum_{j=1}^{r} h_j(\hat{x}(t))GB_j\right)^{-1} \sum_{i=1}^{r} h_i(\hat{x}(t))GL_i(\bar{y}(t_k) - \hat{y}(t)). \tag{2.20}
$$

Substituting equivalent control law (2.20) into state observer (2.11) yields the sliding mode dynamics:

$$\begin{aligned}\dot{\hat{x}}(t) &= \sum_{i=1}^{r}\sum_{j=1}^{r} h_i(\hat{x}(t))h_j(\hat{x}(t))[(A_i + B_i K)\hat{x}(t) - B_i\hat{G}^{-1}(h) \\ &\quad \times GL_j(\bar{y}(t_k) - C\hat{x}(t)) + L_i(\bar{y}(t_k) - C\hat{x}(t))]. \end{aligned} \tag{2.21}$$

After the estimate state arrives at the sliding surface $s(t) = 0$ on the estimate space, the state estimate in (2.11) can be written as the sliding mode dynamics (2.21). Thus, the closed-loop systems consist of sliding mode dynamics (2.21) and estimate error (2.12). In the following theorem, sufficient conditions on the input-to-state practical stability for the closed-loop systems will be given.

Theorem 2.2 *For given positive parameters σ, γ, $\bar{\mu}$, θ, λ, ε, and matrices $M > 0$, $N > 0$, the estimation error system (2.12) and sliding mode dynamics (2.21) can be guaranteed to be ISpSiM, if there exist matrices $Q > 0$, $T > 0$, $\bar{U} > 0$, Y, and $\bar{L}_i$ satisfying the following conditions (for $i = 1, 2, \ldots, r$)*

$$\Gamma = \begin{bmatrix} \Gamma_{11} & \Gamma_{12} & \Gamma_{13} \\ * & \Gamma_{22} & 0 \\ * & * & \Gamma_{33} \end{bmatrix} < 0, \tag{2.22}$$

$$CQ = \bar{U}C, \tag{2.23}$$

with

$$\begin{aligned}
\Gamma_{11} &= \begin{bmatrix} He(A_i Q + B_i Y) & (1-\bar{\mu})(\bar{L}_i C)^T \\ * & He(A_i Q - \bar{\mu}\bar{L}_i C) \end{bmatrix}, \\
\Gamma_{12} &= \begin{bmatrix} 0 & 0 \\ -\bar{\mu}\bar{L}_i C C^T & 0 \end{bmatrix}, \quad \Gamma_{13} = \begin{bmatrix} QC^T & 0 \\ QC^T & I \end{bmatrix}, \\
\Gamma_{22} &= \begin{bmatrix} (1+\lambda)(\varepsilon^2 N^{-1} - 2\varepsilon C Q C^T) & 0 \\ * & (\frac{\lambda}{\theta} - \gamma) I \end{bmatrix}, \\
\Gamma_{33} &= \begin{bmatrix} -\frac{1}{(1+\lambda)\sigma} M^{-1} & 0 \\ * & -T \end{bmatrix}.
\end{aligned}$$

Meanwhile, control gain K and observer gains L_i can be calculated as $K = YQ^{-1}$ and $L_i = \bar{L}_i \bar{U}^{-1}$.

Proof Construct the following Lyapunov function:

$$V(\varsigma(t)) = \hat{x}^T(t)G^T PG\hat{x}(t) + \tilde{x}^T(t)G^T PG\tilde{x}(t) + \eta(t) \tag{2.24}$$

with $\varsigma(t) = [\hat{x}^T(t), \tilde{x}^T(t), \eta(t)]^T$.

Along with expressions (2.4), (2.12) and (2.21), it can be obtained from (2.24) that

$$\begin{aligned}
&\mathbf{E}\{\dot{V}(\varsigma(t))\} \\
&= \mathbf{E}\left\{2\hat{x}^T(t)G^T PG\dot{\hat{x}}(t) + 2\tilde{x}^T(t)G^T PG\dot{\tilde{x}}(t) + \dot{\eta}(t)\right\} \\
&= \sum_{i=1}^{r} h_i(\hat{x}(t))\Big\{2\hat{x}^T(t)G^T PG(A_i + B_i K)\hat{x}(t) + 2\tilde{x}^T(t)G^T PG(A_i - L_i C)\tilde{x}(t) \\
&\quad - 2\bar{\mu}\tilde{x}^T(t)G^T PGL_i e(t) + 2(1-\bar{\mu})\tilde{x}(t)G^T PGL_i C(\tilde{x}(t) + \hat{x}(t))\Big\} \\
&\quad + 2\tilde{x}^T(t)G^T PG\bar{w}(t) - \gamma\eta(t) + \sigma y^T(t)My(t) - e^T(t)Ne(t) + \kappa. \qquad (2.25)
\end{aligned}$$

According to the triggering condition in (2.3), the following expression holds for $t \in [t_k, t_{k+1})$:

$$\frac{1}{\theta}\eta(t) + \sigma y^T(t)My(t) - e^T(t)Ne(t) + \kappa \geq 0. \qquad (2.26)$$

Thus, for any positive scalar λ, one can deduce from (2.25) and (2.26)

$$\begin{aligned}
&\mathbf{E}\{\dot{V}(\varsigma(t))\} \\
&\leq \sum_{i=1}^{r} h_i(\hat{x}(t))\Big\{2\hat{x}^T(t)G^T PG(A_i + B_i K)\hat{x}(t) \\
&+ 2\tilde{x}^T(t)G^T PG(A_i - L_i C)\tilde{x}(t) \\
&- 2\bar{\mu}\tilde{x}^T(t)G^T PGL_i e(t) + 2(1-\bar{\mu})\tilde{x}(t)G^T PGL_i C(\tilde{x}(t) + \hat{x}(t)) \\
&+ \quad 2\tilde{x}^T(t)G^T PG\bar{w}(t)\Big\} + \left(\frac{\lambda}{\theta} - \gamma\right)\eta(t) + (1+\lambda)\sigma y^T(t)My(t) \\
&- (1+\lambda)e^T(t)Ne(t) + (1+\lambda)\kappa. \qquad (2.27)
\end{aligned}$$

According to the measurement output $y(t)$ in IT2 T-S fuzzy systems (2.2), one has

$$(1+\lambda)\sigma y^T(t)My(t) = (1+\lambda)\sigma\zeta^T(t)\bar{C}^T M\bar{C}\zeta(t) \qquad (2.28)$$

with $\zeta(t) \triangleq [\hat{x}^T(t), \tilde{x}^T(t)]^T$ and $\bar{C} \triangleq [C, C]$.

Besides, for any matrix $T > 0$, the following inequality holds

$$2\tilde{x}^T(t)G^T PG\bar{w}(t) \leq \tilde{x}^T(t)(G^T PG)T^{-1}(G^T PG)^T\tilde{x}(t) + \lambda_{\max}(T)||\bar{w}(t)||^2. \qquad (2.29)$$

Substituting (2.29) into (2.27) yields

$$
\begin{aligned}
&\mathbf{E}[\dot{V}(\varsigma(t))] \\
&\leq \sum_{i=1}^{r} h_i(\hat{x}(t))\{\hat{x}^T(t)[2G^T PG(A_i + B_i K) + (1+\lambda)\sigma C^T MC]\hat{x}(t) \\
&+ \tilde{x}^T(t)[2G^T PG(A_i - \bar{\mu}L_i C) + G^T PGT^{-1}(G^T PG)^T \\
&+ (1+\lambda)\sigma C^T MC]\tilde{x}(t) \\
&- 2\tilde{x}^T(t)G^T PG\bar{\mu}L_i e(t) + 2\tilde{x}(t)(1-\bar{\mu})G^T PGL_i C\hat{x}(t)\} \\
&+ 2(1+\lambda)\sigma\hat{x}^T(t)C^T MC\tilde{x}(t) + (\frac{\lambda}{\theta} - \gamma)\eta(t) - (1+\lambda)e^T(t)Ne(t) \\
&+ \lambda_{\max}(T)||\bar{w}(t)||^2 + (1+\lambda)\kappa \\
&\leq \sum_{i=1}^{r} h_i(\hat{x}(t))\bar{x}^T(t)\Pi\bar{x}(t) + \lambda_{\max}(T)||\bar{w}(t)||^2 + (1+\lambda)\kappa,
\end{aligned} \tag{2.30}
$$

with $\bar{x}(t) \triangleq [\hat{x}^T(t), \tilde{x}^T(t), e^T(t), \sqrt{\eta(t)}]^T$, and

$$
\Pi \triangleq \begin{bmatrix} \Pi_{11} & \Pi_{12} & 0 & 0 \\ * & \Pi_{22} & \Pi_{23} & 0 \\ * & * & \Pi_{33} & 0 \\ * & * & * & \Pi_{44} \end{bmatrix}
$$

$$
\begin{aligned}
&\Pi_{11} \triangleq \mathrm{He}(G^T PG(A_i + B_i K)) + (1+\lambda)\sigma C^T MC, \\
&\Pi_{12} \triangleq (1+\lambda)\sigma C^T MC + (1-\bar{\mu})(G^T PGL_i C)^T, \\
&\Pi_{22} \triangleq \mathrm{He}(G^T PG(A_i - \bar{\mu}L_i C)) + (G^T PG)T^{-1}(G^T PG)^T + (1+\lambda)\sigma C^T MC, \\
&\Pi_{23} \triangleq -\bar{\mu}G^T PGL_i,\ \Pi_{33} \triangleq -(1+\lambda)N,\ \Pi_{44} \triangleq (\frac{\lambda}{\theta} - \gamma)I.
\end{aligned}
$$

By left and right multiplying $\mathrm{diag}\{Q^{-1}, Q^{-1}, (CQC^T)^{-1}, I, I, I\}$ to the inequality (2.22) and letting $Q^{-1} \triangleq G^T PG$, $\bar{L}_i \triangleq L_i\bar{U}$, $Y \triangleq KQ$, one has

$$
\begin{bmatrix} \Theta_{11} & \Theta_{12} & 0 & 0 & C^T & 0 \\ * & \Theta_{22} & \Theta_{23} & 0 & C^T & G^T PG \\ * & * & -(1+\lambda)N & 0 & 0 & 0 \\ * & * & * & (\frac{\lambda}{\theta} - \gamma)I & 0 & 0 \\ * & * & * & * & -\frac{1}{(1+\lambda)\sigma M} & 0 \\ * & * & * & * & * & -T \end{bmatrix} < 0 \tag{2.31}
$$

with $\Theta_{11} \triangleq \mathrm{He}(G^T PG(A_i + B_i K))$, $\Theta_{12} \triangleq (1-\bar{\mu})(G^T PGL_i C)^T$, $\Theta_{22} \triangleq \mathrm{He}(G^T PG(A_i - \bar{\mu}L_i C))$, and $\Theta_{23} \triangleq -\bar{\mu}G^T PGL_i$. By adopting Schur's complement, the inequality (2.31) can ensure inequality $\Pi < 0$.

Furthermore, it can be obtained from (2.30) that

$$\begin{aligned}\mathbf{E}[\dot{V}(\varsigma(t))] &\leq -\min_{1\leq i\leq r}\lambda_{\min}(-\Pi)||\bar{x}(t)||^2 + \lambda_{\max}(T)||\bar{w}(t)||^2 + (1+\lambda)\kappa \\ &\leq -\alpha V(\varsigma(t)) + \beta||\bar{w}(t)||^2 + \bar{\kappa}, \end{aligned} \tag{2.32}$$

with $\alpha \triangleq \min_{1\leq i\leq r}\lambda_{\min}(-\Pi)/\lambda_{\max}(\mathrm{diag}\{G^T PG, G^T PG, I\})$, $\beta \triangleq \lambda_{\max}(T)$, and $\bar{\kappa} \triangleq (1+\lambda)\kappa$.

By using the Comparison Lemma in [10], one can obtain

$$\begin{aligned}\mathbf{E}[V(t)] &\leq e^{-\alpha t}V(0) + \int_0^t e^{\alpha(s-t)}(\beta||\bar{w}_t||_\infty^2 + \bar{\kappa})ds \\ &\leq e^{-\alpha t}V(0) + \frac{\beta}{\alpha}||\bar{w}_t||_\infty^2 + \frac{\bar{\kappa}}{\alpha} \\ &\triangleq \gamma_1(V(0), t) + \gamma_2(||\bar{w}_t||_\infty) + \varrho, \end{aligned} \tag{2.33}$$

with $\varrho = \bar{\kappa}/\alpha$, from which it is easily shown that $\gamma_1(\cdot,\cdot) \in \mathcal{KL}$ and $\gamma_2(\cdot) \in \mathcal{K}$. Thus, according to Definition 2.2 and the expression (2.15), it is obtained from (2.30) that the closed-loop systems consisting of estimation error system (2.12) and sliding mode dynamics (2.21) is ISpSiM. □

Here, it should be pointed out that Theorem 2.2 contains an equality constraint (2.23), which is difficulty solved. To circumvent this obstacle, by means of the similar approach in [11], equality constraint (2.23) can be converted into the following matrix inequality:

$$(CQ - \bar{U}C)^T(CQ - \bar{U}C) - \xi I < 0, \tag{2.34}$$

with the variable parameter $\xi > 0$. Apparently, if $\xi = 0$, the inequality (2.34) will imply the equality (2.23), that is, the smaller the parameter ξ, the closer the inequality (2.34) is to the equality (2.23). The inequality (2.34) can be transformed into the following LMI by Schur's complement:

$$\begin{bmatrix} -\xi I & (CQ - \bar{U}C)^T \\ CQ - \bar{U}C & -I \end{bmatrix} < 0. \tag{2.35}$$

Thus, solving the condition (2.22) with equality constraint (2.23) in Theorem 2.2 may be transformed into the minimization problem of parameter ξ:

$$\begin{aligned} &\min_{\mathcal{S}} \quad \xi, \\ &\text{s.t. LMIs (2.22) and (2.35)} \end{aligned} \tag{2.36}$$

with $\mathcal{S} = \{Q,\ T,\ \bar{U},\ Y,\ \bar{L}_i\}$.

Compared with the conventional time-driven case, the ETM is employed to regulate the transmission of measurement signal, effectively reducing the data transmission rate. For the event-triggered control framework, it is critical to avoid the Zeno phenomenon for continuous-time control systems. In subsequent analysis, it will be shown that the inter-event intervals are strictly positive.

2.4.3 Zeno-Free Phenomenon Analysis

Theorem 2.3 *For the continuous-time IT2 T-S fuzzy system (2.1), the dynamic event-triggering condition (2.3) can guarantee that the triggering interval* $\Delta T_k = t_{k+1} - t_k$, $(k \in \mathbb{N})$ *satisfies*

$$\Delta T_k \geq \frac{1}{\phi} \ln \left[1 + \frac{\phi(\eta(t_{k+1}) + \theta \sigma y^T(t_{k+1}) M y(t_{k+1}) + \theta \kappa)}{\varphi \theta \lambda_{\max}(N)} \right] \triangleq T^*, \tag{2.37}$$

where

$$\begin{aligned}
\phi \triangleq & \|C\bar{A}\| + \|C\bar{B}K + C\bar{B}\hat{G}^{-1}(h)G\hat{L}C\| + \|C\bar{B}\hat{G}^{-1}(h)G\hat{L}\| + \rho\|C\bar{B}\| + \|C\bar{D}\|, \\
\varphi \triangleq & \|C\bar{A}\| \|x(t)\|^2 + \|C\bar{B}K + C\bar{B}\hat{G}^{-1}(h)G\hat{L}C\| \|\hat{x}(t)\|^2 \\
& + \|C\bar{B}\hat{G}^{-1}(h)G\hat{L}\| \|\bar{y}(t_k)\|^2 + \rho m\|C\bar{B}\| + \|C\bar{D}\|\varpi^2, \\
\bar{A} \triangleq & \sum_{i=1}^{r} m_i(x(t))A_i, \quad \bar{B} \triangleq \sum_{i=1}^{r} m_i(x(t))B_i, \\
\bar{D} \triangleq & \sum_{i=1}^{r} m_i(x(t))D_i, \quad \hat{L} \triangleq \sum_{i=1}^{r} h_i(\hat{x}(t))L_i.
\end{aligned}$$

That is, there exist a positive scalar T^* *to avoid the Zeno phenomenon.*

Proof For $\forall t \in [t_k,\ t_{k+1})$, one has

$$\begin{aligned}
\dot{e}(t) &= -\sum_{i=1}^{r} h_i(x(t))[CA_i x(t) + CB_i u(t) + CD_i d(t)] \\
&= -\sum_{i=1}^{r} h_i(x(t))[CA_i x(t) + CB_i K\hat{x}(t) - CB_i\hat{G}^{-1}(h)G\hat{L}(\bar{y}(t_k) - \hat{y}(t)) \\
&\quad -\rho CB_i \text{sign}(\hat{G}^T(h)s(t)) + CD_i d(t)] \\
&= -C\bar{A}x(t) - (C\bar{B}K + C\bar{B}\hat{G}^{-1}(m)G\hat{L}C)\hat{x}(t) \\
&\quad +C\bar{B}\hat{G}^{-1}(h)G\hat{L}\bar{y}(t_k) + \rho C\bar{B}\text{sign}(\hat{G}^T(h)s(t)) - C\bar{D}d(t).
\end{aligned} \tag{2.38}$$

Besides, it follows from (2.38)

$$
\begin{aligned}
&\frac{\mathrm{d}}{\mathrm{dt}}\|e(t)\|^2 \\
&\leq 2\|e(t)\|\Big\| - C\bar{A}x(t) - \big(C\bar{B}K + C\bar{B}\hat{G}^{-1}(h)G\hat{L}C\big)\hat{x}(t) \\
&+ C\bar{B}\hat{G}^{-1}(h)G\hat{L}\bar{y}(t_k) + \rho C\bar{B}\mathrm{sign}\big(\hat{G}^T(h)s(t)\big) - C\bar{D}d(t)\Big\| \\
&\leq \big[\|C\bar{A}\| + \|C\bar{B}K + C\bar{B}\hat{G}^{-1}(h)G\hat{L}C\| + \|C\bar{B}\hat{G}^{-1}(h)G\hat{L}\| + \rho||C\bar{B}\| \\
&+ \|C\bar{D}\|\big]\|e(t)\|^2 + \|C\bar{A}\|\|x(t)||^2 + \|C\bar{B}K + C\bar{B}\hat{G}^{-1}(h)G\hat{L}C\|\|\hat{x}(t)\|^2 \\
&+ \|C\bar{B}\hat{G}^{-1}(h)G\hat{L}\|\|\bar{y}(t_k)||^2 + \rho m\|C\bar{B}\| + \|C\bar{D}\|\varpi^2 \\
&\triangleq \phi\|e(t)\|^2 + \varphi.
\end{aligned}
\tag{2.39}
$$

By means of Comparison Lemma in [10] and $e(t_k) = 0$ at the triggering instant, one can obtain from (2.39)

$$||e(t_{k+1})||^2 \leq \frac{\varphi}{\phi}(e^{\phi(t_{k+1}-t_k)} - 1). \tag{2.40}$$

According to the dynamic event-triggering condition (2.3), one has

$$e^T(t_{k+1})Ne(t_{k+1}) \geq \frac{1}{\theta}\eta(t_{k+1}) + \sigma y^T(t_{k+1})My(t_{k+1}), \tag{2.41}$$

thus, it can be obtained from (2.40) and (2.41) that

$$\frac{\eta(t_{k+1}) + \theta\sigma y^T(t_{k+1})My(t_{k+1}) + \theta\kappa}{\theta\lambda_{\max}(N)} \leq ||e(t_{k+1})||^2, \tag{2.42}$$

from (2.40) and (2.42), the triggering interval can be given as

$$t_{k+1} - t_k \geq \frac{1}{\phi}\ln\Big[1 + \frac{\phi(\eta(t_{k+1}) + \theta\sigma y^T(t_{k+1})My(t_{k+1}) + \theta\kappa)}{\varphi\theta\lambda_{\max}(N)}\Big] \triangleq T^*. \tag{2.43}$$

Since the parameters θ and κ are positive constants, which implies that the term $\phi(\eta(t_{k+1}) + \theta\sigma y^T(t_{k+1})My(t_{k+1}) + \theta\kappa)$ is strictly greater than zero. Thus, it can be concluded from (2.43) that a lower bound $T^*(> 0)$ can be found to avoid the Zeno phenomenon. □

2.5 Simulation Example

Consider a 3-rule continuous-time IT-2 T-S fuzzy system with the following parameters:

$$A_1 = \begin{bmatrix} -1.5 & 1 \\ 1.8 & -1 \end{bmatrix}, A_2 = \begin{bmatrix} -2.2 & 1.5 \\ 1.5 & -1 \end{bmatrix}, A_3 = \begin{bmatrix} -1.5 & 2 \\ 1.2 & -1 \end{bmatrix},$$
$$B_1 = \begin{bmatrix} 1 \\ 0.5 \end{bmatrix}, B_2 = \begin{bmatrix} 0 \\ 0.5 \end{bmatrix}, B_3 = \begin{bmatrix} 1 \\ 0.8 \end{bmatrix},$$
$$D_1 = \begin{bmatrix} 0.1 \\ 0.3 \end{bmatrix}, D_2 = \begin{bmatrix} 0.2 \\ 0.1 \end{bmatrix}, D_3 = \begin{bmatrix} 0.1 \\ 0.2 \end{bmatrix}.$$
$$C = \begin{bmatrix} 0.2 & 0.5 \end{bmatrix}, d(t) = 0.4\sin(0.5t)$$

and the upper and lower MFs:

$$\underline{m}_1(x(t)) = 0.95\left(1 - \frac{1}{1 + e^{-(x_1(t)+9/2)/8}}\right),$$
$$\overline{m}_1(x(t)) = 0.95\left(1 - \frac{1}{1 + e^{-(x_1(t)+7/2)/8}}\right),$$
$$\underline{m}_3(x(t)) = \frac{0.925}{1 + e^{-(x_1(t)-9/2)/8}}, \quad \overline{m}_3(x(t)) = \frac{0.925}{1 + e^{-(x_1(t)-7/2)/8}},$$
$$\underline{m}_2(x(t)) = 1 - \overline{m}_1(x(t)) - \overline{m}_3(x(t)),\ \overline{m}_2(x(t)) = 1 - \underline{m}_1(x(t)) - \underline{m}_3(x(t)).$$

The state observer (2.11) is designed with MFs $h_i(\hat{x}(t)) = 0.4\underline{m}_i(\hat{x}(t)) + 0.6\overline{m}_i(\hat{x}(t))$ $(i = 1, 3)$, and $h_2(\hat{x}(t)) = 1 - h_1(\hat{x}(t)) - h_3(\hat{x}(t))$. Choosing the sliding matrix $G = [1, 1]$ and the dynamic event-triggering condition (2.3) and dynamic rule (2.4) with parameters $\sigma = 0.4$, $M = 1$, $N = 20$, $\theta = 0.8$, $\gamma = 0.7$ and $\kappa = 0.001$. Firstly, for the fading channels with difference fading gains $\bar{\mu} = 0.8$, $\bar{\mu} = 0.5$, and $\bar{\mu} = 0.2$, by solving the minimization problem (2.36), the control matrix K, the observer matrices L_1 and L_2 are obtained in Table 2.1.

For executing the simulation, the initial conditions are set as $x(0) = [-1.5, 1.5]^T$, $\hat{x}(0) = [0.5, -1]^T$ and $\eta(0) = 0.2$, and the weighting coefficients of system's upper and lower MFs are chosen as $\underline{\alpha}_1(x(t)) = \underline{\alpha}_3(x(t)) = \sin^2(x_1(t))$ and $\overline{\alpha}_1(x(t)) = \overline{\alpha}_3(x(t)) = 1 - \sin^2(x_1(t))$ to obtain fuzzy system's MFs as $m_i(x(t)) = \underline{\alpha}_i(x(t))\underline{m}_i(x(t)) + \overline{\alpha}_i(x(t))\overline{m}_i(x(t))$ $(i = 1, 3)$, and $m_2(x(t)) = 1 - m_1(x(t)) - m_3(x(t))$.

For the above three cases with different fading channels, the simulation results are given in Figs. 2.2, 2.3 and 2.4, and the triggering numbers are shown in Table 2.1, from which it can be seen that the smaller $\bar{\mu}$, i.e., the worse channel, will cause the slower convergence and the more event triggering. Actually, a smaller parameter $\bar{\mu}$ means that the performance of the network transmission channel is worse, which implies that the transmitted signals are more susceptible to channel fading. Meanwhile, due to the influence of channel fading, the controller cannot obtain accurate signals. Thus,

Table 2.1 The proposed control strategy under different fading channels

Fading channel	K	L_i	Triggering number
$\bar{\mu} = 0.8$	$K =$ $[\,-35.7700\ -59.3604\,]$	$L_1 =$ $[\,1.4395\ 3.5983\,]^T$ $L_2 =$ $[\,1.4537\ 3.6345\,]^T$ $L_3 =$ $[\,1.4452\ 3.6308\,]^T$	19
$\bar{\mu} = 0.5$	$K =$ $[\,-33.6829\ -54.1059\,]$	$L_1 =$ $[\,1.6780\ 4.2026\,]^T$ $L_2 =$ $[\,1.7953\ 4.4900\,]^T$ $L_3 =$ $[\,1.7701\ 4.4230\,]^T$	23
$\bar{\mu} = 0.2$	$K =$ $[\,-11.7308\ -18.8441\,]$	$L_1 =$ $[\,0.0606\ 0.2246\,]^T$ $L_2 =$ $[\,0.0969\ 0.2422\,]^T$ $L_3 =$ $[\,0.1153\ 0.2923\,]^T$	27

as the network deteriorates from $\bar{\mu} = 0.8$ to $\bar{\mu} = 0.2$, more event triggering will be needed to compensate for the loss of performance. This has also been verified in view of the number of event triggering in the simulation results and Table 2.1.

2.6 Conclusion

In this chapter, the dynamic event-triggered SMC problem of IT2 T-S fuzzy systems suffering from the unmeasurable state and the fading channel has been addressed. A dynamic ETM has been utilized in the fading S/C channels to determine the transmission of measurement output, and a fuzzy state observer based on faded triggered output has been designed to estimate unmeasured state. On the state estimate space, an integral sliding function has been constructed, and then, an observer-based sliding mode controller can be designed to ensure the closed-loop systems to be ISpSiM. Besides, the Zeno behavior can be effectively avoided under the proposed event-triggered SMC scheme.

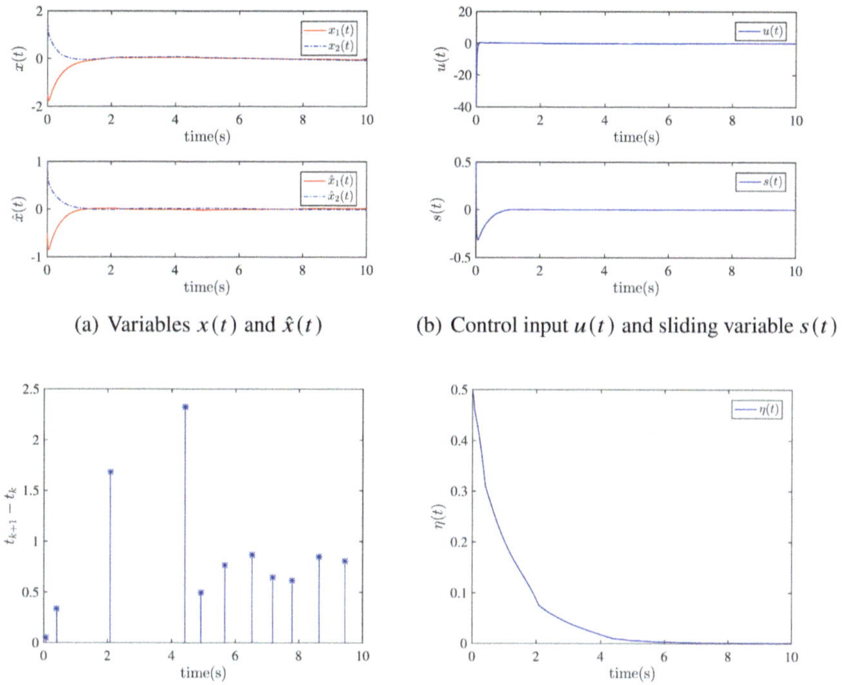

(a) Variables $x(t)$ and $\hat{x}(t)$

(b) Control input $u(t)$ and sliding variable $s(t)$

(c) Triggering interval $t_{k+1} - t_k$

(d) Dynamic variable $\eta(t)$

Fig. 2.2 Simulation results under fading channels with $\bar{\mu} = 0.8$

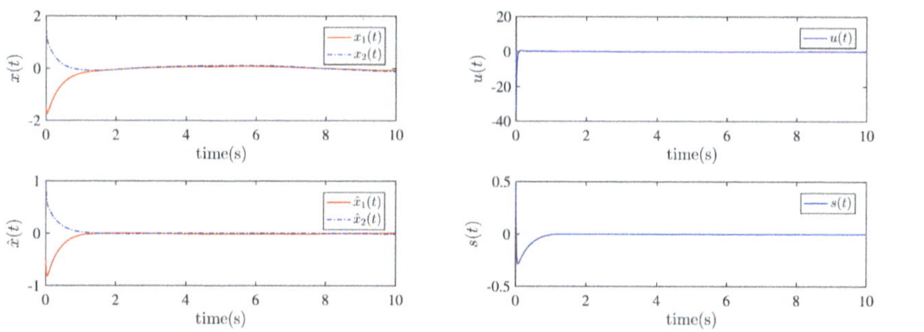

(a) Variables $x(t)$ and $\hat{x}(t)$

(b) Control input $u(t)$ and sliding variable $s(t)$

Fig. 2.3 Simulation results under fading channels with $\bar{\mu} = 0.5$

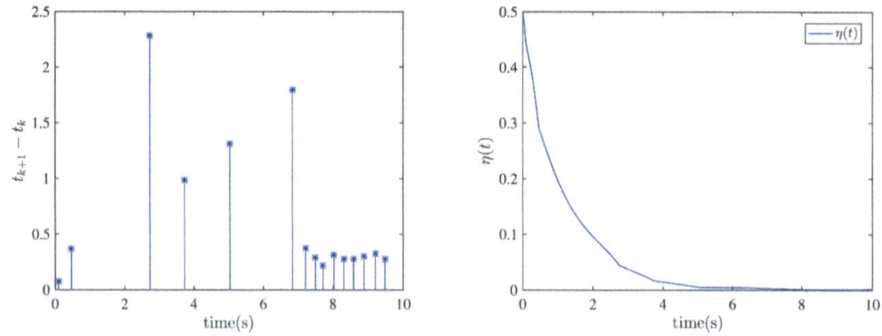

(c) Triggering interval $t_{k+1} - t_k$

(d) Dynamic variable $\eta(t)$

Fig. 2.3 (continued)

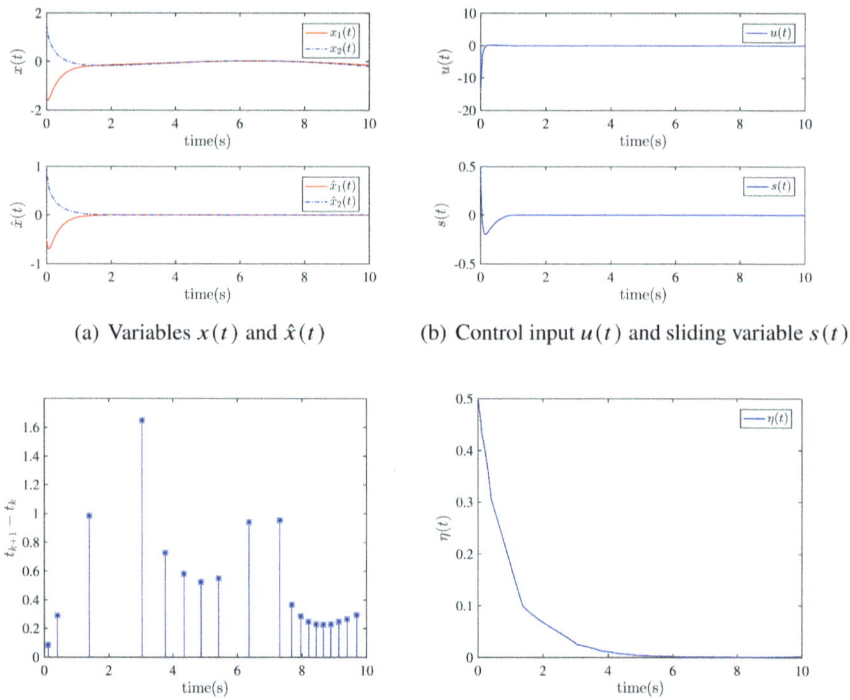

(a) Variables $x(t)$ and $\hat{x}(t)$

(b) Control input $u(t)$ and sliding variable $s(t)$

(c) Triggering interval $t_{k+1} - t_k$

(d) Dynamic variable $\eta(t)$

Fig. 2.4 Simulation results under fading channels with $\bar{\mu} = 0.2$

References

1. Pan, Y., Yang, G.H.: A novel event-based fuzzy control approach for continuous-time fuzzy systems. Neurocomputing **338**, 55–62 (2019)
2. Elia, N.: Remote stabilization over fading channels. Syst. Control Lett. **54**(3), 237–249 (2005)
3. Xiao, N., Niu, Y., Xie, L.: State feedback stabilization over finite-state fading channels. Asian J. Control **18**(3), 1052–1061 (2016)
4. Ding, D., Wang, Z., Lam, J., et al.: Finite-horizon H_∞ control for discrete time-varying systems with randomly occurring nonlinearities and fading measurements. IEEE Trans. Autom. Control **60**(9), 2488–2493 (2014)
5. Song, J., Niu, Y., Lam, J., et al.: Fuzzy remote tracking control for randomly varying local nonlinear models under fading and missing measurements. IEEE Trans. Fuzzy Syst. **26**(3), 1125–1137 (2017)
6. Li, J., Niu, Y.: Sliding mode control subject to rice channel fading. IET Control Theory Appl. **13**(16), 2529–2537 (2019)
7. Zhang, Z., Su, S.F., Niu, Y.: Dynamic event-triggered control for interval type-2 fuzzy systems under fading channel. IEEE Trans. Cybern. **51**(11), 5342–5351 (2021)
8. Zhao, P., Kang, Y., Zhai, D.: On input-to-state stability of stochastic nonlinear systems with Markovian jumping parameters. Int. J. Control **85**(4), 343–349 (2012)
9. Huang, L., Mao, X.: On input-to-state stability of stochastic retarded systems with Markovian switching. IEEE Trans. Autom. Control **54**(8), 1898–1902 (2009)
10. Khalil, H.K.: Nonlinear Systems, 3rd edn. Prentice Hall (2002)
11. Niu, Y., Ho, D., Lam, J.: Robust integral sliding mode control for uncertain stochastic systems with time-varying delay. Automatica **41**(5), 873–880 (2005)

Chapter 3
Component-Based Dynamic Event-Triggered Sliding Mode Control

3.1 Introduction

In Chap. 2, a dynamic ETS based on a common triggering condition is utilized to determine the transmission of measurement output, where all components of measurement output are simultaneously transmitted when the triggering condition is satisfied. This means that the ETM requires to collect signals from all sensors and then these signals are judged and transmitted simultaneously. In practice, sensor nodes may be distributed in different geographical locations, which makes it difficult for the common event triggering mechanism to obtain the data of all sensor nodes at the same time. On the other hand, the data of different sensor nodes have different physical meanings, the design of independent triggering threshold for each sensor node is more in line with the actual situation. Besides, by designing triggering threshold for each sensor node independently, unnecessary transmission of some node signals can be avoided.

Based on the above discussion, this chapter will study component-based event-triggered SMC scheme for a class of discrete-time IT2 fuzzy singularly perturbed systems, where the system involves two time scales, i.e., fast and slow states. To address the communication constraints, independent triggering conditions are designed for each state component to determine the transmission of each sensor node, by which the *component-based dynamic ETS* is proposed. At the controller side, the triggering sequence of all sensor nodes is mapped into a triggered state updating sequence, and a fuzzy sliding mode controller is designed by using the triggered state signal. By analyzing the relationship between the MFs in fuzzy model and controller, sufficient conditions based on MFD method are obtained, which ensures the asymptotic stability of the closed-loop systems and the reachability of sliding domain. Meanwhile, an optimization algorithm based on minimizing sliding domain is presented to improve the control performance.

Y. Yang et al., *Analysis and Synthesis of Interval Type-2 Fuzzy Systems Under Communication Protocols*, Studies in Systems, Decision and Control 636,
https://doi.org/10.1007/978-981-95-2355-9_3

3.2 Problem Description

3.2.1 System Model

Consider the discrete-time IT2 fuzzy singularly perturbed systems with the following rule:

Rule i: IF $f_1(x(k))$ is $\mathscr{F}_1^i$ and ... and $f_\hbar(x(k))$ is $\mathscr{F}_\hbar^i$, THEN

$$\begin{cases} x_1(k+1) = A_{11i}x_1(k) + \varepsilon A_{12i}x_2(k) + B_{1i}(u(k) + d(x(k), k)), \\ x_2(k+1) = A_{21i}x_1(k) + \varepsilon A_{22i}x_2(k) + B_{2i}(u(k) + d(x(k), k)), \end{cases} \tag{3.1}$$

where $x_1(k) \in \mathbb{R}^{n_1}$ and $x_2(k) \in \mathbb{R}^{n_2}$ $(n_1 + n_2 = n)$ are the slow and fast state vectors, respectively. $u(k) \in \mathbb{R}^m$ represents the control input. The known small positive scalar ε denotes the singular perturbation parameter. $f_\iota(x(k))$ $(\iota = 1, \ldots, \hbar)$ is the premise variable, $\mathscr{F}_\iota^i$ $(i = 1, \ldots, r)$ is the IT2 fuzzy set in i-th IF-THEN rule. Matrices A_{11i}, A_{12i}, A_{21i}, A_{22i}, B_{1i}, and B_{2i} are known. The external disturbance $d(x(k), k)$ is assumed to satisfy $||d(x(k), k)|| \leq \varpi ||x(k)||$ with a known constant $\varpi \geq 0$.

The firing strength of the i-th rule is obtained into the following interval:

$$\overrightarrow{m}_i(x(k)) = [\underline{m}_i(x(k)), \overline{m}_i(x(k))],$$

where $\overline{m}_i(x(k)) = \prod_{\iota=1}^{\hbar} \overline{\mu}_{\mathscr{F}_\iota^i}(f_\iota(x(k)))$ and $\underline{m}_i(x(k)) = \prod_{\iota=1}^{\hbar} \underline{\mu}_{\mathscr{F}_\iota^i}(f_\iota(x(k)))$ are upper and lower MFs, and $\underline{\mu}_{\mathscr{F}_\iota^i}(f_\iota(x(k)))$ and $\overline{\mu}_{\mathscr{F}_\iota^i}(f_\iota(x(k)))$ are the lower and upper grades of the IT2 MFs, satisfying $0 \leq \underline{\mu}_{\mathscr{F}_\iota^i}(f_\iota(x(k))) \leq \overline{\mu}_{\mathscr{F}_\iota^i}(f_\iota(x(k))) \leq 1$, and then one has $0 \leq \underline{m}_i(x(k)) \leq \overline{m}_i(x(k)) \leq 1$.

Blending the fuzzy models with the MFs, the following IT2 fuzzy singularly perturbed systems can be obtained:

$$x(k+1) = \sum_{i=1}^{r} m_i(x(k))[A_i E_\varepsilon x(k) + B_i(u(k) + d(x(k), k))] \tag{3.2}$$

with

$$A_i = \begin{bmatrix} A_{11i} & A_{12i} \\ A_{21i} & A_{22i} \end{bmatrix}, E_\varepsilon = \begin{bmatrix} I_{n_1} & 0 \\ 0 & \varepsilon I_{n_2} \end{bmatrix}, B_i = \begin{bmatrix} B_{1i} \\ B_{2i} \end{bmatrix},$$

and the membership function can be represented as

$$m_i(x(k)) = \underline{m}_i(x(k))\underline{\alpha}_i(x(k)) + \overline{m}_i(x(k))\overline{\alpha}_i(x(k))$$

where $\overline{m}_i(x(k))$ and $\underline{m}_i(x(k))$ are known upper and lower MFs, and the weighting coefficients $\underline{\alpha}_i(x(k)), \overline{\alpha}_i(x(k)) \in [0, 1]$ satisfy $\underline{\alpha}_i(x(k)) + \overline{\alpha}_i(x(k)) = 1$.

3.2.2 Component-Based Dynamic Event-Triggered Scheme

To improve the utilization of the limited communication network resources and reduce unnecessary transmission, the event-triggered protocol is an effective method. However, the conventional approach in existing works is to release all components of the state vector $x(k) = [x_1(k), \ldots, x_n(k)]^T$, if the sum of the difference between the latest released signal and the current value is greater than a predefined threshold, e.g., $\|x(k) - x(k_l)\|^2 \geq \rho\|x(k)\|^2$. Apparently, under the kind of ETS, some state components with smaller difference may be unnecessarily transmitted, which will increase the communication burden. Aiming at this issue, a component-based dynamic event-triggered protocol will be proposed to determine the transmission of the state signals, where each individual state component will be judged by different triggering thresholds (as shown in Fig. 3.1). Under this case, only the components with greater error will be transmitted at each trigging instant.

For the measurement state $x(k) = [x_1(k), \ldots, x_n(k)]^T$, the individual triggering function $\psi_\iota(k)$ for the state component $x_\iota(k)$ is designed as

$$\psi_\iota(k) = |e_i(k)|^2 - \lambda_\iota|x_\iota(k)|^2 - \frac{1}{\theta_\iota}\eta_\iota(k), \tag{3.3}$$

where $e_\iota(k) \triangleq x_\iota(l_p^\iota) - x_\iota(k)$ is the difference between the previous triggered value $x_\iota(l_p^\iota)$ and the current one $x_\iota(k)$, $\lambda_\iota \in (0, 1)$ and θ_ι are two given positive scalars, and $\eta_\iota(k)$ is the internal dynamic variable. The sequence $\{l_p^\iota\}_{p\geq 0}$ denotes the triggering instant for the ι-th state component $x_\iota(k)$.

Meanwhile, the dynamic variable $\eta_\iota(k)$ in (3.3) will be updated as

$$\eta_\iota(k+1) = \gamma_\iota\eta_\iota(k) - |e_\iota(k)|^2 + \lambda_\iota|x_\iota(k)|^2 \tag{3.4}$$

with $\eta_\iota(0) > 0$, $\gamma_\iota > 0$, and $\gamma_\iota\theta_\iota > 1$.

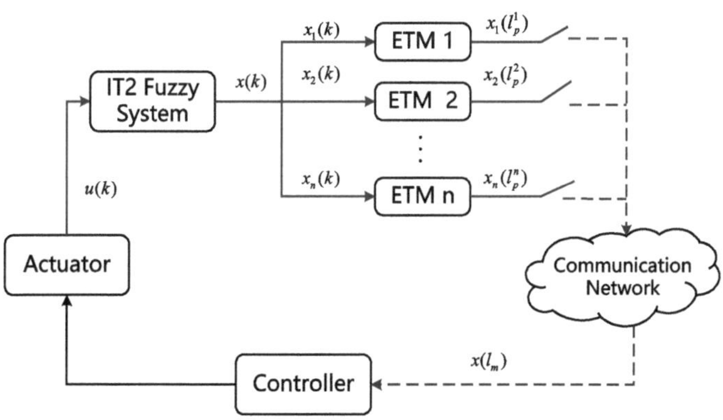

Fig. 3.1 Control system under the communication network

The next triggering instant for the component state $x_\iota(k)$ will be determined via the following dynamic ETM

$$l^\iota_{p+1} = \inf\left\{k \in \mathbb{N} \mid k > l^\iota_p,\, \psi_\iota(k) \geq 0\right\}. \tag{3.5}$$

For the controller, the available state signals will be updated once there exist state component is triggered according to the rule (3.5). To facilitate the subsequent control design and stability analysis, the triggering sequence in the controller side can be represented as $\{l_m\}_{m\geq 0} = \bigcup_{\iota=1}^{n}\{l^\iota_p\}$. That is, the triggering sequence of all components will be mapped into a common one in the controller side. For $k \in [l_m,\ l_{m+1})$, the signal received by the controller is $x(l_m) = [x_1(l^1_p), x_2(l^2_p), \ldots, x_n(l^n_p)]^T$, where l^ι_p is the latest triggering instant for the ι-th state component $x_\iota(k)$ and the unreleased component will hold the previous released value via the ZOH strategy.

As an example, the triggering sequence with two sensor nodes is shown in Fig. 3.2, from which it can be seen that the triggering instant in controller side is decided by the triggering instant of all state components, e.g., $x(l_0) = [x_1(l^1_0), x_2(l^2_0)]^T$ at $k = l_0$ and $x(l_1) = [x_1(l^1_1), x_2(l^2_0)]^T$ at $k = l_1$.

Remark 3.1 Different from the conventional ETS, the component-based dynamic ETS composed of (3.3)–(3.5) only determines the next triggering instant of the ι-th state component $x_\iota(k)$. The individual triggering threshold eliminates the influence of different state components. Moreover, it is known that the errors of those transmitted states are greater than those not transmitted states. Thus, the unnecessary transmissions of those states with smaller errors are avoided.

For the subsequent analysis, the following lemmas are provided.

Lemma 3.1 *For given positive scalars θ_ι and γ_ι satisfying $\theta_\iota\gamma_\iota > 0$, dynamic rule (3.4), and $\eta_\iota(0) = \eta_{\iota 0} > 0$, the dynamic variable $\eta_\iota(k)$ satisfies*

$$\eta_\iota(k) \geq 0, \forall k \in Z^+. \tag{3.6}$$

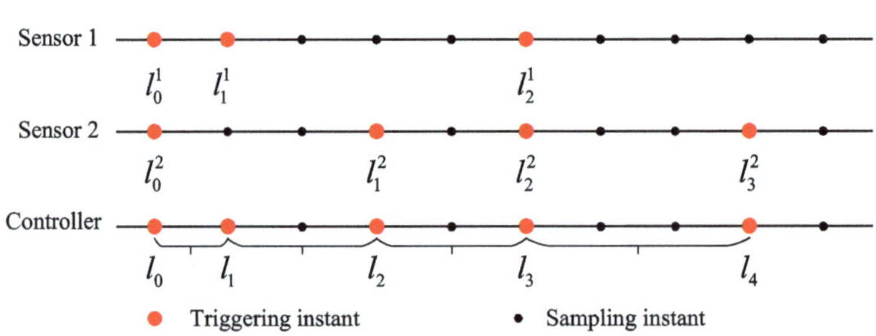

Fig. 3.2 The triggering sequence in the sensor and controller sides

Proof For $\forall k \in [l^{\iota}_{p}, l^{\iota}_{p+1})$, one has

$$|e_{\iota}(k)|^2 - \lambda_{\iota}|x_{\iota}(k)|^2 - \frac{1}{\theta_{\iota}}\eta_{\iota}(k) \leq 0. \tag{3.7}$$

It follows from (3.4) and (3.7) that

$$\begin{aligned}
\eta_{\iota}(k+1) &\geq (\gamma_{\iota} - \frac{1}{\theta_{\iota}})\eta_{i\iota}(k) \\
&\geq (\gamma_{\iota} - \frac{1}{\theta_{\iota}})^2\eta_{\iota}(k-1) \\
&\geq \cdots \\
&\geq (\gamma_{\iota} - \frac{1}{\theta_{\iota}})^{k+1}\eta_{\iota}(0).
\end{aligned}$$

It can be obtained from $\gamma_{\iota}\theta_{\iota} > 1$ and $\eta_{\iota}(0) = \eta_{\iota 0} > 0$ that

$$\eta_{\iota}(k) \geq (\gamma_{\iota} - \frac{1}{\theta_{\iota}})^k\eta_{\iota 0} \geq 0.$$

Thus, the proof is completed. □

Lemma 3.2 [1] *For a scalar* $\varepsilon > 0$ *and symmetric matrices* T_1 *and* T_2*, if* $T_1 \leq 0$*, and* $T_1 + \bar{\varepsilon}T_2 < 0$*, then the following inequality holds:*

$$T_1 + \varepsilon T_2 < 0, \quad \forall \varepsilon \in (0, \bar{\varepsilon}]. \tag{3.8}$$

3.3 Event-Triggered Sliding Mode Control Design

In this section, the component-based event-triggered sliding mode controller will be designed for the IT2 fuzzy singularly perturbed system (3.2).

Firstly, the following linear sliding surface is constructed

$$s(k) = Gx(k) \tag{3.9}$$

where G is the sliding matrix to be determined later.

Under the component-based dynamic event-triggered protocol (3.5), the controller only receives the transmitted signal $x(l_m)$ for $\forall k \in [l_m,\ l_{m+1})$. Thus, the IT2 fuzzy sliding mode controller is designed as

$$u(k) = \sum_{i=1}^{r} h_i(x(l_m))K_iE_{\varepsilon}x(l_m) - \varpi||x(l_m)||\text{sign}(\bar{s}(k)),\ k \in [l_m,\ l_{m+1}) \tag{3.10}$$

where K_i is the control matrix to be determined later, ϖ is the known bound of the external disturbance defined in (3.1). The MFs are reconstructed as $h_i(x(l_m)) = \underline{\beta}_i(x(l_m))\underline{m}_i(x(l_m)) + \overline{\beta}_i(x(l_m))\overline{m}_i(x(l_m))$ with the coefficients $\underline{\beta}_i(x(l_m))$ and $\overline{\beta}_i(x(l_m))$, and the sliding variable $s(k)$ in (3.9) is replaced as $\bar{s}(k) = Gx(l_m)$. Here, by means of the available states $x(l_m)$, the sliding variable $\bar{s}(k)$ and the MFs $h_i(x(l_m))$ are redesigned so that the sliding mode controller (3.10) is implementable in practice.

Remark 3.2 It can be observed that the fuzzy sliding mode controller only depends on the received signals $x(l_m)$. On the other hand, since the MFs $m_i(x(k))$ in IT2 fuzzy model are unknown, the MFs $h_i(x(l_m))$ in controller (3.10) are reconstructed. Furthermore, the controller is related to the singularly perturbed matrix E_ε, this reflects the effect of the singular perturbation parameter ε.

Substituting the fuzzy sliding mode controller (3.10) into the IT2 fuzzy systems (3.2) yields the closed-loop systems as follows:

$$\begin{aligned} x(k+1) = \sum_{i=1}^{r}\sum_{j=1}^{r} m_i(x(k))h_j(x(l_m))\Big[&(A_i + B_iK_j)E_\varepsilon x(l_m) \\ &+ B_i(d(x(k),k) - \varpi||x(l_m)||\text{sign}(\bar{s}(k)))\Big], \end{aligned} \tag{3.11}$$

for $\forall k \in [l_m,\ l_{m+1})$.

Remark 3.3 Notice that the MFs $h_i(x(l_m))$ in the controller (3.10) are distinct from the ones $m_i(x(k))$ in the IT2 fuzzy systems (3.2), which is called as mismatched MFs. Consequently, the mismatched MFs are involved in the closed-loop system (3.11). By establishing the relation $h_i(x(l_m)) - \delta_i m_i(x(k)) \geq 0$ (in Theorem 3.3) between the MFs in the fuzzy system and the controller, the information of MFs and some slack matrices are introduced to facilitate the stability analysis and reduce the conservatism of stability conditions.

3.4 Performance Analysis Under Component-Based Dynamic Event-Triggered Scheme

In this section, it will be proved that the proposed component-based event-triggered SMC law (3.44) can guarantee the stability of the closed-loop system (3.11). Furthermore, it will ensure that the system state trajectories converge to a small neighborhood around the predefined sliding surface.

3.4.1 Stability of Closed-Loop Systems

Theorem 3.1 *For given positive parameters ν_1, $\lambda_\iota \in (0, 1)$, γ_ι, and θ_ι satisfying $\gamma_\iota \theta_\iota > 1$ in dynamic ETS (3.3)–(3.4), if there exist matrices $P > 0$, G, K_j, and scalars $\xi_1 > 0$, $\varphi > 0$ satisfying the following conditions (for $i, j = 1, \ldots, r$, and $\iota = 1, \ldots, n$)*

$$B_i^T P B_i \leq \xi_1 I, \tag{3.12}$$

and

$$\sum_{i=1}^{r} \sum_{j=1}^{r} m_i(x(k)) h_j(x(l_m)) \Omega_{ij\varepsilon} < 0, \tag{3.13}$$

with

$$\Omega = \begin{bmatrix} \Omega_{11} & * & * \\ \Omega_{21} & \Omega_{22} & * \\ 0 & 0 & \Omega_{33} \end{bmatrix} < 0,$$

and

$$\begin{aligned}
\Omega_{11} &= (1+\nu_1)(A_i E_\varepsilon + B_i K_j E_\varepsilon)^T P (A_i E_\varepsilon + B_i K_j E_\varepsilon) + 2(1+\nu_1^{-1})(1+m)\varpi^2 \xi_1 I \\
&\quad + (1+\varphi)\Lambda - P, \\
\Omega_{21} &= (1+\nu_1)(B_i K_j E_\varepsilon)^T P (A_i E_\varepsilon + B_i K_j E_\varepsilon) + 2(1+\nu_1^{-1}) m \varpi^2 \xi_1 I, \\
\Omega_{22} &= (1+\nu_1)(B_i K_j E_\varepsilon)^T P B_i K_j E_\varepsilon - (1+\varphi) I + 2(1+\nu_1^{-1}) m \varpi^2 \xi_1 I, \\
\Omega_{33} &= -\Upsilon + \varphi \Theta, \\
\Lambda &= \mathrm{diag}\{\lambda_1, \lambda_2, \ldots, \lambda_n\}, \\
\Theta &= \mathrm{diag}\left\{\frac{1}{\theta_1}, \frac{1}{\theta_2}, \ldots, \frac{1}{\theta_n}\right\}, \\
\Upsilon &= \mathrm{diag}\{1-\gamma_1, 1-\gamma_2, \ldots, 1-\gamma_n\},
\end{aligned}$$

then, the closed-loop system (3.11) under the component-based dynamic event-triggering protocol is asymptotically stable.

Proof Choose the Lyapunov function as

$$V(\rho(k)) = x^T(k) P x(k) + \sum_{\iota=1}^{n} \eta_\iota(k) \tag{3.14}$$

with $\rho(k) = [x^T(k), \eta_1(k), \ldots, \eta_n(k)]^T$. It can be obtained from Lemma 3.1 (i.e., $\eta_\iota(k) \geq 0$) that the Lyapunov function $V(\rho(k))$ is positive.

For $\forall k \in [l_m,\ l_{m+1})$, one has from (3.4), (3.11)–(3.12) and (3.14) that

$$
\begin{aligned}
&\Delta V(\rho(k)) \\
&= V(\rho(k+1)) - V(\rho(k)) \\
&= x^T(k+1)Px(k+1) + \sum_{\iota=1}^{n}\eta_\iota(k+1) - x^T(k)Px(k) - \sum_{\iota=1}^{n}\eta_\iota(k) \\
&\le \sum_{i=1}^{r}\sum_{j=1}^{r} m_i(x(k))h_j(x(l_m))[(A_i + B_iK_j)E_\varepsilon x(k) + B_iK_jE_\varepsilon e(k) + B_i(d(x(k),k) \\
&- \varpi||x(l_m)||\text{sign}(\bar{s}(k)))]^T P[(A_i + B_iK_j)E_\varepsilon x(k) + B_iK_jE_\varepsilon e(k) \\
&+ B_i(d(x(k),k) - \varpi||x(l_m)||\text{sign}(\bar{s}(k)))] + \sum_{\iota=1}^{n}(\gamma_\iota\eta_\iota(k) - |e_\iota(k)|^2 + \lambda_\iota|x_\iota(k)|^2) \\
&- x^T(k)Px(k) - \sum_{\iota=1}^{n}\eta_\iota(k) \\
&\le \sum_{i=1}^{r}\sum_{j=1}^{r} m_i(x(k))h_j(x(l_m))\Big\{(1+\nu_1)[(A_i + B_iK_j)E_\varepsilon x(k) + B_iK_jE_\varepsilon e(k)]^T \\
&P[(A_i + B_iK_j)E_\varepsilon x(k) + B_iK_jE_\varepsilon e(k)] + (1+\nu_1^{-1})\xi_1\beta^T(k)\beta(k)\Big\} \\
&+ \sum_{\iota=1}^{n}[(\gamma_\iota - 1)\eta_\iota(k) - e_\iota^T(k)e_\iota(k) + \lambda_\iota x_\iota^T(k)x_\iota(k)] - x^T(k)Px(k)
\end{aligned} \tag{3.15}
$$

with $e(k) \triangleq [e_1(k), e_2(k), \ldots, e_n(k)]^T$, $\beta(k) \triangleq d(x(k),k) - \varpi||x(l_m)||\text{sign}(\bar{s}(k))$.

We can further obtain

$$
\begin{aligned}
\beta^T(k)\beta(k) &\le 2\varpi^2||x(k)||^2 + 2m\varpi^2||x(l_m)||^2 \\
&= 2\varpi^2||x(k)||^2 + 2m\varpi^2[x(k)+e(k)]^T[x(k)+e(k)].
\end{aligned} \tag{3.16}
$$

Substituting (3.16) into (3.15) yields

$$
\begin{aligned}
&\Delta V(\rho(k)) \\
&\le \sum_{i=1}^{r}\sum_{j=1}^{r} m_i(x(k))h_j(x(l_m))\Big\{(1+\nu_1)[(A_i + B_iK_j)E_\varepsilon x(k) + B_iK_jE_\varepsilon e(k)]^T \\
&\times P[(A_i + B_iK_j)E_\varepsilon x(k) + B_iK_jE_\varepsilon e(k)] + 2(1+\nu_1^{-1})\xi_1\varpi^2||x(k)||^2 \\
&+ 2m(1+\nu_1^{-1})\xi_1\varpi^2[x(k)+e(k)]^T[x(k)+e(k)]\Big\} \\
&- \sqrt{\eta(k)}^T\Upsilon\sqrt{\eta(k)} + e^T(x)e(k) + x^T(k)\Lambda x(k) - x^T(k)Px(k)
\end{aligned} \tag{3.17}
$$

with $\sqrt{\eta(k)} = [\sqrt{\eta_1(k)}, \ldots, \sqrt{\eta_n(k)}]^T$.

Considering the condition (3.7) for $\forall k \in [l_m,\ l_{m+1})$, one has

$$\sum_{\iota=1}^{n}\left[\lambda_\iota |x_\iota(k)|^2 - |e_\iota(k)|^2 + \frac{1}{\theta_\iota}\eta_\iota(k)\right] \geq 0, \tag{3.18}$$

which can be represented as the following compact form

$$x^T(k)\Lambda x(k) + \sqrt{\eta(k)}^T\Theta\sqrt{\eta(k)} - e^T(k)e(k) \geq 0. \tag{3.19}$$

From the expressions (3.17) and (3.19), there exists a positive scalar φ so that the following expression holds:

$$\begin{aligned}
&\Delta V(\rho(k)) \\
&\leq \Delta V(\rho(k)) + \varphi[x^T(k)\Lambda x(k) + \sqrt{\eta(k)}^T\Theta\sqrt{\eta(k)} - e^T(t)e(t)] \\
&\leq \sum_{i=1}^{r}\sum_{j=1}^{r} m_i(x(k))h_j(x(l_m))\Big\{(1+\nu_1)[(A_i + B_iK_j)E_\varepsilon x(k) + B_iK_jE_\varepsilon e(k)]^T \\
&\times P[(A_i + B_iK_j)E_\varepsilon x(k) + B_iK_jE_\varepsilon e(k)] + 2(1+\nu_1^{-1})\varpi^2\xi_1 x^T(k)x(k) \\
&+ 2(1+\nu_1^{-1})m\varpi^2\xi_1[x(k)+e(k)]^T[x(k)+e(k)]\Big\} + (1+\varphi)x^T(k)\Lambda x(k) \\
&- (1+\varphi)e^T(k)e(k) + \sqrt{\eta(k)}^T(-\Upsilon + \varphi\Theta)\sqrt{\eta(k)} - x^T(k)Px(k) \\
&\leq \varrho^T(k)\sum_{i=1}^{r}\sum_{j=1}^{r} m_i(x(k))h_j(x(l_m))\Omega_{ij\varepsilon}\varrho(k)
\end{aligned} \tag{3.20}$$

with $\varrho(k) = [x^T(k), e^T(k), \sqrt{\eta(k)}^T]^T$.

It can be obtained from the conditions (3.12) and (3.13) that $\Delta V(\rho(k)) < 0$ holds for $\varrho(k) \neq 0$, which means that the closed-loop system (3.11) under the component-based event-triggered protocol is asymptotically stable. □

3.4.2 Reachability of Sliding Surface

Under the proposed event-triggered SMC scheme, the reachability of the specified sliding surface $s(k) = 0$ can be ensured under the following conditions.

Theorem 3.2 *For given positive scalars ν_1, ν_2, $\lambda_i \in (0, 1)$, θ_i, and γ_ι satisfying $\gamma_i\theta_i > 1$, if there exist matrices $P > 0$, $Q > 0$, G, K_j, and scalars $\xi_1 > 0$, $\xi_2 > 0$, $\varphi > 0$ satisfying condition (3.12) and the following conditions (for $i, j = 1, \ldots, r$, and $\iota = 1, \ldots, n$)*

$$B_i^T G^T Q G B_i \leq \xi_2 I, \tag{3.21}$$

and

$$\sum_{i=1}^{r}\sum_{j=1}^{r} m_i(x(k))h_j(x(l_m))\Gamma_{ij\varepsilon} < 0, \tag{3.22}$$

with

$$\Gamma_{ij\varepsilon} = \Omega_{ij\varepsilon} + (1+\nu_2)\mathscr{G}_{ij}^T Q\mathscr{G}_{ij} + 2(1+\nu_2^{-1})m\mu^2\xi_2 I,$$

and

$$\mathscr{G}_{ij} = \begin{bmatrix} G(A_i + B_iK_j)E_\varepsilon & GB_iK_jE_\varepsilon \end{bmatrix},$$

then, the state trajectories will enter into the following domain

$$\mathcal{D} = \left\{ s(k) \mid ||s(k)|| \leq \sqrt{\frac{2(1+\nu_2^{-1})\varpi^2\xi_2}{\lambda_{\min}(Q)}}||x(k)|| \right\}. \tag{3.23}$$

Proof Substituting (3.10) into (3.9), one has

$$\begin{aligned} s(k+1) = \sum_{i=1}^{r}\sum_{j=1}^{r} m_i(x(k))h_j(x(l_m))\big[&G(A_i + B_iK_j)E_\varepsilon x(k) + GB_iK_jE_\varepsilon e(k) \\ &+ GB_i(d(x(k),k) - \varpi||x(l_m)||\text{sign}(\bar{s}(k)))\big]. \end{aligned} \tag{3.24}$$

Choose the Lyapunov function as

$$U(\rho(k), s(k)) = V(\rho(k)) + V_s(k) \tag{3.25}$$

with $V(\rho(k))$ defined in (3.14), and $V_s(k) \triangleq s^T(k)Qs(k)$.

Along with the state trajectories of closed-loop system (3.24), we can obtain

$$\begin{aligned} &\Delta V_s(k) \\ &\quad = s^T(k+1)Qs(k+1) - s^T(k)Qs(k) \\ &\quad \leq \sum_{i=1}^{r}\sum_{j=1}^{r} m_i(x(k))h_j(x(l_m))\Big\{(1+\nu_2)\big[G(A_i+B_iK_j)E_\varepsilon x(k) + GB_iK_jE_\varepsilon e(k)\big]^T Q \\ &\times \big[G(A_i+B_iK_j)E_\varepsilon x(k) + GB_iK_jE_\varepsilon e(k)\big] + 2(1+\nu_2^{-1})\varpi^2\xi_2||x(k)||^2 \\ &\quad + 2(1+\nu_2^{-1})m\varpi^2\xi_2 x^T(l_m)x(l_m)\Big\} - s^T(k)Qs(k) \\ &\leq \sum_{i=1}^{r}\sum_{j=1}^{r} m_i(x(k))h_j(x(l_m))\Big\{(1+\nu_2)\big[G(A_i+B_iK_j)E_\varepsilon x(k) + GB_iK_jE_\varepsilon e(k)\big]^T \\ &\quad \times Q\big[G(A_i+B_iK_j)E_\varepsilon x(k) + GB_iK_jE_\varepsilon e(k)\big]\Big\} \\ &\quad + 2(1+\nu_2^{-1})m\varpi^2\xi_2||(x(k)+e(k))||^2 \\ &- \big[\lambda_{\min}(Q)||s(k)||^2 - 2(1+\nu_2^{-1})\varpi^2\xi_2||x(k)||^2\big] \end{aligned} \tag{3.26}$$

for $\forall k \in [l_m,\ l_{m+1})$.

Combining (3.20) with (3.26), one has

$$\begin{aligned}
&\Delta U(\rho(k), s(k)) \\
&\quad \le \Delta V(\rho(k)) + \Delta V_s(k) + \varphi\left[x^T(k)\Lambda\Phi x(k) + \sqrt{\eta(k)}\Theta\sqrt{\eta(k)} - e^T(t)\Phi e(t)\right] \\
&\quad \le \varrho^T(k)\sum_{i=1}^{r}\sum_{j=1}^{r} m_i(x(k))h_j(x(l_m))\Gamma_{ij\varepsilon}\varrho(k) - \left[\lambda_{\min}(Q)||s(k)||^2\right. \\
&\quad \left. - 2(1+\nu_2^{-1})\varpi^2\xi_2||x(k)||^2\right].
\end{aligned} \tag{3.27}$$

Thus, if the state trajectories lie outside the domain $\mathcal{D}$, i.e.,

$$||s(k)|| \ge \sqrt{\frac{2(1+\nu_2^{-1})\varpi^2\xi_2}{\lambda_{\min}(Q)}}||x(k)||, \tag{3.28}$$

one has from the condition (3.22)

$$\Delta U(\rho(k), s(k)) \le \varrho^T(k)\sum_{i=1}^{r}\sum_{j=1}^{r} m_i(x(k))h_j(x(l_m))\Gamma_{ij\varepsilon}\varrho(k) < 0, \tag{3.29}$$

which implies that the state trajectories will be driven into the domain $\mathcal{D}$. □

3.5 Stability Conditions and Optimization Solving Algorithm

In this section, we shall address the mismatched MFs and obtain sufficient conditions on the stability and reachability. Furthermore, an optimization algorithm will be given to solve these sufficient conditions and improve system performance.

3.5.1 Membership-Function-Dependent Sufficient Conditions

It should be noted that two issues need to be addressed, i.e., ① the conditions (3.13) and (3.22) in Theorems 3.1 and 3.2 involve the unknown MFs $m_i(x(k))$ and $h_i(x(l_m))$. ② the singular perturbation parameter ε may cause ill-conditioned result of the conditions (3.13) and (3.22). To tackle the above two problems, we further give the ε-independent stability criterions in the following theorem by means of the relation between $m_i(x(k))$ and $h_i(x(l_m))$.

Theorem 3.3 *For given positive scalars ν_1, ν_2, $\lambda_i \in (0, 1)$, θ_ι and γ_ι satisfying $\gamma_\iota\theta_\iota > 1$, and scalar $\delta_i > 0$ satisfying $h_i(x(l_m)) - \delta_i m_i(x(k)) \geq 0$, if there exist matrices $P > 0$, $\bar{Q} > 0$, $Y_i = Y_i^T$, G, K_j, and scalars $\xi_1 > 0$, $\xi_2 > 0$, $\varphi > 0$ satisfying the following conditions (for $i, j = 1, \ldots, r$, and $\iota = 1, \ldots, n$)*

$$\begin{bmatrix} -\xi_1 I & * \\ PB_i & -P \end{bmatrix} \leq 0, \quad \begin{bmatrix} -\xi_2 I & * \\ GB_i & -\bar{Q} \end{bmatrix} \leq 0, \tag{3.30}$$

$$\Xi_{ij1} - Y_i \leq 0, \quad \Xi_{ij\bar{\varepsilon}} - Y_i < 0, \tag{3.31}$$
$$\delta_i \Xi_{ii1} + Y_i - \delta_i Y_i \leq 0, \quad \Xi_{ii\bar{\varepsilon}} + Y_i - \delta_i Y_i < 0, \tag{3.32}$$
$$\delta_j \Xi_{ij1} + Y_i - \delta_j Y_i + \delta_i \Xi_{ji1} + Y_j - \delta_i Y_j \leq 0, i < j, \tag{3.33}$$
$$\delta_j \Xi_{ij\bar{\varepsilon}} + Y_i - \delta_j Y_i + \delta_i \Xi_{ji\bar{\varepsilon}} + Y_j - \delta_i Y_j < 0, i < j \tag{3.34}$$

with

$$\Xi_{ij1} = \begin{bmatrix} \Xi^{11} & * & * \\ \hat{\Xi}^{21} & \Xi^{22} & * \\ \hat{\Xi}^{31} & 0 & \Xi^{33} \end{bmatrix}, \quad \Xi_{ij\bar{\varepsilon}} = \begin{bmatrix} \Xi^{11} & * & * \\ \bar{\Xi}^{21} & \Xi^{22} & * \\ \bar{\Xi}^{31} & 0 & \Xi^{33} \end{bmatrix},$$

and

$$\Xi^{11} = \mathrm{diag}\{-P + (1+\varphi)\Lambda + 2(1+\nu_1^{-1})\varpi^2\xi_1 I, -(1+\varphi)I, -\Upsilon + \varphi\Theta\},$$
$$\Xi^{22} = \mathrm{diag}\left\{-\frac{1}{1+\nu_1}P, -\frac{1}{1+\nu_1^{-1}}\xi_1\right\}, \quad \Xi^{33} = \mathrm{diag}\left\{-\frac{1}{1+\nu_2}\bar{Q}, -\frac{1}{1+\nu_2^{-1}}\xi_2\right\},$$
$$\hat{\Xi}^{21} = \begin{bmatrix} P(A_i + B_iK_j)E_1 & PB_iK_jE_1 & 0 \\ \sqrt{2m}\varpi\xi_1 I & \sqrt{2m}\varpi\xi_1 I & 0 \end{bmatrix},$$
$$\hat{\Xi}^{31} = \begin{bmatrix} G(A_i + B_iK_j)E_1 & GB_iK_jE_1 & 0 \\ \sqrt{2m}\varpi\xi_2 I & \sqrt{2m}\varpi\xi_2 I & 0 \end{bmatrix},$$
$$\bar{\Xi}^{21} = \begin{bmatrix} P(A_i + B_iK_j)E_{\bar{\varepsilon}} & PB_iK_jE_{\bar{\varepsilon}} & 0 \\ \sqrt{2m}\varpi\xi_1 I & \sqrt{2m}\varpi\xi_1 I & 0 \end{bmatrix},$$
$$\bar{\Xi}^{31} = \begin{bmatrix} G(A_i + B_iK_j)E_{\bar{\varepsilon}} & GB_iK_jE_{\bar{\varepsilon}} & 0 \\ \sqrt{2m}\varpi\xi_2 I & \sqrt{2m}\varpi\xi_2 I & 0 \end{bmatrix},$$

and

$$E_1 = \begin{bmatrix} I_{n_1} & 0 \\ 0 & 0 \end{bmatrix}, E_{\bar{\varepsilon}} = \begin{bmatrix} I_{n_1} & 0 \\ 0 & \bar{\varepsilon} I_{n_2} \end{bmatrix},$$

then, both the stability of the closed-loop system (3.11) and the reachability of the sliding domain $\mathcal{D}$ can be attained.

Proof It is noted that the condition (3.13) in Theorem 3.1 can be ensured by the condition (3.22) in Theorem 3.2.

Denote $Q^{-1} \triangleq \bar{Q}$. By virtue of the Schur Complement, the condition (3.30) in Theorem 3.3 can ensure the conditions (3.12) and (3.21).

Meanwhile, the condition (3.22) can be ensured by the following inequality

$$\sum_{i=1}^{r}\sum_{j=1}^{r} m_i(x(k))h_j(x(l_m))\Xi_{ij\varepsilon} < 0 \tag{3.35}$$

with

$$\Xi_{ij\varepsilon} = \begin{bmatrix} \Xi^{11} & * & * \\ \Xi^{21} & \Xi^{22} & * \\ \Xi^{31} & 0 & \Xi^{33} \end{bmatrix},$$

and

$$\Xi^{21} = \begin{bmatrix} P(A_i + B_iK_j)E_\varepsilon & PB_iK_jE_\varepsilon & 0 \\ \sqrt{2m}\mu\xi_1 I & \sqrt{2m}\mu\xi_1 I & 0 \end{bmatrix},$$

$$\Xi^{31} = \begin{bmatrix} G(A_i + B_iK_j)E_\varepsilon & GB_iK_jE_\varepsilon & 0 \\ \sqrt{2m}\mu\xi_2 I & \sqrt{2m}\mu\xi_2 I & 0 \end{bmatrix}.$$

Note that

$$\begin{aligned} &\sum_{i=1}^{r}\sum_{j=1}^{r} m_i(x(k))(m_j(x(k)) - h_j(x(l_m)))Y_i \\ &= \sum_{i=1}^{r} m_i(x(k))\left(\sum_{j=1}^{r} m_j(x(k)) - \sum_{j=1}^{r} h_j(x(l_m))\right)Y_i = 0. \end{aligned} \tag{3.36}$$

We have from the expressions (3.35) and (3.36)

$$\begin{aligned} &\sum_{i=1}^{r}\sum_{j=1}^{r} m_i(x(k))h_j(x(l_m))\Xi_{ij\varepsilon} \\ &= \sum_{i=1}^{r}\sum_{j=1}^{r} m_i(x(k))\left(h_j(x(l_m)) - \delta_j m_j(x(k)) + \delta_j m_j(x(k))\right)\Xi_{ij\varepsilon} \\ &+ \sum_{i=1}^{r}\sum_{j=1}^{r} m_i(x(k))\left(m_j(x(k)) - h_j(x(l_m)) + \delta_j m_j(x(k)) - \delta_j m_j(x(k))\right)Y_i \\ &= \sum_{i=1}^{r}\sum_{j=1}^{r} m_i(x(k))\left(h_j(x(l_m)) - \delta_j m_j(x(k))\right)\left[\Xi_{ij\varepsilon} - Y_i\right] \\ &+ \sum_{i=1}^{r} m_i^2(x(k))\left[\delta_i \Xi_{ii\varepsilon} + Y_i - \delta_i Y_i\right] \\ &+ \sum_{i=1}^{r}\sum_{j>i} m_i(x(k))m_j(x(k))\left[\delta_j \Xi_{ij\varepsilon} + Y_i - \delta_j Y_i + \delta_i \Xi_{ji\varepsilon} + Y_j - \delta_i Y_j\right]. \end{aligned} \tag{3.37}$$

By virtue of Lemma 3.2, the ε-independent condition (3.31) can guarantee the expression (3.37). Thus, the proof is completed. □

It is noted the conditions (3.30)–(3.34) are nonlinear matrix inequalities on account of the coupling terms $P(A_i + B_i K_j)$ and $G(A_i + B_i K_j)$. Apparently, if the matrices P and G are determined in advance, the nonlinear conditions (3.30)–(3.34) will become the linear ones. Thus, a feasible strategy is to check the solvability of conditions (3.30)–(3.34) for a series of given matrices P and G. Based on this idea, we shall propose an optimization algorithm in the following subsection to search appropriate matrices P and G via minimizing the sliding domain $\mathcal{D}$ in (3.23) around the sliding surface.

3.5.2 PSO-Based Solving Algorithm

As already known, the PSO algorithm method can obtain the optimized solutions via performing the random search. In this subsection, the dynamic PSO algorithm is firstly introduced to enhance the information exchanging ability of the entire particle swarm. And then, a dynamic PSO-based optimization algorithm is proposed to solve the conditions (3.30)–(3.34) in Theorem 3.3 by searching feasible matrices P and G.

In the standard PSO algorithm, each particle is characterized by velocity, position, and fitness values, where the fitness value reflects the quality of the each particle. During each iteration, the position z_i and velocity v_i of i-th particle are updated according to the particle's local best position $pbest$ and the global best position $gbest$ as follows:

$$v_i^{l+1} = wv_i^l + c_1 rand_1(pbest_i^l - z_i^l) + c_2 rand_2(gbest^l - z_i^l), \tag{3.38}$$

$$z_i^{l+1} = z_i^l + v_i^l, \tag{3.39}$$

where w denotes the inertia weighting factor, c_1 and c_2 represent the acceleration parameters, $rand_1$ and $rand_2 \in [0, 1]$ are two uniformly distributed random numbers. For the previous l iterations, the best position of the i-th particle is denoted as $pbest_i^l$, and the best position among the entire particle swarm is denoted as $gbest^l$.

Remark 3.4 It can be observed that the movement of particles is only affected by its best information $pbest$ and the global best information $gbest$. The information from the neighbouring particles is neglected in the updating process, which weakens the interaction ability of the particle swarm.

For enhancing the information exchange of the entire swarm, the dynamic learning approach was proposed in [2] to make the particle learn the optimal information from itself or its neighbors. Thus, historical best information from the entire swarm is involved for updating the particle's velocity and the optimized solutions can be searched by dynamically exchanging the better searched information among all particles. The dynamic PSO updating rules are designed as

$$v_i^{l+1} = wv_i^l + c_1 rand_1(pbest_{di}^l - z_i^l) + c_2 rand_2(gbest^l - z_i^l), \tag{3.40}$$
$$z_i^{l+1} = z_i^l + v_i^l, \tag{3.41}$$

where $pbest_{di}^l$ denotes the best fitness value of the di-th particle determined by the learning probability

$$L_i = a_1 + a_2 \frac{1}{\sqrt{R_i}}, \tag{3.42}$$

where R_i is the ranking number of the i-th particle according to the fitness value, a_1 and a_2 are the weighting coefficients within [0, 1]. If a random number $rand > L_i$, the i-th particle will learn from itself, i.e., $pbest_{di} = pbest_i$. Otherwise, the i-th particle learns from its neighbor's best information around N_p particles, i.e., $di = Mod(i + rands * N_p, N)$, where N is the number of the particle and $rands$ is a random number within the interval $[-1, 1]$.

In this chapter, the dynamic PSO with the updating rules (3.40), (3.41), and (3.42) is utilized to search the feasible matrices P and G. To this end, choose P and G as a particle encoded as

$$z = [p_{11}, \ldots, p_{1n}, p_{22}, \ldots, p_{nn}, g_{11}, \ldots, g_{1n}, g_{21}, \ldots, g_{mn}],$$

whose position and velocity will be updated according to the dynamic updating rules (3.40) and (3.41). Meanwhile, the search space of each particle is defined as $\mathbb{S} \triangleq \{z_i, v_i \mid z_i \in [\underline{z}_i, \overline{z}_i],\ v_i \in [\underline{v}_i, \overline{v}_i],\ i = 1, \ldots, N\}$, where the $\underline{v}_i$ and $\overline{v}_i$ are the bound of the velocity, $\underline{z}_i$ and $\overline{z}_i$ are the search scope of the position for i-th particle.

By using the above dynamic PSO algorithm to search matrices P and G, the PSO-based optimization algorithm is proposed to solve stability conditions (3.30)–(3.34) by minimizing the sliding domain $\mathcal{D}$ with the objective function as $F_{obj} = \sqrt{2(1 + \nu_2^{-1})\varpi^2 \xi_2 / \lambda_{\min}(Q)}$.

Thus, the control design problem can be transformed into the following optimization problem:

$$\begin{aligned} \min_{P,G} \quad & F_{obj} \\ \text{s.t.} \quad & \text{conditions (3.30)}-\text{(3.34)}. \end{aligned} \tag{3.43}$$

For the above minimization problem, the objective function F_{obj} shall be minimized via the following dynamic PSO-based optimization algorithm:

It should be noted that some generated matrices G and P may result in the infeasibility of stability conditions (3.30)–(3.34). In this case, the value of the objective function is assigned as a huge value, e.g., 10^5.

Algorithm 3.1 Dynamic PSO-based optimization algorithm

Initialization

1: Give system parameters: A_i, B_i, $\bar{\varepsilon}$, and design parameters: λ_ι, γ_ι, θ_ι.

2: Randomly generate the initial particles z_i in the given search space $\mathbb{S}$, and initialize the objective function F_{obj}.

Evaluation

3: Solve the conditions (3.30)–(3.34) with the obtained matrices P and G via particles z_i, and calculate the objective function F_{obj}.

4: Compare the current objective function with the previous best value. If the current value is less than the previous best value, the best value is updated. Otherwise, the best value is not updated. Go to the **Improvement** step.

Improvement

5: The P and G are improved according to the current values and the updating rules (3.40)–(3.41).

6: If the terminal condition is reached, i.e., the number of iterative attains the maximum M, stop and return the optimized solutions. Otherwise, go to the **Evaluation** step.

7: **return** G and K_j.

3.6 Simulation Example

Consider the IT2 fuzzy singularly perturbed systems (3.2) with the following parameters:

$$A_1 = \begin{bmatrix} 0.8 & 0.3 \\ 0.1 & 1.0 \end{bmatrix}, \ A_2 = \begin{bmatrix} 1.1 & 0.4 \\ 0.2 & 1.0 \end{bmatrix}, \ B_1 = \begin{bmatrix} 0.1 \\ 0.2 \end{bmatrix}, \ B_2 = \begin{bmatrix} 0.1 \\ 0.5 \end{bmatrix},$$
$$E_\varepsilon = \text{diag}\{1,\ 0.01\},\ d(x(k), k) = \mu\sqrt{\|2x_1(k)x_2(k)\|}.$$

The upper and lower MFs are chosen as

$$\overline{m}_1(x(k)) = \frac{0.2(3x_1^2(k)+2)}{3},\ \underline{m}_1(x(k)) = \frac{0.2(x_1^2(k)+2)}{3},$$
$$\overline{m}_2(x(k)) = 1 - \underline{m}_1(x_1(k)),\ \underline{m}_2(x(k)) = 1 - \overline{m}_1(x_1(k)).$$

According to the proposed IT2 fuzzy SMC law (3.10), one has

$$u(k) = \sum_{i=1}^{2} h_i(x(l_m))K_i E_\varepsilon x(l_m) - \mu||x(l_m)||\text{sign}(\bar{s}(k)), \tag{3.44}$$

with MFs $h_1(x(l_m)) = 0.6\underline{m}_1(x(l_m)) + 0.4\overline{m}_1(x(l_m))$ and $h_2(x(l_m)) = 1 - h_1(x(l_m))$. Assuming the system state $x_1(k) \in [-2,\ 2]$, we can choose the parameters $\delta_1 = 0.2708$ and $\delta_2 = 0.4462$ satisfying $h_i(x(l_m)) - \delta_i m_i(x(k)) \geq 0$.

To perform the proposed PSO-based solving algorithm to solve stability conditions (3.30)–(3.34), the matrices P and G with $P = \begin{bmatrix} p_{11} & p_{12} \\ p_{12} & p_{22} \end{bmatrix}$ and $G = \begin{bmatrix} g_1 & g_2 \end{bmatrix}$ are encoded as $z_i = [p_{11}, p_{12}, p_{22}, g_1, g_2]$, where the search scopes are assigned as $p_{11}, p_{22} \in [0,\ 5]$, $p_{12}, g_1, g_2 \in [-1,\ 1]$, the velocities are bounded within $v_i \in [-0.5, 0.5]$. The parameters of the optimization algorithm are set as: $N = 40$, $M = 60$, $w = 0.8$, $c_1 = c_2 = 2$, $a_1 = 0.1$, $a_2 = 0.5$, and $N_p = 2$. Besides, suppose that the upper bound of external disturbance is $\varpi = 0.1$, and the design parameters are selected as $\nu_1 = \nu_2 = 0.2$, $\bar{\varepsilon} = 0.01$, $\lambda_1 = 0.2$, $\lambda_2 = 0.1$, $\theta_1 = 10$, $\theta_2 = 8$, $\gamma_1 = 0.4$, and $\gamma_2 = 0.5$. By performing the proposed dynamic PSO-based optimization algorithm 3.1, we obtain the following design parameters:

$$P = \begin{bmatrix} 3.5974 & -0.1546 \\ -0.1546 & 1.4556 \end{bmatrix}, \ G = \begin{bmatrix} -0.1160 & -0.5921 \end{bmatrix},$$
$$K_1 = \begin{bmatrix} -1.1706 & -4.0553 \end{bmatrix}, \ \ K_2 = \begin{bmatrix} -1.1703 & -3.5647 \end{bmatrix}.$$

Moreover, the convergence process of fitness values in the PSO-based optimization algorithm is illustrated in Fig. 3.3, where it is evident that the objective function F_{obj} successfully converges to the optimized solutions.

For the initial conditions $x(0) = [-2.0,\ -1.5]^T$ and $[\eta_1(0),\ \eta_2(0)] = [0.1,\ 0.2]$, the simulation results under the proposed SMC scheme with component-based dynamic event-triggered strategy are shown in Figs. 3.4a–f. As depicted in Fig. 3.4a, the stability of the closed-loop system is ensured. Furthermore, Fig. 3.4c demonstrates that the reachability of the sliding surface is guaranteed. Additionally, the control input illustrated in Fig. 3.4b is a piecewise function, which is updated only at the triggering instants. In Fig. 3.4f, the number of events for the states $x_1(k)$ and $x_2(k)$ are 27 and 29, respectively. Moreover, it can be observed from Fig. 3.4f that the triggering instants for different state components are often asynchronous.

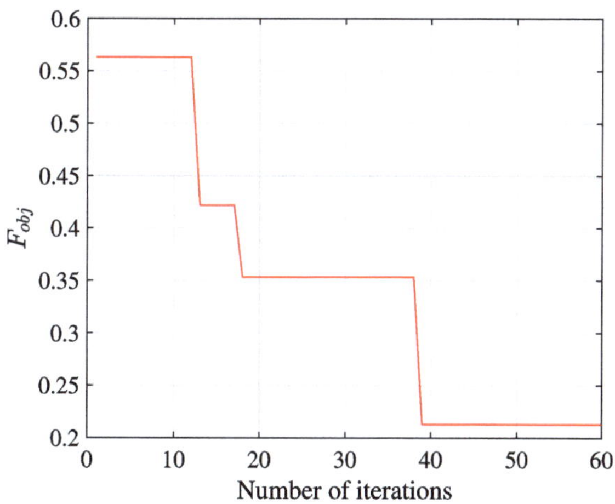

Fig. 3.3 The evolution of F_{obj} under the PSO-based optimization algorithm

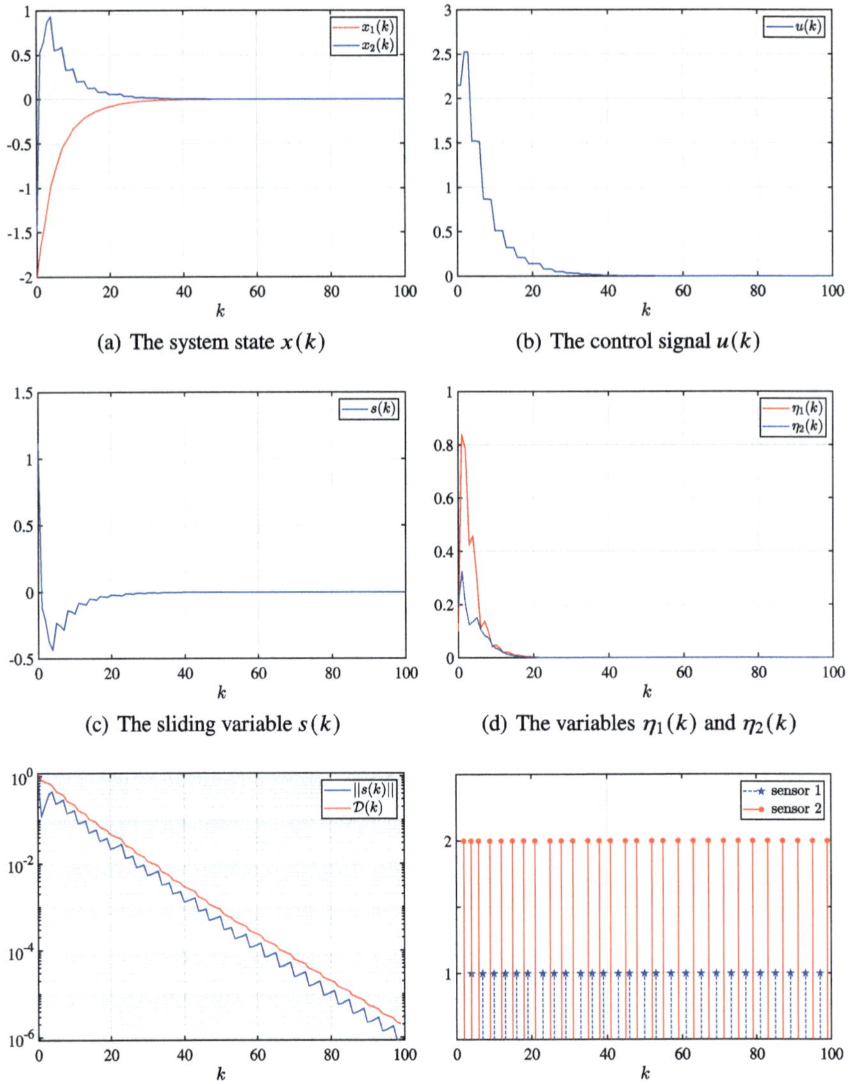

(a) The system state $x(k)$

(b) The control signal $u(k)$

(c) The sliding variable $s(k)$

(d) The variables $\eta_1(k)$ and $\eta_2(k)$

(e) The sliding domain $\mathcal{D}(k)$ and $||s(k)||$

(f) The triggering instants

Fig. 3.4 The simulation results under the component-based dynamic ETS

3.7 Conclusion

In this chapter, the component-based dynamic event-triggered SMC problem has been studied for IT2 fuzzy singularly perturbed systems. An individual ETM is designed for each sensor node. A fuzzy sliding mode controller based on triggered state has been designed, and sufficient conditions have been derived so that the stability of closed-loop systems and the reachability of sliding domain can be guaranteed. Besides, a dynamic PSO-based optimization algorithm has been developed to minimize the sliding domain and improve the control performance.

References

1. Wan, X., Wang, Z., Wu, M., Liu, X.: H_∞ state estimation for discrete-time nonlinear singularly perturbed complex networks under the Round-Robin protocol. IEEE Trans. Neural Netw. Learn. Syst. **30**(2), 415–426 (2019)
2. Liu, Z.-H., Wei, H.-L., Li, X.-H., Liu, K., Zhong, Q.-C.: Global identification of electrical and mechanical parameters in PMSM drive based on dynamic self-learning PSO. IEEE Trans. Power Electron. **33**(12), 10858–10871 (2018)

Chapter 4
Sliding Mode Quantized Control Under Membership-Function-Based Dynamic Event-Triggered Scheme

4.1 Introduction

Quantization has been demonstrated to be an effective approach for conserving communication resources in networked control systems. By converting analog signals into digital signals with finite levels, it reduces data transmission while maintaining control performance. Building on the dynamic ETSs examined in the previous two chapters, which utilizes adjustable triggering thresholds to balance communication efficiency and system performance, this chapter will investigate the integration of quantization with event-triggered SMC for IT2 T-S fuzzy systems under communication constraints. While static quantization strategies quantize signals using fixed quantizer parameters [1, 2], dynamic quantization offers superior adaptability by adjusting quantization levels in real time to reduce quantization errors, as highlighted in prior works [2, 3]. In [4], a dynamic quantization parameter was online adjusted according to the system state, and transmitted to the controller. Following the similar lines, the input quantization were also concerned in [5, 6]. However, in the context of event-triggered control, the interplay between quantization and triggering mechanisms remains underexplored, particularly for IT2 T-S fuzzy systems with inherent uncertainties.

This chapter focuses on the co-design of an event-triggered mechanism and a dynamic quantizer. By integrating the characteristics of IT2 fuzzy systems, a membership-function-based dynamic event-triggered scheme (DETS) is developed to regulate the transmission of system states. The proposed ETS incorporates two triggering conditions: one dependent on the system state and the other on the controller's MFs. It can judge the deviation of controller's MFs for preventing large fluctuations in MFs, and reduce unnecessary communication. Then, an event-triggered SMC law is designed based on the received triggered states. In order to facilitate the transmission of the control signal, a dynamic quantizer is introduced in C/A channels. Meanwhile, the dynamic quantizer's parameters can be on-line adjusted depending on the control signal and are transmitted via the network channels. Since

Y. Yang et al., *Analysis and Synthesis of Interval Type-2 Fuzzy Systems Under Communication Protocols*, Studies in Systems, Decision and Control 636,
https://doi.org/10.1007/978-981-95-2355-9_4

the proposed event-triggered fuzzy sliding mode controller contains different MFs from the system, and the controller's MFs are only updated at triggering instants, the above two factors result in the mismatching phenomenon between the system's and the controller's MFs. By virtue of the triggering conditions, the mismatching MFs between the fuzzy controller and the controlled system are handled and the *local* information of controller's and system's MFs can be employed to analyze the system performance. By leveraging the local MF information from both the system and controller, *local*-boundary-information-dependent approach is utilized to derive sufficient conditions for ensuring the control performance of the closed-loop dynamic system and the sliding mode dynamics. Additionally, a trade-off between communication resource conservation and control performance is formulated as an optimization problem, resolved via the PSO algorithm to reduce triggering frequency and convergence domain. The proposed framework effectively balances resource utilization and system performance under communication constraints and mismatching MFs.

4.2 Problem Description

4.2.1 System Model

Consider the following IT2 T-S fuzzy system with the i-th rule:

Rule i: IF $f_1(x(k))$ is $\mathscr{F}_1^i$ and ... and $f_\hbar(x(k))$ is $\mathscr{F}_\hbar^i$, THEN

$$x(k+1) = A_i x(k) + B_i(u(k) + d(k)), \tag{4.1}$$

where $f_\iota(x(k))$ $(\tau = 1, \ldots, \hbar)$ is the premise variable, $\mathscr{F}_\iota^i$ $(i = 1, \ldots, r)$ is the IT2 fuzzy set, $x(k) \in \mathbb{R}^n$ is the system state, and $u(k) \in \mathbb{R}^m$ is the control input. $d(k)$ is the external disturbance with a upper bound $\varpi \geq 0$, i.e., $||d(k)|| \leq \varpi$. Matrix A_i and B_i are known with appropriate dimensions. The firing strength in fuzzy rule i satisfies

$$\vec{m}_i(x(k)) = [\underline{m}_i(x(k)), \overline{m}_i(x(k))]$$

with the upper MFs $\overline{m}_i(x(k)) = \prod_{\iota=1}^{\hbar} \overline{\mu}_{\mathscr{F}_\iota^i}(f_\iota(x(k))) \geq 0$, and lower MFs $\underline{m}_i(x(k)) = \prod_{\iota=1}^{\hbar} \underline{\mu}_{\mathscr{F}_\iota^i}(f_\iota(x(k))) \geq 0$, in which the upper and lower membership grades $\overline{\mu}_{\mathscr{F}_\iota^i}(f_\tau(x(k)))$ and $\underline{\mu}_{\mathscr{F}_\iota^i}(f_\iota(x(k)))$ satisfy $0 \leq \underline{\mu}_{\mathscr{F}_\iota^i}(f_\iota(x(k))) \leq \overline{\mu}_{\mathscr{F}_\iota^i}(f_\iota(x(k))) \leq 1$. It is easy to deduce from the above relations that $0 \leq \underline{m}_i(x(k)) \leq \overline{m}_i(x(k)) \leq 1$.

According to the fuzzy model (4.1), the IT2 fuzzy system can be obtained as

$$x(k+1) = \sum_{i=1}^{r} m_i(x(k))[A_i x(k) + B_i(u(k) + d(k))], \tag{4.2}$$

where $m_i(x(k))$ is the membership function with

$$m_i(x(k)) = \overline{m}_i(x(k))\overline{\alpha}_i(x(k)) + \underline{m}_i(x(k))\underline{\alpha}_i(x(k)), \tag{4.3}$$

and $\sum_{i=1}^{r} m_i(x(k)) = 1$ with weighting coefficients $\overline{\alpha}_i(x(k)), \underline{\alpha}_i(x(k)) \in [0, 1]$ satisfying $\overline{\alpha}_i(x(k)) + \underline{\alpha}_i(x(k)) = 1$. In the IT2 fuzzy systems (4.2), only the upper and lower MFs (i.e., $\overline{m}_i(x(k))$ and $\underline{m}_i(x(k))$) are known, and their weighting coefficients $\overline{\alpha}_i(x(k))$ and $\underline{\alpha}_i(x(k))$ are unknown nonlinear functions related to uncertainties.

4.2.2 Analysis on Conventional Fuzzy Controller

Consider the conventional state-feedback fuzzy controller with the c-rule satisfying the following rules:

Rule j: IF $g_1(x(k))$ is $\mathscr{M}_1^j$ and ... and $g_{\wp}(x(k))$ is $\mathscr{M}_{\wp}^j$, THEN

$$v(k) = K_j x(k) \tag{4.4}$$

with $g_\tau(x(k))$ and $\mathscr{M}_\tau^j$ ($\tau = 1, \ldots, \wp,\ j = 1, \ldots, c$) being the premise variables and fuzzy sets, and control gains K_j to be designed later.

For j-th rule, its firing strength is defined as

$$\overrightarrow{h}_j(x(k)) = [\underline{h}_j(x(k)),\ \overline{h}_j(x(k))] \tag{4.5}$$

with $\overline{h}_j(x(k)) = \prod_{\tau=1}^{\wp} \overline{\mu}_{\mathscr{M}_\tau^j}(g_\tau(x(k))) \geq 0$, and $\underline{h}_j(x(k)) = \prod_{\tau=1}^{\wp} \underline{\mu}_{\mathscr{M}_\tau^j}(g_\tau(x(k))) \geq 0$, where $\overline{\mu}_{\mathscr{M}_\tau^j}(g_\tau(x(k)))$ and $\underline{\mu}_{\mathscr{M}_\tau^j}(g_\tau(x(k))) \in [0, 1]$ are the upper and lower grades of the MFs and satisfy $\underline{\mu}_{\mathscr{M}_\tau^j}(g_\tau(x(k))) \leq \overline{\mu}_{\mathscr{M}_\tau^j}(g_\tau(x(k)))$.

Then, one can design the following fuzzy controller

$$v(k) = \sum_{j=1}^{c} h_j(x(k))K_j x(k), \tag{4.6}$$

where the membership function $h_j(x(k))$ may be designed via the known upper MFs of $\overline{m}_j(x(k))$ and lower MFs $\underline{m}_j(x(k))$ of fuzzy system (4.2).

Remark 4.1 When system states $x(k)$ are successfully transmitted from sensors to the controller, the premise variables in fuzzy controller (4.4) and the fuzzy system (4.1) may be synchronously updated, which is called as the synchronous premise variables. As pointed out in [7], according to the varying range of system state $x(k)$ in each sub-domain, the local boundary information of MFs $h_j(x(k))$ in (4.10) and $m_i(x(k))$ in (4.2) can be obtained in local sub-domains of state $x(k)$. This is named as local-boundary-information-dependent (LBID) approach, which can gather more useful and exact information to reduce the conservatism in the stability analysis. For

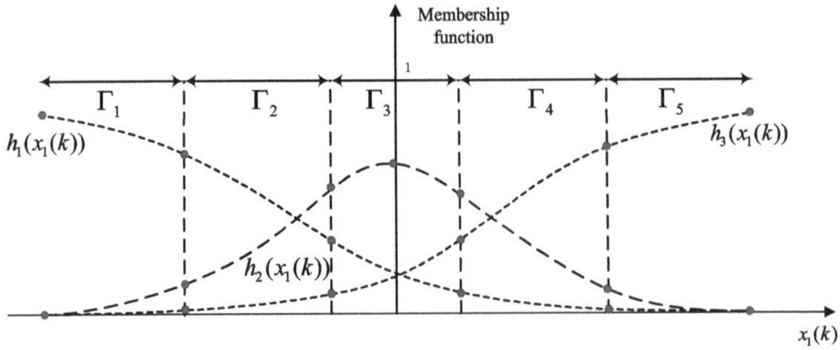

Fig. 4.1 Local upper and lower bounds of MFs

example, as shown in Fig. 4.1, according to the sub-domains $\Gamma_1, \ldots, \Gamma_5$ of system state $x_1(k)$, local upper and lower bounds (i.e., values at red dots) of MFs in (4.9) can be calculated and used in stability analysis for reducing the conservatism of the derived stability conditions.

4.2.3 Event-Triggered Communication

To relieve the transmission burden and avoid the network congestion under the limited resources, a conventional ETS as in (4.7) is usually used to regulate the transmission of state signal $x(k)$ from sensors to the controller (as shown in Fig. 4.2).

$$k_{l+1} = \inf\left\{k \in \mathbb{N} \mid k > k_l, e^T(k)\Omega e(k) \geq \rho x^T(k)\Omega x(k)\right\}, \tag{4.7}$$

where $e(k) \triangleq x(k_l) - x(k)$ is the difference between the current state and the latest triggered state, $\Omega > 0$ is the weighting matrix, $\rho \in (0, 1)$ is the triggering parameter, and $\{k_0, k_1, \ldots, k_l, \ldots\}(l \in \mathbb{N})$ is the triggering sequence.

It should be pointed out that under the ETS (4.7), the fuzzy controller can only utilize the triggered state $x(k_l)$, that is, controller's MFs will keep constant (i.e., $h_j(x(k_l)))$ for $k \in [k_l, k_{l+1})$, which will lead to the asynchronism of premise variables between the fuzzy controller and the controlled system (4.1) for the triggering interval (k_l, k_{l+1}). Due to the asynchronous MFs between the controller and system, the local information (i.e., the lower and upper boundaries during each local sub-domain) of the controller's MFs $h_j(x(k_l))$ cannot be obtained by using the system state $x(k)$, and even the relation between the system state $x(k)$ and its triggered value $x(k_l)$ is also difficult to establish. Thus, the existing MFD approach in [7] cannot be utilized directly. In addition, the conventional ETS (4.7) only focuses on the deviation of the system state $x(k)$, but it ignores the feature and change of MFs within the triggering interval. Hence, it is necessary to consider the derivation of MFs in ETS and facilitate the MFD analysis approach.

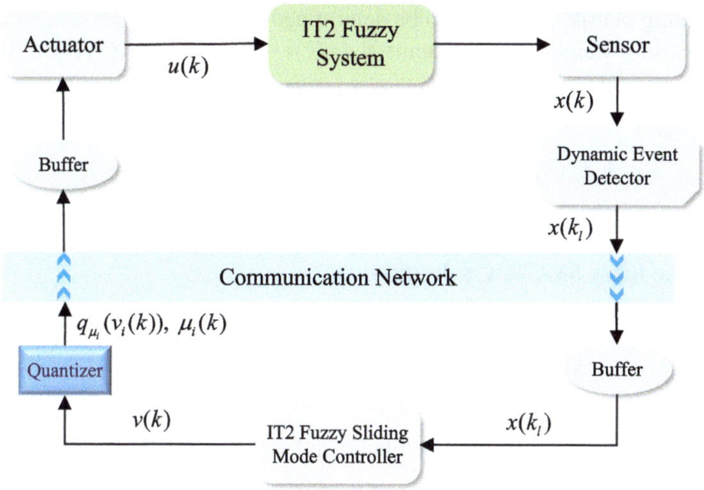

Fig. 4.2 Control system under DETS and dynamic quantization scheme

Based on the above discussion, the following objective is to design a novel ETS involving the deviation of MFs to govern the communication process of system state, and an effective fuzzy control scheme is proposed while the corresponding MFD stability conditions are derived via the MFs' local information.

4.3 Event-Triggered SMC Design and Performance Analysis

In this section, an SMC law is designed firstly for the IT2 T-S fuzzy system (4.1) suffering from the external disturbance $d(k)$. Then, a dynamic ETS involving the controller's MFs will be designed to determine the transmission of system states. Besides, the event-based control signal will be quantized via a dynamic quantizer to further improve the utilization of limited network resources in the C/A channels. Meanwhile, sufficient conditions will be derived for ensuring a desired performance of the IT2 fuzzy systems under the ETS and dynamic quantization.

4.3.1 IT2 Fuzzy Sliding Mode Control

Firstly, choose a sliding function as

$$s(k) = Gx(k), \tag{4.8}$$

with a sliding matrix $G \in \mathbb{R}^{m\times n}$ to be determined later.

By choosing weighting coefficients $\overline{\beta}_j(x(k)) \in [0, 1]$ and $\underline{\beta}_j(x(k)) \in [0, 1]$ satisfying $\underline{\beta}_j(x(k)) + \overline{\beta}_j(x(k)) = 1$, the following controller's MFs can be designed:

$$h_j(x(k)) = \overline{\beta}_j(x(k))\overline{h}_j(x(k)) + \underline{\beta}_j(x(k))\underline{h}_j(x(k)) \tag{4.9}$$

with $h_j(x(k)) \in [0, 1]$ and $\sum_{j=1}^{c} h_j(x(k)) = 1$.

Then, the fuzzy SMC law is designed as

$$v(k) = \sum_{j=1}^{c} h_j(x(k))K_j x(k) - \varpi \operatorname{sign}(s(k)), \tag{4.10}$$

with control gains K_j to be designed later, and ϖ being the known upper bound of the external disturbance $d(k)$.

4.3.2 *Membership-Function-Based Dynamic Event-Triggered Scheme*

To reduce the communication resource of network channels, the ETS is always utilized to regulate the transmission of system state. In this chapter, by considering the feature of the IT2 T-S fuzzy systems, a membership-function-based ETS will be proposed, which contains the state-based triggering condition and the membership-function-based triggering condition.

(i) *State-based triggering condition:* Firstly, we design the following state-based dynamic event-triggered condition as

$$e(k)^T \Omega e(k) \geq \rho x^T(k)\Omega x(k) + \frac{1}{\theta}\eta(k) \tag{4.11}$$

with the following updating rule of dynamic variable $\eta(k)$ as:

$$\eta(k+1) = \gamma\eta(k) - e^T(k)\Omega e(k) + \rho x^T(k)\Omega x(k), \tag{4.12}$$

where $e(k) \triangleq x(k_l) - x(k)$ is the difference between the current state and the latest triggered state, $\Omega > 0$ is the triggering weighting metric to be designed later, and $0 < \rho < 1$, $0 < \gamma < 1$, and $\theta > 0$ are the triggering parameters satisfying $\gamma\theta > 1$.

As discussed in Sect. 4.2.3, the triggered-based transmission manner of system state $x(k)$ in (4.11) will also result in the asynchronism between the controller's MFs and the system's MFs, which brings challenges on the local MFD analysis method.

Furthermore, the triggering condition (4.11) only focuses on the deviation of system state $x(k)$ over the triggering interval, ignoring the changes of MFs during triggering interval. This motivates us to further design the following membership-function-based triggering condition.

(ii) *Membership-function-based triggering condition:* Additionally, the following triggering condition with respect to MFs is formulated to judge the deviation of MFs:

$$|h_j(x(k)) - h_j(x(k_l))| \geq \tau_j, \tag{4.13}$$

with the triggering parameter $0 < \tau_j < 1$, which can also be given or optimized by combining the system performance index.

The membership-function-based triggering condition (4.13) implies that the triggered-state-based controller's MFs $h_j(x(k_l))$ can maintain within a small range around state-based MFs $h_j(x(k))$ in (4.9) without an ETS, i.e., $\max\{0, h_j(x(k)) - \tau_j\} \leq h_j(x(k_l)) \leq \min\{h_j(x(k)) + \tau_j, 1\}$. Taking the membership function $h_2(x_1(k))$ in Fig. 4.1 as an example, the membership function $h_2(x_1(k_l))$ under event-triggering condition (4.13) can be ensured within the range of red dotted lines. Obviously, the triggering condition (4.13) focuses on the variation of MFs, differing from the triggering condition (4.11) focusing on the variation of state signal $x(k)$. Hence, the introduction of condition (4.13) can avoid the situation that slight fluctuations of the state signal $x(k)$ lead to drastic changes in MFs.

By integrating triggering conditions (4.11) and (4.13), the membership-function-based dynamic event-triggered scheme (MFD-DETS) is obtained as

$$k_{l+1} = \inf\Big\{k \mid k > k_l,\ \psi_1(x(k), x(k_l)) \geq 0 \ \vee \ \psi_2(x(k), x(k_l)) \geq 0\Big\}, \tag{4.14}$$

with triggering conditions

$$\psi_1(x(k), x(k_l)) \triangleq e(k)^T \Omega e(k) - \rho x^T(k) \Omega x(k) - \frac{1}{\theta}\eta(k),$$

$$\psi_2(x(k), x(k_l)) \triangleq \max_{1 \leq j \leq r} \{|h_j(x(k)) - h_j(x(k_l))| - \tau_j\}.$$

Similar to Lemma 3.1 in Chap. 3, under the proposed dynamic event-triggering conditions in (4.11) with the updating rule (4.12), the following lemma can guarantee that $\eta(k) \geq 0$ holds for all $k \geq 0$.

Lemma 4.1 *For given $\eta(0) > 0$, parameters $0 < \gamma < 1$ and $\theta > 0$ satisfying $\gamma\theta > 1$, the dynamic parameter $\eta(k) \geq 0$ holds for all $k \geq 0$.*

Under the proposed DETS (4.14), we further redesign the following event-triggered SMC law (4.10) for $k \in [k_l, k_{l+1})$:

$$v(k) = \sum_{j=1}^{c} h_j(x(k_l))K_j x(k_l) - \varpi \operatorname{sign}(s(k_l)), \tag{4.15}$$

where the MFs $h_j(x(k_l))$ and $s(k_l)$ are obtained by replacing $x(k)$ with $x(k_l)$ in $h_j(x(k))$ and $s(k)$. Since the fuzzy controller (4.15) can only receive the latest triggered state signal $x(k_l)$ at the triggering instants to update the control input $v(k)$, the control signal $v(k)$ is only sent to actuator at the triggering instants. Obviously, the introduction of ETS can simultaneously improve the utilization of the limited communication resources in the S/C and C/A channels.

Remark 4.2 In the MFD-DETS (4.14), except for a conventional dynamic event-triggering condition $\psi_1(x(k), x(k_l)) \geq 0$, another condition $\psi_2(x(k), x(k_l)) \geq 0$ involving in controller's MFs is introduced for preventing a large deviation between MFs $h_j(x(k))$ and $h_j(x(k_l))$ due to the noncontinuous transmission. By means of the MFD triggering function $\psi_2(x(k), x(k_l))$, the range of the controller's MFs $h_j(x(k_l))$ can be explicitly represented via the system state $x(k)$ and controller's MFs $h_j(x(k))$, which is beneficial to analyze the local information of controller's MFs $h_j(x(k_l))$. Specifically, the controller's MFs satisfy $|h_j(x(k)) - h_j(x(k_l))| \leq \tau_j$ for $k \in [k_l, k_{l+1})$, from which we have $h_j(x(k_l)) \leq h_j(x(k)) + \tau_j$. Hence, we can utilize MFs $h_j(x(k))$ and the constant threshold τ_j to estimate the range of the MFs $h_j(x(k_l))$ under ETS, then the asynchronous problem between the controller and fuzzy system can be effectively handled while the *local* information of MFs $h_j(x(k))$ can be used in the stability analysis later. Recalling Fig. 4.3, we can find that under this triggering condition (4.13), the *local* upper and lower bounds of $h_2(x_1(k_l))$ in different sub-domains (i.e., $\Gamma_1, \ldots, \Gamma_5$) can be obtained by calculating and comparing those blue dots.

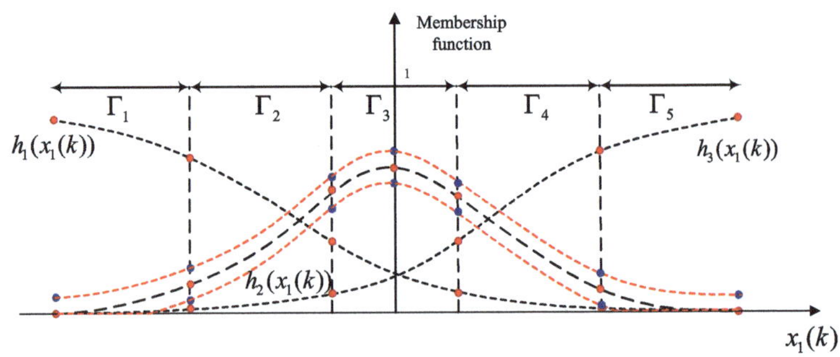

Fig. 4.3 Local boundaries of MFs under ETS (4.13)

4.3.3 Dynamic Quantization Scheme

As shown in Fig. 4.2, the control input $v(k) \triangleq [v_1(k), v_2(k), \ldots, v_m(k)]^T$ is firstly quantized before being transmitted to the actuator.

For a static uniform quantizer, the control input $v_\iota(k)$ $(\iota = 1, 2, \ldots, m)$ is quantized according to the following conditions:

$$\begin{aligned}&\text{if } |v_\iota(k)| \le M_\iota, \text{ then } |q_\iota(v_\iota(k)) - v_\iota(k)| \le \Delta_\iota,\\&\text{if } |v_\iota(k)| > M_\iota, \text{ then } |q_\iota(v_\iota(k))| > M_\iota - \Delta_\iota,\end{aligned}$$

where M_ι is the quantization level, $q_\iota(v_\iota(k))$ is quantized input signal, and Δ_ι is the quantization error.

For the purpose of enhancing the quantization procedure of the above static quantizer, we further adopt the following dynamic quantizer similar to [4]:

$$q_{\mu_\iota}(v_\iota(k)) = q_\iota\left(\frac{v_\iota(k)}{\mu_\iota(k)}\right)\mu_\iota(k), \tag{4.16}$$

which satisfies the following conditions:

$$\begin{aligned}&\text{if } |v_\iota(k)| \le M_\iota\mu_\iota(k), \text{ then } |q_{\mu_\iota}(v_\iota(k)) - v_\iota(k)| \le \Delta_\iota\mu_\iota(k),\\&\text{if } |v_\iota(k)| > M_\iota\mu_\iota(k), \text{ then } |q_{\mu_\iota}(v_\iota(k))| > (M_\iota - \Delta_\iota)\mu_\iota(k),\end{aligned}$$

where $\mu_\iota(k)$ is the dynamic quantization parameter to be designed.

Define the quantization error $\tilde{v}_\iota(k) \triangleq q_{\mu_\iota}(v_\iota(k)) - v_\iota(k)$, and dynamic quantization parameter $\mu(k) = [\mu_1(k), \mu_2(k), \ldots, \mu_m(k)]^T$. The quantized control input can be obtained as

$$u(k) = q_\mu(v(k)) = q\left(\frac{v(k)}{\mu(k)}\right)\mu(k) \tag{4.17}$$

with $q(\frac{v(k)}{\mu(k)}) = \text{diag}\left\{q_1(\frac{v_1(k)}{\mu_1(k)}), q_2(\frac{v_2(k)}{\mu_2(k)}), \ldots, q_m(\frac{v_m(k)}{\mu_m(k)})\right\}$.

By substituting quantized SMC law (4.15) and (4.17) into the IT2 fuzzy systems (4.2), the closed-loop control systems can be obtained as

$$\begin{aligned}x(k+1) = \sum_{i=1}^{r}\sum_{j=1}^{c} m_i(x(k))h_j(x(k_l))\Big\{&\mathcal{A}_{ij}x(k) + B_iK_je(k)\\&+ B_i\tilde{v}(k) + B_i(d(k) - \varpi\,\text{sign}(s(k_l)))\Big\},\end{aligned} \tag{4.18}$$

with $\mathcal{A}_{ij} \triangleq A_i + B_iK_j$ and $\tilde{v}(k) \triangleq [\tilde{v}_1(k), \ldots, \tilde{v}_m(k)]^T$.

The following definition is given for subsequent analysis.

Definition 4.1 The IT2 fuzzy system (4.18) is said to be ultimately bounded if there exist a constant $\chi > 0$ such that the system state $x(k)$ satisfies

$$\lim_{k \to \infty} ||x(k)|| \leq \chi, \tag{4.19}$$

where χ is an ultimate upper bound.

4.3.4 Analysis of Stability and Reachability

In this section, we will analyze the stability of the closed-loop systems (4.18) and the reachability of the specified sliding surface $s(k) = 0$, and the corresponding sufficient conditions will be derived.

4.3.4.1 Stability Analysis

Under the event-triggered fuzzy sliding mode controller (4.15) with dynamic quantization (4.17), the ultimate boundedness of the closed-loop system (4.18) can be attained under the following conditions.

Theorem 4.1 *Given parameter $\kappa_1 > 0$, quantization range $M_\iota > 0$ with $\Delta_\iota > 0$, triggering parameters $\rho > 0$, $\theta > 0$ and $\gamma \in (0,\ 1)$ satisfying $\gamma\theta > 1$, if there exist matrices $P > 0$, $\Omega > 0$, G, K_j, and scalars $\lambda > 0$, $\zeta_\iota > 0$ satisfying the following conditions (for $i = 1, \ldots, r$, $j = 1, \ldots, c$, and $\iota = 1, \ldots, m$)*

$$\sqrt{\zeta_\iota} \geq \frac{1}{M_\iota}, \tag{4.20}$$

$$\sum_{i=1}^{r}\sum_{j=1}^{c} m_i(x(k))h_j(x(k_l))\Pi_{ij} < 0 \tag{4.21}$$

with

$$\begin{aligned}
\Pi_{ij} &\triangleq \mathrm{diag}\{-P + \bar{\theta}\rho\Omega, -\bar{\theta}\Omega, -\lambda I, -\bar{\theta}(1-\gamma)I\} \\
&\quad + (1+\kappa_1)\mathcal{M}^T P \mathcal{M} + 8\lambda\zeta\mathcal{N}^T\mathcal{N}, \\
\mathcal{M} &\triangleq [A_i + B_i K_j,\ B_i K_j,\ B_i,\ 0],\ \mathcal{N} \triangleq [\Delta K_j, \Delta K_j, 0, 0], \\
\zeta &\triangleq \mathrm{diag}\{\zeta_1, \zeta_2, \ldots, \zeta_m\}, \Delta \triangleq \mathrm{diag}\{\Delta_1, \Delta_2, \ldots, \Delta_m\},
\end{aligned}$$

and $\bar{\theta} \triangleq 1/\theta$, then the IT2 T-S fuzzy system (4.18) under the MFD-DETS (4.14) and input quantization (4.17) is ultimately bounded, and the system state $x(k)$ can ultimately enter the following domain:

$$\mathcal{B} = \left\{ x(k) \mid \|x(k)\| \le \sqrt{\frac{\bar{\varphi}}{\vartheta}} \right\}, \tag{4.22}$$

with $\bar{\varphi} \triangleq (1+\kappa_1^{-1})(1+\sqrt{m})^2 \max_{1\le i\le r}\{\lambda_{\max}(B_i^T P B_i)\}\varpi^2 + 8\lambda\varpi^2 \sum_{\iota=1}^{m} \Delta_\iota^2 \zeta_\iota$, $\vartheta \triangleq \min\{\bar{\vartheta}(k)\}$, *and* $\bar{\vartheta}(k) \triangleq \lambda_{\min}\Big(-\sum_{i=1}^{r}\sum_{j=1}^{c} m_i(x(k))h_j(x(k_l))\Pi_{ij}\Big)$.

Proof Choose a Lyapunov function as

$$V(x(k), \eta(k)) = x^T(k)Px(k) + \frac{1}{\theta}\eta(k) \tag{4.23}$$

with $\eta(k) > 0$ according to Lemma 4.1.

One has

$$\Delta V(x(k), \eta(k)) = x^T(k+1)Px(k+1) + \frac{1}{\theta}\eta(k+1) - V(x(k), \eta(k)). \tag{4.24}$$

Along the closed-loop system (4.18) under the event-based fuzzy SMC law (4.15), one has

$$\begin{aligned}
&\Delta V(x(k), \eta(k)) \\
&\le \sum_{i=1}^{r}\sum_{j=1}^{c} m_i(x(k))h_j(x(k_l))\Big\{\Big[\mathcal{A}_{ij}x(k) + B_iK_je(k) + B_i\tilde{v}(k) + B_iF(k)\Big]^T \\
&\times P\Big[\mathcal{A}_{ij}x(k) + B_iK_je(k) + B_i\tilde{v}(k) + B_iF(k)\Big] + \frac{1}{\theta}[\gamma\eta(k) - e(k)^T\Omega e(k) \\
&+ \rho x(k)^T\Omega x(k)] - x^T(k)Px(k) - \frac{1}{\theta}\eta(k)\Big\} \\
&\le \sum_{i=1}^{r}\sum_{j=1}^{c} m_i(x(k))h_j(x(k_l))\Big\{(1+\kappa_1)\Big[\mathcal{A}_{ij}x(k) + B_i(K_je(k) + \tilde{v}(k))\Big]^T \\
&\times P\Big[\mathcal{A}_{ij}x(k) + B_i(K_je(k) + \tilde{v}(k))\Big] + (1+\kappa_1^{-1})F^T(k)B_i^TPB_iF(k) \\
&- x^T(k)Px(k) + \bar{\theta}\rho x^T(k)\Omega x(k) - \bar{\theta}e^T(k)\Omega e(k) - \bar{\theta}(1-\gamma)\eta(k)\Big\}
\end{aligned} \tag{4.25}$$

with $\kappa_1 > 0$, $\bar{\theta} \triangleq 1/\theta$ and $F(k) \triangleq d(k) - \varpi\,\mathrm{sign}(s(k_l))$.

In addition, according to the property of quantization, the condition $\sqrt{\zeta_\iota}|v_\iota(k)| \le \mu_\iota(k)$ in (4.51) implies that input quantizer is not saturate. Thus, we can obtain $|\tilde{v}_\iota(k)| = |q_{\mu_\iota}(v_\iota(k)) - v_\iota(k)| \le \Delta_\iota\mu_\iota(k)$. By means of (4.51), one has $|\tilde{v}_\iota(k)| \le 2\Delta_\iota\sqrt{\zeta_\iota}|v_\iota(k)|$.

Furthermore, by means of $||\tilde{v}(k)|| \leq 2||\Delta\sqrt{\zeta}v(k)||$, one has

$$\begin{aligned}
\tilde{v}^T(k)\tilde{v}(k) &\leq 4v^T(k)\Delta^2\zeta v(k)\\
&\leq 4\Big(\sum_{j=1}^{c} h_j(x(k_l))K_jx(k_l) - \varpi\,\text{sign}(s(k_l))\Big)^T\Delta^2\zeta\\
&\times\Big(\sum_{j=1}^{c} h_j(x(k_l))K_jx(k_l) - \varpi\,\text{sign}(s(k_l))\Big)\\
&\leq 8\sum_{j=1}^{c} h_j(x(k_l))(x(k)+e(k))^TK_j^T\Delta^2\zeta K_j(x(k)+e(k))\\
&+ 8\varpi^2\sum_{l=1}^{m}\Delta_l^2\zeta_l.
\end{aligned} \tag{4.26}$$

Due to $||F(k)|| \leq (1+\sqrt{m})\varpi$, it can be obtained from conditions (4.25) and (4.26) that

$$\begin{aligned}
&\Delta V(x(k),\eta(k))\\
&\leq \sum_{i=1}^{r}\sum_{j=1}^{c} m_i(x(k))h_j(x(k_l))\Big\{(1+\kappa_1)\Big[\mathcal{A}_{ij}x(k) + B_i(K_je(k)+\tilde{v}(k))\Big]^T\\
&\times P\Big[\mathcal{A}_{ij}x(k) + B_i(K_je(k)+\tilde{v}(k))\Big] - x^T(k)Px(k) + \bar{\theta}\rho x^T(k)\Omega x(k)\\
&- \bar{\theta}e^T(k)\Omega e(k) - \bar{\theta}(1-\gamma)\eta(k) + (1+\kappa_1^{-1})(1+\sqrt{m})^2\lambda_{\max}(B_i^TPB_i)\varpi^2\\
&+ \lambda\Big[8(x(k)+e(k))^TK_j^T\Delta^2\zeta K_j(x(k)+e(k)) + 8\varpi^2\sum_{l=1}^{m}\Delta_l^2\zeta_l - \tilde{v}^T(k)\tilde{v}(k)\Big]\Big\}\\
&\leq \sum_{i=1}^{r}\sum_{j=1}^{c} m_i(x(k))h_j(x(k_l))\phi^T(k)\Pi_{ij}\phi(k) + \bar{\varphi}
\end{aligned} \tag{4.27}$$

with $\phi(k) \triangleq \Big[x^T(k), e^T(k), \tilde{v}^T(k), \sqrt{\eta(k)}\Big]^T$.

If the condition (4.21) holds, we have from (4.27)

$$\Delta V(k) \leq -\vartheta||\phi(k)||^2 + \bar{\varphi} \leq 0, \tag{4.28}$$

with a positive scalar $\vartheta = \min\{\bar{\vartheta}(k)\}$ and $\bar{\vartheta}(k) \triangleq \lambda_{\min}\Big(-\sum_{i=1}^{r}\sum_{j=1}^{c} m_i(x(k))h_j(x(k_l))\Pi_{ij}\Big)$.

Due to $||x(k)|| \leq ||\phi(k)||$, if $||x(k)|| > \sqrt{\frac{\bar{\varphi}}{\vartheta}}$, one has

$$\Delta V(x(k), \eta(k)) \leq -\vartheta\left(||\phi(k)||^2 - \frac{\bar{\varphi}}{\vartheta}\right) < 0. \tag{4.29}$$

According to Definition 4.1, the expression (4.29) means that system state $x(k)$ of the closed-loop system (4.18) can ultimately converge into the domain $\mathcal{B}$. □

From the convergence domain $\mathcal{B}$ in (4.22), it can be seen that both the quantization error Δ_ι and the parameter ζ_ι related to quantization range M_ι will affect the upper bound of convergence domain of system state $x(k)$. In addition, the event-triggering parameters ρ, θ, and Ω affect the solvability of sufficient conditions (4.20) and (4.21), and have an implicit effect on the convergence domain. In other words, the influence of ETS and dynamic quantization strategy on control performance is reflected via the convergence domain. A similar phenomenon can be found in Theorem 4.2.

4.3.4.2 Reachability Analysis

In what follows, the reachability of the sliding surface $s(k) = 0$ will be analyzed.

Theorem 4.2 *Given parameters $\kappa_1 > 0$, $\kappa_2 > 0$, quantization range $M_\iota > 0$ with $\Delta_\iota > 0$, triggering parameters $\rho > 0$, $\theta \in (0,\ 1)$ and $\gamma > 0$ satisfying $\theta\gamma > 1$, if there exist matrices $P > 0$, $R > 0$, $\Omega > 0$, G, K_j, and scalars $\zeta_\iota > 0$, $\lambda > 0$ that satisfy conditions (4.20)–(4.21) and the following condition (for $i = 1, \ldots, r$, $j = 1, \ldots, c$, and $\iota = 1, \ldots, m$)*

$$\sum_{i=1}^{r}\sum_{j=1}^{c} m_i(x(k))h_j(x(k_l))\Phi_{ij} < 0 \tag{4.30}$$

with $\Phi_{ij} \triangleq \Pi_{ij} + (1+\kappa_2)\mathcal{H}^T G^T R G \mathcal{H}$ and $\mathcal{H} \triangleq [A_i + B_i K_j,\ B_i K_j,\ B_i,\ 0]$, then the event-triggered fuzzy SMC law (4.15) can guarantee the system state of the closed-loop system (4.18) to converge into the following sliding domain:

$$\mathcal{D} = \left\{s(k) \mid \|s(k)\| \leq \hat{\varphi}\right\} \tag{4.31}$$

with $\hat{\varphi} \triangleq \max\limits_{1\leq i\leq r} \sqrt{\frac{\varphi_i}{\lambda_{\min}(R)}}$, $\varphi_i \triangleq \bar{\varphi} + (1+\kappa_2^{-1})(1+\sqrt{m})^2\lambda_{\max}((GB_i)^T RGB_i)\varpi^2$ and $\bar{\varphi}$ defined in (4.22).

Proof Choose the Lyapunov function $U(x(k), \eta(k))$ as

$$U(x(k), \eta(k)) = V(x(k), \eta(k)) + s^T(k)Rs(k) \tag{4.32}$$

with $V(x(k), \eta(k))$ defined in (4.23).

One has

$$\Delta U(x(k), \eta(k)) = \Delta V(x(k), \eta(k)) + s^T(k+1)Rs(k+1) - s^T(k)Rs(k). \quad (4.33)$$

In the light of the expressions (4.8) and (4.18), one has

$$\begin{aligned} s(k+1) = \sum_{i=1}^{r}\sum_{j=1}^{c} m_i(x(k))h_j(x(k_l))\Big\{G[\mathcal{A}_{ij}x(k) \\ + B_i(K_j e(k) + \tilde{v}(k) + F(k))]\Big\}, \end{aligned} \quad (4.34)$$

from which we can obtain

$$\begin{aligned} & s^T(k+1)Rs(k+1) \\ & \leq \sum_{i=1}^{r}\sum_{j=1}^{c} m_i(x(k))h_j(x(k_l))\Big\{(1+\kappa_2)\Big[\mathcal{A}_{ij}x(k) + B_i(K_j e(k) + \tilde{v}(k))\Big]^T G^T \\ & \times RG\Big[\mathcal{A}_{ij}x(k) + B_i(K_j e(k) + \tilde{v}(k))\Big] + (1+\kappa_2^{-1})(1+\sqrt{m})^2 \\ & \times \lambda_{\max}(B_i^T G^T RGB_i)\varpi^2\Big\}, \end{aligned} \quad (4.35)$$

with a scalar $\kappa_2 > 0$ and $F(k) \triangleq d(k) - \varpi \operatorname{sign}(s(k_l))$.
Substituting (4.35) and (4.25) into (4.33) yields

$$\begin{aligned} & \Delta U(x(k), \eta(k)) \\ & \leq \sum_{i=1}^{r}\sum_{j=1}^{c} m_i(x(k))h_j(x(k_l))\Big\{\phi^T(k)\Pi_{ijl}\phi(k) + (1+\kappa_2^{-1})\Big[\mathcal{A}_{ij}x(k) + B_i(K_j e(k) \\ & + \tilde{v}(k))\Big]^T G^T RG\Big[\mathcal{A}_{ij}x(k) + B_i(K_j e(k) + \tilde{v}(k))\Big] - [\lambda_{\min}(R)||s(k)||^2 - \bar{\varphi} \\ & - (1+\kappa_2^{-1})(1+\sqrt{m})^2\lambda_{\max}(B_i^T G^T RGB_i)\varpi^2]\Big\} \\ & \leq \sum_{i=1}^{r}\sum_{j=1}^{c} m_i(x(k))h_j(x(k_l))\Big[\phi^T(k)\Phi_{ij}\phi(k) - (\lambda_{\min}(R)||s(k)||^2 - \varphi_i)\Big]. \end{aligned} \quad (4.36)$$

If the state trajectories remain outside the domain $\mathcal{D}$, i.e., $||s(k)|| > \hat{\varphi}$, ($\hat{\varphi}$ is defined in (4.31)), the condition (4.30) can ensure

$$\Delta U(x(k), \eta(k)) < 0, \quad (4.37)$$

which implies that the state trajectories of the closed-loop system (4.18) can be finally driven into the sliding domain $\mathcal{D}$. □

According to Theorems 4.1 and 4.2, the state trajectories under the proposed event-triggered quantized SMC control scheme can be driven into the sliding domain $\mathcal{D}$, then go along this domain into the bounded neighborhood $\mathcal{B}$ around the origin. This is the feature of discrete-time SMC. However, in Theorems 4.1 and 4.2, sufficient conditions involving the controller's and system's MFs are not linear and difficult to solve. To remove the obstacle, these conditions will be further transformed into some solvable results while utilizing enough useful information about MFs for reducing the conservatism, meanwhile, the relation between control performance and communication resource consumption will also be evaluated in the following section.

4.4 Sufficient Conditions and Performance Optimization

In this section, by means of the LBID method, sufficient conditions for stability and reachability will be given. Additionally, the design parameters will be optimized to enhance control performance. Furthermore, an online updating rule for implementing the dynamic quantizer in practical applications will be presented.

4.4.1 LBID Sufficient Conditions

Since the condition (4.21) in Theorem 4.1 and the condition (4.30) in Theorem 4.2 involve time-varying MFs $m_i(x(k))$ and $h_j(x(k_l))$, they cannot be directly solved. Besides, under the ETS (4.14), the mismatched MFs $m_i(x(k))$ and $h_j(x(k_l))$ are dependent on the state $x(k)$ and the triggered value $x(k_l)$, respectively. Thus, it is difficult to obtain the local information of the controller' MFs $h_j(x(k_l))$ according to system state $x(k)$, so that the existing local-boundary-information-dependent methods in [7] are inapplicable for establishing the MFD conditions with less conservatism. Apparently, if we directly ignore the MFs and obtain the MFI results, i.e., $\Phi_{ij} < 0$, it will increase more conservatism in stability conditions. To circumvent this dilemma, we will develop the MFD stability results to reduce the conservatism by using the relation in the triggering condition $\psi_2(x(k), x(k_l))$ in (4.14). Specifically, by partitioning the modeling domain of the system state $x(k)$ into multiple subdomains, the local boundaries of MFs $h_j(x(k))$ can be determined. Leveraging these boundaries and the upper bound τ_j in membership-function-based triggering condition (4.14), the local boundaries of MFs $h_j(x(k_l))$ for the event-triggered fuzzy controller can then be estimated. Based on the above analysis, we can utilize the local information of both system's MFs and controller's ones to derive the MFD results and reduce the conservatism. To this end, the modeling domain of the controlled system, Γ, is divided into $\Gamma = \bigcup_{d=1}^{D} \Gamma_d$, in which D represents the divided subdomains. It is assumed that the lower and upper boundaries of $m_i(x(k))h_j(x(k))$ over d-th sub-domain are denoted as $\underline{h}_{ijd}$ and $\bar{h}_{ijd}$ that satisfy $\underline{h}_{ijd} \le m_i(x(k))h_j(x(k)) \le \bar{h}_{ijd}$, which can be used to obtain the following LBID conditions.

Theorem 4.3 *Given parameters $\kappa_1 > 0$, $\kappa_2 > 0$, quantization range $M_\iota > 0$ with $\Delta_\iota > 0$, triggering parameters $\rho > 0$, $\tau_j \in (0, 1)$, $\theta \in (0, 1)$ and $\gamma > 0$ satisfying $\theta\gamma > 1$, if there exist matrices $X > 0$, $\hat{R} > 0$, $\hat{\Omega} > 0$, $\bar{\Lambda}_{ijd} \ge 0$, $\hat{\Upsilon}_{ijd} \ge 0$, $\hat{\Xi}_{ijd} \ge 0$, G, Y_j, and scalars $\xi_\iota > 0$, $\hat{\lambda} > 0$ satisfying the following conditions (for $i = 1, \ldots, r$, $j = 1, \ldots, c$, and $\iota = 1, \ldots, m$)*

$$\xi_\iota - \hat{\lambda} M_\iota^2 \le 0, \tag{4.38}$$

$$\Psi_{ij} - \hat{\Lambda}_{ijd} \le 0, \tag{4.39}$$

$$\Psi_{ij} + \sum_{j=1}^{p} \tau_j \hat{\Lambda}_{ijd} - \hat{\Upsilon}_{ijd} + \hat{\Xi}_{ijd} + \sum_{a=1}^{r}\sum_{b=1}^{p} (\bar{h}_{abd}\hat{\Upsilon}_{abd} - \underline{h}_{abd}\hat{\Xi}_{abd}) < 0, \tag{4.40}$$

with

$$\Psi_{ij} = \begin{bmatrix} \Psi^{11} & * & * & * \\ \Psi^{21} & -\frac{1}{1+\kappa_1}X & * & * \\ \Psi^{31} & 0 & -\frac{1}{2}\xi & * \\ \Psi^{41} & 0 & 0 & -\frac{1}{1+\kappa_2}\hat{R} \end{bmatrix}$$

$$\begin{aligned}
\Psi^{11} &= \mathrm{diag}\{-X + \bar{\theta}\rho\hat{\Omega}, -\bar{\theta}\hat{\Omega}, -\hat{\lambda}I, -\bar{\theta}(1-\gamma)I\}, \\
\Psi^{21} &= \begin{bmatrix} A_iX + B_iY_j & B_iY_j & \hat{\lambda}B_i & 0 \end{bmatrix}, \\
\Psi^{31} &= \begin{bmatrix} 2\Delta Y_j & 2\Delta Y_j & 0 & 0 \end{bmatrix}, \\
\Psi^{41} &= \begin{bmatrix} GA_iX + GB_iY_j & GB_iY_j & \hat{\lambda}GB_i & 0 \end{bmatrix}, \\
\xi &= \mathrm{diag}\{\xi_1, \xi_2, \ldots, \xi_m\}, \Delta = \mathrm{diag}\{\Delta_1, \Delta_2, \ldots, \Delta_m\}, \\
\hat{\Lambda}_{ijd} &= \mathrm{diag}\{\bar{\Lambda}_{ijd}, 0, 0, 0\}, \bar{\theta} = 1/\theta,
\end{aligned}$$

then, the results in Theorems 4.1 and 4.2 can be guaranteed, simultaneously. Meanwhile, we can obtain the control gains $K_j = Y_jX^{-1}$, the weighting matrix $\Omega = X^{-1}\hat{\Omega}X^{-1}$, and $\zeta_\iota = \hat{\lambda}/\xi_\iota$.

Proof When system state lies in d-th sub-domain, i.e., $x(k) \in \Gamma_d$, by resorting to the triggering condition $\psi_2(x(k), x(k_l)) \le 0$ for $k \in [k_l, k_{l+1})$, we can obtain from the condition (4.30) in Theorem 4.2 that

$$\begin{aligned}
&\sum_{i=1}^{r}\sum_{j=1}^{c} m_i(x(k))h_j(x(k_l))\Phi_{ij} \\
&= \sum_{i=1}^{r}\sum_{j=1}^{c} m_i(x(k))(h_j(x(k)) + \Delta h_j(k))\Phi_{ij} \\
&\leq \sum_{i=1}^{r}\sum_{j=1}^{c} m_i(x(k))h_j(x(k))\Phi_{ij} + \sum_{i=1}^{r}\sum_{j=1}^{c} m_i(x(k))\tau_j\Lambda_{ijd} \\
&= \sum_{i=1}^{r}\sum_{j=1}^{c} m_i(x(k))h_j(x(k))(\Phi_{ij} + \sum_{j=1}^{c}\tau_j\Lambda_{ijd}),
\end{aligned} \tag{4.41}$$

with $\Delta h_j(k) \triangleq h_j(x(k_l)) - h_j(x(k))$ and slack matrices $\Lambda_{ijd} = \Lambda_{ijd}^T > 0$ satisfying $\Lambda_{ijd} \geq \Phi_{ij}$.

For $\Upsilon_{ijd} = \Upsilon_{ijd}^T \geq 0$ and $\Xi_{ijd} = \Xi_{ijd}^T \geq 0$, by using the property of $\underline{h}_{ijd} \leq m_i(x(k))h_j(x(k)) \leq \bar{h}_{ijd}$, one has

$$\sum_{i=1}^{r}\sum_{j=1}^{c}(\bar{h}_{ijd} - m_i(x(k))h_j(x(k)))\Upsilon_{ijd} \geq 0, \tag{4.42}$$

$$\sum_{i=1}^{r}\sum_{j=1}^{c}(m_i(x(k))h_j(x(k)) - \underline{h}_{ijd})\Xi_{ijd} \geq 0. \tag{4.43}$$

Then, it can be deduced for $x(k) \in \Gamma_d$ that

$$\begin{aligned}
&\sum_{i=1}^{r}\sum_{j=1}^{c} m_i(x(k))h_j(x(k_l))\Phi_{ij} \\
&\leq \sum_{i=1}^{r}\sum_{j=1}^{c}\Bigg\{ m_i(x(k))h_j(x(k))\left(\Phi_{ij} + \sum_{j=1}^{c}\tau_j\Lambda_{ijd}\right) \\
&+ (\bar{h}_{ijd} - m_i(x(k))h_j(x(k)))\Upsilon_{ijd} \\
&+ (m_i(x(k))h_j(x(k)) - \underline{h}_{ijd})\Xi_{ijd}\Bigg\} \\
&= \sum_{i=1}^{r}\sum_{j=1}^{c} m_i(x(k))h_j(x(k))\Bigg[\Phi_{ij} + \sum_{j=1}^{c}\tau_j\Lambda_{ijd} - \Upsilon_{ijd} + \Xi_{ijd} \\
&+ \sum_{a=1}^{r}\sum_{b=1}^{c}(\bar{h}_{abd}\Upsilon_{abd} - \underline{h}_{abd}\Xi_{abd})\Bigg].
\end{aligned} \tag{4.44}$$

Thus, the following condition can ensure condition (4.30)

$$\Phi_{ij}+\sum_{j=1}^{c}\tau_j\Lambda_{ijd}-\Upsilon_{ijd}+\Xi_{ijd}+\sum_{a=1}^{r}\sum_{b=1}^{c}(\bar{h}_{abd}\Upsilon_{abd}-\underline{h}_{abd}\Xi_{abd})<0 \quad (4.45)$$

for $d=1,\ldots,D$.

By Schur complement, the condition (4.45) is equivalent to

$$\boldsymbol{\Psi}_{ij}+\sum_{j=1}^{c}\tau_j\boldsymbol{\Lambda}_{ijd}-\boldsymbol{\Upsilon}_{ij}-\boldsymbol{\Xi}_{ij}+\sum_{a=1}^{r}\sum_{b=1}^{c}(\bar{h}_{abd}\boldsymbol{\Upsilon}_{abd}-\underline{h}_{abd}\boldsymbol{\Xi}_{abd})<0 \quad (4.46)$$

with

$$\boldsymbol{\Psi}_{ij}=\begin{bmatrix}\boldsymbol{\Psi}^{11} & * & * & * \\ \boldsymbol{\Psi}^{21} & -\frac{1}{1+\kappa_1}P^{-1} & * & * \\ \boldsymbol{\Psi}^{31} & 0 & -\frac{\lambda^{-1}\zeta^{-1}}{2}I & * \\ \boldsymbol{\Psi}^{41} & 0 & 0 & -\frac{1}{1+\kappa_2}R^{-1}\end{bmatrix}$$

$$\begin{aligned}
\boldsymbol{\Psi}^{11}&=\mathrm{diag}\{-P+\bar{\theta}\rho\Omega,-\bar{\theta}\Omega,-\lambda I,-\bar{\theta}(1-\gamma)I\},\\
\boldsymbol{\Psi}^{21}&=\begin{bmatrix}A_i+B_iK_j & B_iK_j & B_i & 0\end{bmatrix},\\
\boldsymbol{\Psi}^{31}&=\begin{bmatrix}2\Delta K_j & 2\Delta K_j & 0 & 0\end{bmatrix}\\
\boldsymbol{\Psi}^{41}&=\begin{bmatrix}GA_i+GB_iK_j & GB_iK_j & GB_i & 0\end{bmatrix},\\
\boldsymbol{\Lambda}_{ijd}&=\mathrm{diag}\{\Lambda_{ijd},0,0,0\}.
\end{aligned}$$

Pre- and post-multiplying $\mathrm{diag}\{P^{-1},P^{-1},\lambda^{-1},I,I,I,I\}$, and letting $X=P^{-1}$, $\hat{\lambda}=\lambda^{-1}$, $Y_j=K_jX$, $\hat{\Omega}=X\Omega X$, $\xi_l=\hat{\lambda}\zeta_l^{-1}$, $\bar{\Lambda}_{ijd}=\mathrm{diag}\{X,X,\hat{\lambda},I\}\Lambda_{ijd}\mathrm{diag}\{X,X,\hat{\lambda},I\}$, it is seen that condition (4.40) can ensure condition (4.30). Similarly, the condition (4.39) can ensure $\Phi_{ij}\leq\Lambda_{ijd}$. □

Remark 4.3 In the expression (4.41), the membership function $h_j(x(k_l))$ of the event-triggered fuzzy controller is expressed as a combination of the real-time state-based membership function $h_j(x(k))$ and the difference $\Delta h_j(k)$. According to the triggering condition $\psi_2(x(k),x(k_l))$ in (4.14), the difference $\Delta h_j(k)$ is bounded by a constant τ_j. Based on this relationship, it can calculate the local boundary information of the membership function $m_i(x(k))h_j(x(k))$, (i.e., $\underline{h}_{ijd}$ and $\bar{h}_{ijd}$ satisfying $\underline{h}_{ijd}\leq m_i(x(k))h_j(x(k))\leq\bar{h}_{ijd}$), and then incorporate the triggering threshold τ_i, thereby enabling to derive the MFD stability condition (4.40). This approach addresses skillfully the difficulty that the local boundaries of mismatching membership function $m_i(x(k))h_j(x(k_l))$ cannot be directly calculated due to the event-triggering transmission. It should be pointed out that the term $\sum_{j=1}^{c}\tau_j\hat{\Lambda}_{ijd}$ in (4.40) arises from the ETS (4.14). If the triggering thresholds $\tau_j=0$ for all $j=1,\ldots,p$, the conditions in Theorem 4.3 reduce to the results under periodic sampling.

In what follows, a tradeoff issue with respect to the system performance and the communication resource consumption can be dealt with by establishing an optimization problem. Meanwhile, an online adjustment strategy will be given so that the dynamic quantizer's parameters are available for the controller and the actuator.

4.4.2 Optimization of Triggering Threshold and Control Performance

It should be pointed out that the nonlinear terms GA_iX and GB_iY_j in Theorem 4.3 make conditions (4.39) and (4.40) are difficult to solve. If the sliding matrix G is given aforehand, conditions (4.39) and (4.40) will become linear. However, the choice of matrix G may affect the sliding domain $\mathcal{D}$ and the feasibility of stability conditions. To improve the SMC performance, we can minimize $\hat{\varphi}$ in (4.31) to obtain a smaller sliding domain $\mathcal{D}$. However, a smaller $\hat{\varphi}$ may also require more resource consumption and triggering.

Notice that two triggering functions $\psi_1(x(k), x(k_l)) \geq 0$ and $\psi_2(x(k), x(k_l)) \geq 0$ in MFD-DETS (4.14) are used to conjointly determine the triggering instants. For the first event-triggering condition $\psi_1(x(k), x(k_l))$, a larger parameter ρ may bring a less triggering number for reducing the transmission frequency. For the second event-triggering condition $\psi_2(x(k), x(k_l))$, a set of larger parameters τ_j is also beneficial to reduce the triggering number. Thus, if the triggering parameters ρ and τ_j are maximized, a less communication burden can be achieved.

By comprehensively considering both control performance and communication resource consumption, the following multi-objective optimization function is formulated:

$$F_{obj} = \varepsilon_1 \frac{1}{\rho} + \varepsilon_2 \sum_{j=1}^{c} \frac{1}{\tau_j} + \varepsilon_3 \hat{\varphi} \tag{4.47}$$

where ε_1, ε_2, and ε_3 are adjustable weighting parameters satisfying $\varepsilon_1, \varepsilon_2, \varepsilon_3 \in [0, 1]$ and $\sum_{t=1}^{3} \varepsilon_t = 1$.

Furthermore, the following optimization problem is established:

$$\begin{aligned} &\min_{\rho, \tau_j, G} F_{obj} \\ &\text{s.t. conditions (4.38)–(4.40).} \end{aligned} \tag{4.48}$$

The above optimization problem (4.48) can be addressed by using various smart optimization approaches, e.g., genetic algorithm [8] and PSO algorithm used in Chaps. 2 and 3. Here, the PSO algorithm is employed to solve the optimization problem (4.48) for obtaining the optimal sliding matrix G, parameters ρ, and τ_j. To this end, the following PSO-based sliding mode quantized control design algorithm is given.

Algorithm 1 PSO-based sliding mode quantized control design algorithm

Initialize Particle

1: Give parameters: A_i, B_i, M_ι, parameters Δ_ι, κ_1, κ_2, θ and γ.

2: For the matrix $G \in \mathbb{R}^{m\times n}$ and triggering parameters ρ, τ_j $(j \in \mathbb{L})$ containing $\bar{n} \triangleq m \times n + p + 1$ independent elements, they can be encoded as a particle:

$$\mathrm{w} \triangleq [g_{11}, \ldots, g_{1n}, \ g_{21}, \ldots, g_{mn}, \tau_1, \ldots, \tau_p, \rho].$$

3: Denoting the initial step as $l = 1$, and randomly generate an initial swarm with N particles as $\mathrm{w}^l = \{\mathrm{w}_1^l, \mathrm{w}_2^l, \ldots, \mathrm{w}_N^l\}$, where its position and velocity of t-th particle are denoted as $\mathrm{w}_t^l = [\mathrm{w}_{1t}^l, \ldots, \mathrm{w}_{\bar{n}t}^l]$ and $\mathrm{v}_t^l = [\mathrm{v}_{1t}^l, \ldots, \mathrm{v}_{\bar{n}t}^l]$. Given the search scope of position and velocity as $S = \{\mathrm{w}_{\wp t}, \mathrm{v}_{\wp t} \mid \mathrm{w}_{\wp t} \in [\underline{\mathrm{w}}_\wp, \overline{\mathrm{w}}_\wp], \mathrm{v}_{\wp t} \in [\underline{\mathrm{v}}_\wp, \overline{\mathrm{v}}_\wp], \wp = 1, 2, \ldots, \bar{n}\}$.

Evaluate Objective Function

4: For each particle $\mathrm{w}_{\wp t}^l$, by solving the conditions (4.38)–(4.40) and comparing the objective function F_{obj} with the best individual value pB_t^l and the best global value gB^l can be generated among the entire particle swarm.

Update Position and Velocity

5: According to the individual best value and the global best value, the particle's position and velocity will be updated as follows:

$$\mathrm{v}_t^{l+1} = c_0\mathrm{v}_t^l + c_1 r_1(pB_t^l - \mathrm{w}_t^l) + c_2 r_2(gB^l - \mathrm{w}_t^l), \quad (4.49)$$

$$\mathrm{w}_t^{l+1} = \mathrm{w}_t^l + \mathrm{v}_t^{l+1}, \quad (4.50)$$

where $c_0 \in (0, 1)$ denotes forgetting factor, c_1, c_2 are two adjustable parameters, r_1 and $r_2 \in [0, 1]$ are two random numbers.

6: If the termination condition (i.e., the maximum iterations $T_{\max}$) is attained, stop the algorithm and output the optimal solution. Otherwise, go back the **Evaluate Objective Function** step.

7: **return** K_j, G, Ω, τ_j, and ρ.

Remark 4.4 For achieving the trade-off between the control performance of the controlled system (4.18) and the utilization of communication resources, a multi-objective optimization problem (4.48) via the linear combination approach is proposed to reduce the sliding domain (4.31) and increase the triggering thresholds in (4.14). For the optimization objective F_{obj} in (4.47), the first two terms reflect the communication ratio of ETS, and the last term represents the control performance. In addition, three weighting factors (ε_1, ε_2, and ε_3) imply the importance of control performance and communication resources, that is, larger ε_1 and ε_2 are more conducive to reducing communication resources. On the contrary, a larger value of ε_3 is more beneficial for reducing the sliding mode convergence region $\mathcal{D}$.

4.4.3 Implementation of Dynamic Quantizer

It should be pointed out that the dynamic quantized input $u(k) = q_\mu(v(k))$ is composed of the quantized value $q_{\mu_\iota}(v_\iota(k))$ and the corresponding dynamic quantizer's parameter $\mu_\iota(k)$. However, the parameter $\mu_\iota(k)$ dependent on the designed control input $v_\iota(k)$ is unavailable for the actuator side.

In the following, we shall design an adjustment rule on the time-varying parameter $\mu_\iota(k)$ for obtaining $u_\iota(k) = q_{\mu_\iota}(v_\iota(k))\mu_\iota(k)$ at the actuator side. Inspired by [4], the dynamic quantizer's parameter $\mu_\iota(k)$ can be designed to satisfy the following rule:

$$\sqrt{\zeta_\iota}|v_\iota(k)| < \mu_\iota(k) \leq 2\sqrt{\zeta_\iota}|v_\iota(k)|, \tag{4.51}$$

where parameter ζ_ι can be obtained in Theorem 4.1.

The dynamic quantizer's parameter $\mu_\iota(k)$ in (4.51) can be adjusted online according to the following rules:

Case 1: if $2\sqrt{\zeta_\iota}|v_\iota(k)| \geq 1$, then

$$\mu_\iota(k) = 2^{\varsigma_{1\iota}}, \tag{4.52}$$

where $\varsigma_{1\iota} = \max\{\varsigma_{1\iota} \mid \varsigma_{1\iota} \in \mathbb{N}, 2\sqrt{\zeta_\iota}|v_\iota(k)| \times 2^{-\varsigma_{1\iota}} \geq 1\}$.

Case 2: if $2\sqrt{\zeta_\iota}|v_\iota(k)| < 1$, then

$$\mu_\iota(k) = 2^{-\varsigma_{2\iota}}, \tag{4.53}$$

where $\varsigma_{2\iota} = \min\{\varsigma_{2\iota} \mid \varsigma_{2\iota} \in \mathbb{N}^+, 2\sqrt{\zeta_\iota}|v_\iota(k)| \times 2^{\varsigma_{2\iota}} \geq 1\}$.

Furthermore, we define the following index

$$\bar{\varsigma}_\iota = \begin{cases} -\varsigma_{1\iota}, & \text{for } 2\sqrt{\zeta_\iota}|v_\iota(k)| \geq 1, \\ \varsigma_{2\iota}, & \text{for } 2\sqrt{\zeta_\iota}|v_\iota(k)| < 1, \end{cases} \tag{4.54}$$

which can be transmitted via network channels so that it is available for the controller and actuator sides. According to the index $\bar{\varsigma}_\iota$, we can reconstruct the quantization parameter $\mu_\iota(k)$ at the actuator sides. Thus, the dynamic quantization scheme is practically implementable.

4.5 Simulation Example

Consider a 3-rule fuzzy system (4.2) with the following parameters:

$$\begin{aligned} A_1 &= \begin{bmatrix} 1.0 & 0.1 \\ 0.5 & 0.6 \end{bmatrix}, A_2 = \begin{bmatrix} 1.1 & 0.2 \\ 0.3 & 0.4 \end{bmatrix}, A_3 = \begin{bmatrix} 1.0 & 0.3 \\ 0.4 & 0.2 \end{bmatrix}, \\ B_1 &= \begin{bmatrix} 0.3 \\ 0.2 \end{bmatrix}, B_2 = \begin{bmatrix} 0.3 \\ 0.4 \end{bmatrix}, B_3 = \begin{bmatrix} 0.2 \\ 0.3 \end{bmatrix}, \\ d(k) &= 0.1\sin(2k), \end{aligned}$$

and the upper and lower MFs

$$\overline{m}_1(x(k)) = \frac{e^{-(x_1(k)+3)}}{1+e^{-(x_1(k)+3)}}, \underline{m}_1(x(k)) = \frac{e^{-(x_1(k)+5)}}{1+e^{-(x_1(k)+5)}},$$
$$\overline{m}_3(x(k)) = \frac{1}{1+e^{-1(x_1(k)-3)}}, \underline{m}_3(x(k)) = \frac{1}{1+e^{-1(x_1(k)-5)}},$$

and $\overline{m}_2(x(k)) = 1 - \underline{m}_1(x(k)) - \underline{m}_3(x(k)), \underline{m}_2(x(k)) = 1 - \overline{m}_1(x(k)) - \overline{m}_3(x(k))$.

Different from the system's fuzzy rules and its MFs, we design a 2-rule event-triggered SMC law (4.15) with the following MFs

$$\overline{h}_1(x(k_l)) = \frac{1}{1+e^{-(2x_1(k_l)+1)}}, \quad \underline{h}_1(x(k_l)) = \frac{1}{1+e^{(1-2x_1(k_l))}}$$

with the weighting coefficients $\overline{\beta}_1(x(k_l)) = 0.5$ for obtaining the controller's MFs $h_1(x(k_l)) = 0.5\overline{h}_1(x(k_l)) + 0.5\underline{h}_1(x(k_l))$ and $h_2(x(k_l)) = 1 - h_1(x(k_l))$. Assuming $x_1(k) \in [-5, 5]$, the modeling domain of the controlled system is divided into 5 sub-domains, i.e., $\Gamma_d = [-7+2d, -5+2d]$ and $d = 1, \ldots, 5$.

Given scalars $\kappa_1 = 0.1, \kappa_2 = 1$, quantization parameters $M_1 = 1$, $\Delta_1 = 0.1$, triggering parameters $\theta = 10$, and $\gamma = 0.2$ satisfying $\gamma\theta > 1$. Denote sliding mode gain $G = [g_1, g_2]$ and the particle $\mathrm{w} = [g_1, g_2, \tau_1, \tau_2, \rho]$. The search scope is assigned as $S = \{\mathrm{w}_{\wp t}, \mathrm{v}_{\wp t} \mid g_1, g_2 \in [-2, 2], \tau_1, \tau_2, \rho \in [0.001, 0.999], \mathrm{v}_1, \mathrm{v}_2 \in [-0.4, 0.4], \mathrm{v}_3, \mathrm{v}_4, \mathrm{v}_5 \in [-0.1, 0.1]\}$. The parameters of the PSO-based solving algorithm are set as: $T_{\max} = 50$, $N = 100$, $c_0 = 1.2$, $c_1 = 1.6$ and $c_2 = 1.8$. By solving the optimization problem (4.48) via Algorithm 1 with the weighting values $\varepsilon_1 = \varepsilon_2 = 0.3$ and $\varepsilon_3 = 0.4$, we can obtain parameters $\rho = 0.1958$, $\tau_1 = 0.5027$, $\tau_2 = 0.7534$, $\hat{\lambda} = 2.0387$, $\xi_1 = 2.0339$, and matrices

$$G = \begin{bmatrix} -0.7415 & -0.8813 \end{bmatrix}, \ \Omega = \begin{bmatrix} 13.6286 & 3.7375 \\ 3.7375 & 1.3601 \end{bmatrix},$$
$$K_1 = \begin{bmatrix} -2.0638 & -0.5802 \end{bmatrix}, \ K_2 = \begin{bmatrix} -2.0592 & -0.5853 \end{bmatrix}.$$

Suppose the weightings of system's MFs as $\overline{\alpha}_1(x(k)) = 0.5\sin(x_1^2(k))$, $\underline{\alpha}_1(x(k)) = 1 - \overline{\alpha}_1(x(k))$, $\overline{\alpha}_2(x(k)) = 0.5\cos(x_1^2(k))$, and $\underline{\alpha}_2(x(k)) = 1 - \overline{\alpha}_2(x(k))$. Under the initial condition $x(0) = [-3, 8]^T$ and $\eta(0) = 10$, the simulation results are shown in Fig. 4.4a–f. As shown in Fig. 4.4a, system states $x_1(k)$ and $x_2(k)$ can converge into a small neighborhood around the equilibrium point, which means that the closed-loop system can be ensured to be ultimately bounded. The triggering interval under the proposed DETS (4.14) is shown in Fig. 4.4b, which manifests that the DETS can save the communication resources compared with the periodic sampling scheme. From control input $v(k)$ and quantized input signal $u(k)$ shown in Fig. 4.4c, the quantization error will decrease with the control input decreases. Sliding variable $s(k)$ and the sliding domain $\hat{\varphi}$ are plotted in Fig. 4.4d, from which it can be seen that state trajectories can converge into sliding domain

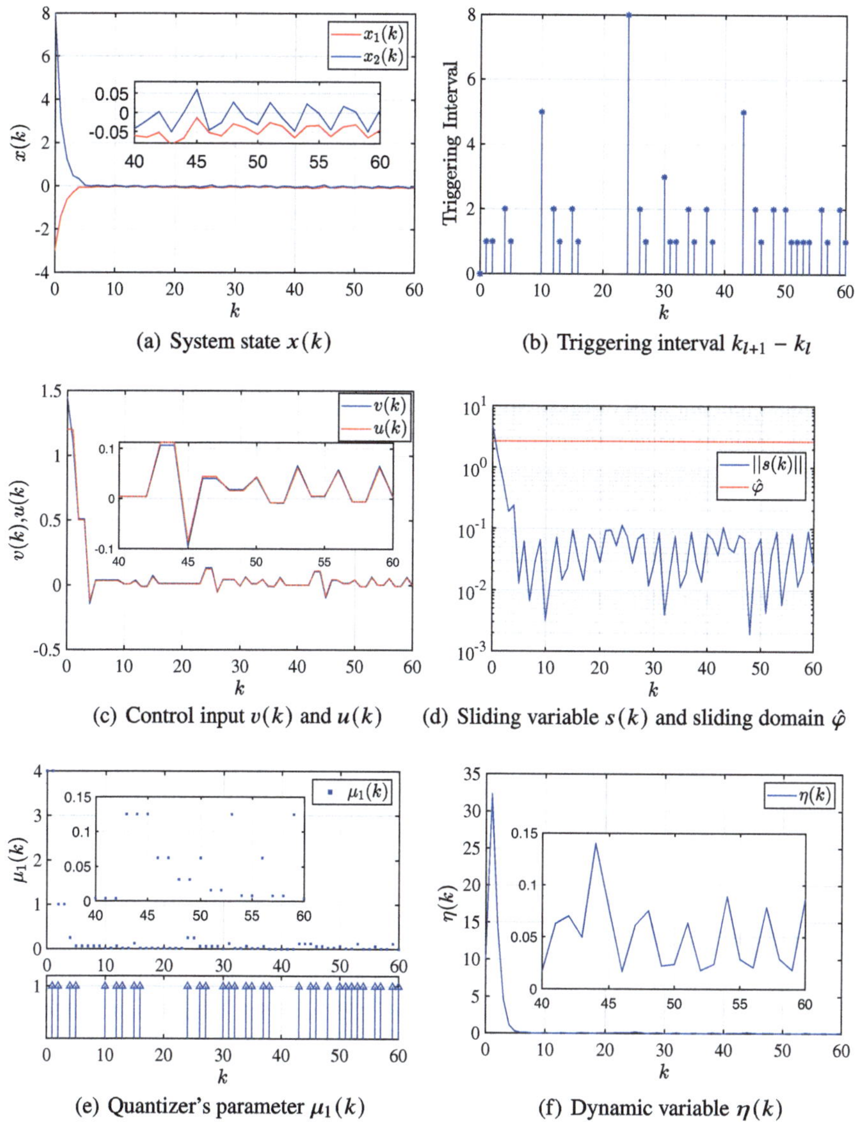

(a) System state $x(k)$

(b) Triggering interval $k_{l+1} - k_l$

(c) Control input $v(k)$ and $u(k)$

(d) Sliding variable $s(k)$ and sliding domain $\hat{\varphi}$

(e) Quantizer's parameter $\mu_1(k)$

(f) Dynamic variable $\eta(k)$

Fig. 4.4 Simulation results under MFD-DETS and quantization

and remain in it. In addition, the dynamic quantization parameter $\mu_1(k)$ is shown in Fig. 4.4e, which will remain constant during the triggering interval. Meanwhile, the dynamic event-triggering parameter $\eta(k)$ as described in Fig. 4.4f is dynamically adjusted and keep to be positive.

To conquer the difficulty of asynchronous MFs, the membership-function-based triggering condition $\psi_2(x(k), x(k_l))$ is introduced in (4.14), which enables the MFs' local information to be available and be conducive to reduce conservatism. Compared with the existing strategy in [9], only the first triggering condition $\psi_1(x(k), x(k_l))$ is adopted, under which the global information of MFs can be obtained. Following this similar approach and keeping the same design parameters as above, no feasible solution can be found, which manifests that the MFs' local information can enlarge the feasible solution space.

In addition, the MFs' local information of asynchronous was also used in [10] under the ETS for continuous-time IT2 fuzzy systems, where the deviation between state $x(t)$ and triggered state $x(t_k)$ was estimated under the assumption that there exist a maximum of the derivative of state $x(t)$ and a linear relation between each components of $x(t)$ and its triggered value $x(t_k)$. In this chapter, the above assumptions are removed, while no constant upper bound on the triggering threshold is required. Hence, the developed MFD-DETS can enlarge the application of a fuzzy control frame in the face of restricted communication capability.

4.6 Conclusion

This chapter has studied the IT2 fuzzy event-triggered SMC issue, and the dynamic quantizer has been used to improve the transmission of the control input. By constructing a novel triggering condition, an event-triggered controller has been designed while the MFs' local information can be used in stability analysis. To achieve the trade-off between the communication rate and the control performance, an optimization problem has been provided to solve design parameters. Finally, the dynamic quantizer's parameter can be adjusted online.

References

1. Coutinho, D.F., Fu, M., de Souza, C.E.: Input and output quantized feedback linear systems. IEEE Trans. Autom. Control **55**(3), 761–766 (2010)
2. Fu, M., Xie, L.: The sector bound approach to quantized feedback control. IEEE Trans. Autom. Control **50**(11), 1698–1711 (2005)
3. Liberzon, D.: Hybrid feedback stabilization of systems with quantized signals. Automatica **39**(9), 1543–1554 (2003)
4. Niu, Y., Ho, D.W.: Control strategy with adaptive quantizer's parameters under digital communication channels. Automatica **50**(10), 2665–2671 (2014)
5. Chang, X.H., Yang, C., Xiong, J.: Quantized fuzzy output feedback H_∞ control for nonlinear systems with adjustment of dynamic parameters. IEEE Trans. Syst. Man Cybern. Syst. **49**(10), 2005–2015 (2019)
6. Zheng, Q., Xu, S., Du, B.: Quantized guaranteed cost output feedback control for nonlinear networked control systems and its applications. IEEE Trans. Fuzzy Syst. **30**(7), 2402–2411 (2022)
7. Lam, H.K.: A review on stability analysis of continuous-time fuzzy-model-based control systems: from membership-function-independent to membership-function-dependent analysis. Eng. Appl. Artif. Intell. **67**, 390–408 (2018)

8. Song, J., Niu, Y., Lam, H.K., Zou, Y.: Asynchronous sliding mode control of singularly perturbed semi-Markovian jump systems: application to an operational amplifier circuit. Automatica **118**, 109026 (2020)
9. Zhang, Z., Su, S.F., Niu, Y.: Dynamic event-triggered control for interval type-2 fuzzy systems under fading channel. IEEE Trans. Cybern. **51**(11), 5342–5351 (2021)
10. Xiao, B., Lam, H.K., Zhong, Z., Wen, S.: Membership-function-dependent stabilization of event-triggered interval type-2 polynomial fuzzy-model-based networked control systems. IEEE Trans. Fuzzy Syst. **28**(12), 3171–3180 (2020)

Chapter 5
Sliding Mode Security Control via Probability-Dependent Adaptive Event-Triggered Scheme

5.1 Introduction

Chapters 2–4 have investigated the dynamic ETSs by introducing the dynamic variable to adjust the triggering threshold. In this chapter, an adaptive event-triggered scheme (AETS) will be proposed for the IT2 T-S fuzzy systems. Compared with the ETSs in the previous chapters with the constant triggering parameters, the triggering parameters of AETS can be adaptively adjusted to improve the flexibility of the triggering threshold for adapting to the changes of the network environment and the signal transmission fluctuations.

On the other hand, over the open communication environments, network security is a significant concern. When information and data are exchanged over the shared communication networks, they become highly vulnerable to malicious cyber-attacks, e.g., DoS attacks [1, 2] and deception attacks [3, 4]. Among them, DoS attacks can block the flow of information and cause the unavailability of data, which may have more serious adverse effect on event-triggered control system since only a small amount of necessary information is transmitted. Recently, some works on ETSs subject to cyber-attacks have been carried out in [5]. In particular, security control under AETS has also been investigated for fuzzy systems subject to deception attacks in [6, 7] and DoS attacks in [8, 9], respectively. In [8], an adaptive event-triggered fuzzy controller was employed for the IT2 fuzzy systems under nonperiodic DoS attacks, where the triggering parameter was switched only between two given constants. In [9], a resilient controller under AETS was developed for the IT2 fuzzy systems suffering from DoS attacks, however, whose adaptive rule may be unimplementable due to its dependence on the unknown system's MFs. In addition, the aforementioned works ignored the effect of network attacks on adaptive rule of triggering parameters such that greater conservatism may be brought. Therefore, how to develop suitable adaptive rules in AETS has practical significance for resisting the adverse effect of cyber-attacks.

Y. Yang et al., *Analysis and Synthesis of Interval Type-2 Fuzzy Systems Under Communication Protocols*, Studies in Systems, Decision and Control 636,
https://doi.org/10.1007/978-981-95-2355-9_5

This chapter focuses on the event-triggered sliding mode security control for the IT2 fuzzy systems subject to randomly occurring DoS attacks. Under the resource-limited and unreliable network, a probability-dependent AETS is proposed, whose triggering parameter can be adaptively adjusted in accordance with the convergence trend of system states and attack probability. By constructing a sliding function with one-step memorized state information, a fuzzy sliding mode resilient controller is designed, whose MFs are dependent on the compromised triggered state, and do not require the same fuzzy rules as the fuzzy system. Furthermore, by employing the global bounds of mismatched MFs and the MFD approach, the co-design scheme of the AETS and the fuzzy controller is proposed to ensure the stability of closed-loop systems and the reachability of specified sliding surface.

5.2 Problem Description

5.2.1 System Model

Consider the following r-rule IT2 fuzzy system:

Rule i: IF $f_1(x(k))$ is $\mathscr{F}_1^i$ and ... and $f_\hbar(x(k))$ is $\mathscr{F}_\hbar^i$, THEN

$$x(k+1) = A_i x(k) + B_i(u(k) + d(x(k), k)) \tag{5.1}$$

where $f_\iota(x(k))$ $(\iota = 1, \ldots, \hbar)$ is the premise variable, $\mathscr{F}_\iota^i$ $(i = 1, \ldots, r)$ is the IT2 fuzzy set, $x(k) \in \mathbb{R}^n$ is the system state, and $u(k) \in \mathbb{R}^m$ is the control input. $d(x(k), k)$ is the external disturbance with a known upper bound $\varpi > 0$ satisfying $||d(x(k), k)|| \leq \varpi ||x(k)||$. The matrices A_i and B_i are known. The firing strength is within the following interval set:

$$\vec{m}_i(x(k)) = [\underline{m}_i(x(k)), \overline{m}_i(x(k))]$$

with the following upper and lower MFs

$$\overline{m}_i(x(k)) = \prod_{\iota=1}^{\hbar} \overline{\mu}_{\mathscr{F}_\iota^i}(f_\iota(x(k))) \geq 0, \tag{5.2}$$

$$\underline{m}_i(x(k)) = \prod_{\iota=1}^{\hbar} \underline{\mu}_{\mathscr{F}_\iota^i}(f_\iota(x(k))) \geq 0, \tag{5.3}$$

in which the upper and lower membership grades $\overline{\mu}_{\mathscr{F}_\iota^i}(f_\iota(x(k)))$ and $\underline{\mu}_{\mathscr{F}_\iota^i}(f_\iota(x(k)))$ satisfy $0 \leq \underline{\mu}_{\mathscr{F}_\iota^i}(f_\iota(x(k))) \leq \overline{\mu}_{\mathscr{F}_\iota^i}(f_\iota(x(k))) \leq 1$. It is easily obtained from (5.2) and (5.3) that $0 \leq \underline{m}_i(x(k)) \leq \overline{m}_i(x(k)) \leq 1$. Furthermore, by means of the upper and lower MFs $\overline{m}_i(x(k))$ and $\underline{m}_i(x(k))$, system's MFs $m_i(x(k))$ can be represented as

$$m_i(x(k)) = \overline{m}_i(x(k))\overline{\alpha}_i(x(k)) + \underline{m}_i(x(k))\underline{\alpha}_i(x(k)),$$

with the weighting coefficients $\underline{\alpha}_i(x(k))$ and $\overline{\alpha}_i(x(k))$ satisfying $\underline{\alpha}_i(x(k)), \overline{\alpha}_i(x(k)) \in [0, 1]$, and $\underline{\alpha}_i(x(k)) + \overline{\alpha}_i(x(k)) = 1$.

By combining the fuzzy model (5.1) with MFs $m_i(x(k))$, the IT2 T-S fuzzy system can be obtained as

$$x(k+1) = \sum_{i=1}^{r} m_i(x(k))[A_i x(k) + B_i(u(k) + d(x(k), k))] \tag{5.4}$$

where $m_i(x(k))$ is the membership function satisfying $\sum_{i=1}^{r} m_i(x(k)) = 1$.

5.2.2 Event-Triggered Communication Under DoS Attacks

Under the resource-limited communication network, ETS is an effective approach to alleviate the communication load by reducing the transmission frequency of data. Generally, an ETS can determine releasing instants of the system information via the following triggering condition:

$$k_{\ell+1} = \inf\{k \mid k > k_\ell,\ \phi(x(k), x(k_\ell)) \geq 0\}, \tag{5.5}$$

where $\phi(x(k), x(k_\ell))$ is a triggering function of system state $x(k)$ and its triggered state $x(k_\ell)$, and k_ℓ denotes the latest triggering instant.

In addition, the open and shared communication environment is vulnerable to malicious network attacks. Typically, DoS attacks will interfere network channels so that the remote controller cannot receive the triggered state. To compensate the effect of attacks, the remote controller only can utilize the previous transmitted data stored in Buffer (shown in Fig. 5.1) when DoS attacks occur. Thus, the available signal $\bar{x}(k)$ during triggering interval $[k_\ell, k_{\ell+1})$ for the controller can be represented as follows:

$$\bar{x}(k_\ell) = (1 - \theta(k_\ell))x(k_\ell) + \theta(k_\ell)\bar{x}(k_\ell - 1), \tag{5.6}$$

$$\bar{x}(k) = \bar{x}(k_\ell),\quad k \in [k_\ell, k_{\ell+1}), \tag{5.7}$$

where $\theta(k_\ell)$ is a stochastic variable obeying Bernoulli distribution with $\mathbf{Pr}\{\theta(k_\ell) = 1\} = \bar{\theta}$, $\mathbf{Pr}\{\theta(k_\ell) = 0\} = 1 - \bar{\theta}$, and a known constant $\bar{\theta} \in [0, 1)$. $\theta(k_\ell) = 1$ represents that DoS attacks occur at the current triggering instant k_ℓ. $\theta(k_\ell) = 0$ denotes the successful transmission of triggered state to the controller. Here, denote $\bar{x}(-1) = 0$. In addition, at the triggering instant, since the transmitted state information may be attacked by DoS attacker, the available state signal for the controller will keep constant during the triggering interval [as given in (5.6)].

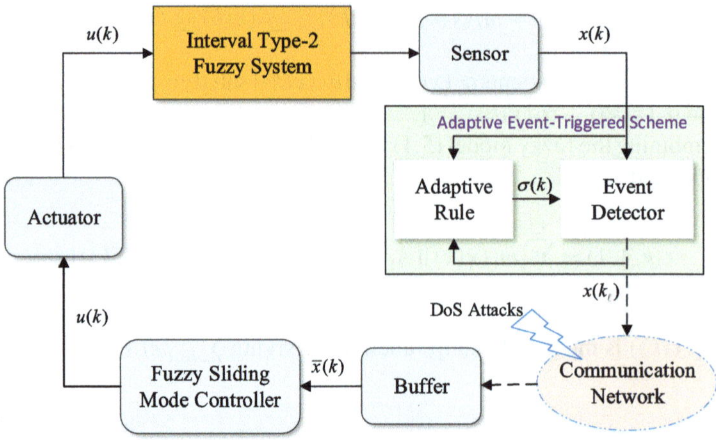

Fig. 5.1 Control system under AETS (5.9) and DoS attacks

By means of the relations (5.6) and (5.7), the available state for the controller can be further represented as

$$\bar{x}(k) = (1 - \theta(k_\ell))x(k_\ell) + \theta(k_\ell)\bar{x}(k-1) \tag{5.8}$$

for $k \in [k_\ell, k_{\ell+1})$.

It should be pointed out that no system state is sent to the controller during the triggering interval. This means that when the controller does not receive new state information, it cannot distinguish whether a DoS attack has occurred or an event has not been triggered. Therefore, the DoS attack cannot be detected for the controller, which brings difficulty and challenges for designing an efficient controller to achieve the desired control performance.

Under the above background, the following objective is to propose an AETS and design a fuzzy sliding mode security controller to guarantee the stability of the IT2 T-S fuzzy system (5.4) subject to DoS attacks.

5.2.3 Probability-Dependent Adaptive Event-Triggered Scheme

Under the unreliable communication network shown in Fig. 5.1, the DoS attacks may block the triggered state $x(k_\ell)$ so that the control performance of system is degraded. Hence, taking the changing trend of the system state and the occurring probability of DoS attacks into the ETS will be beneficial to withstand the adverse effect of attacks. To this end, we design the following AETS as

$$k_{\ell+1} = \inf\left\{k \in \mathbb{N} \mid k > k_\ell, e^T(k)\Omega e(k) \geq \sigma(k)x^T(k_\ell)\Omega x(k_\ell)\right\} \tag{5.9}$$

with $e(k) \triangleq x(k_\ell) - x(k)$ as the variation between the current state $x(k)$ and the previous triggered one $x(k_\ell)$, the triggering matrix $\Omega = \Omega^T > 0$ to be designed later, and the triggering parameter $\sigma(k)$ satisfying the following attack-probability-dependent adaptive rule:

$$\sigma(k+1) = \sigma(k) + \begin{cases} (\sigma(k) - \underline{\sigma}) \tanh(-\varrho(1+\bar{\theta})||e(k)||^2), \text{ if } L(k) \leq 0, \\ (\overline{\sigma} - \sigma(k)) \tanh(\varrho(1-\bar{\theta})||e(k)||^2), \text{ if } L(k) > 0, \end{cases} \tag{5.10}$$

where ϱ, $\underline{\sigma}$, and $\overline{\sigma}$ are positive adjustable constants satisfying $0 < \underline{\sigma} < \overline{\sigma} < 1$, and $L(k) \triangleq ||x(k_\ell)|| - ||x(k)||$ denotes the changing trend of the norm of system state $x(k)$.

The following lemma can ensure that the adaptive triggering parameter $\sigma(k)$ in (5.10) lies the interval $[\underline{\sigma}, \overline{\sigma}]$.

Lemma 5.1 *For any $\sigma(0) \in [\underline{\sigma}, \overline{\sigma}]$, the adaptive rule (5.10) ensures $\sigma(k) \in [\underline{\sigma}, \overline{\sigma}]$ for $\forall k \geq 0$.*

Proof It is easily obtained from (5.10) that for $L(k) \leq 0$, we have $\sigma(k+1) \in [\underline{\sigma}, \sigma(k)]$, and for $L(k) > 0$, we have $\sigma(k+1) \in [\sigma(k), \overline{\sigma}]$. That is, we have $\sigma(k+1) \in [\underline{\sigma}, \overline{\sigma}]$ for $\sigma(k) \in [\underline{\sigma}, \overline{\sigma}]$. Similarly, by means of the recursion process from k to 0, we can obtain that if $\sigma(0) \in [\underline{\sigma}, \overline{\sigma}]$, $\sigma(k) \in [\underline{\sigma}, \overline{\sigma}]$ can be guaranteed. □

Remark 5.1 According to the variation trend $L(k)$ of the norm of system state $x(k)$, the above adaptive rule (5.10) can adaptively adjust the triggering variable $\sigma(k)$ in the increasing or decreasing direction. For $L(k) > 0$, the triggering variable $\sigma(k)$ increases towards $\bar{\sigma}$, otherwise, the triggering variable $\sigma(k)$ decreases towards $\underline{\sigma}$. Moreover, by introducing two parameters $\underline{\sigma}$ and $\overline{\sigma}$ in (5.10), the triggering variable $\sigma(k)$ can be adjusted within a bounded range $[\underline{\sigma}, \overline{\sigma}]$, which is different from the existing AETS in [10, 11] that can only ensure the existence of a lower bound of $\sigma(k)$. The upper bound $\overline{\sigma}$ can prevent $\sigma(k)$ from increasing continuously without bound, and compel the triggering condition (5.9) to be violated when the system state fluctuates greatly. Meanwhile, the lower bound $\underline{\sigma}$ can ensure events to not occur frequently for achieving the purpose of reducing the communication resources.

Remark 5.2 It can be seen from the adaptive rule (5.10) that when the probability $\bar{\theta}$ of DoS attacks increases, the speed of $\sigma(k+1)$ approaching $\underline{\sigma}$ will increase as $L(k) \leq 0$, meanwhile, its speed approaching $\overline{\sigma}$ will slow down as $L(k) > 0$ so that a smaller triggering threshold in (5.9) is generated. Conversely, the speed of $\sigma(k+1)$ approaching $\underline{\sigma}$ will decrease while its speed approaching $\overline{\sigma}$ will increase. Furthermore, a scalar $\varrho > 0$ can be chosen to adjust the effect of the probability $\bar{\theta}$ on the updating speed of $\sigma(k)$ according to the changing trend of the difference $L(k)$. Thus, the introduction of $\bar{\theta}$ and ϱ can flexibly adjust the triggering threshold in (5.9) via $\sigma(k)$ for resisting the adverse effect of DoS attacks.

5.3 Sliding Mode Security Control Design

In this section, we shall design a desired fuzzy sliding mode resilient controller. First, the sliding function is chosen as

$$s(k) = (1-\bar{\theta})Gx(k) + \bar{\theta}Gx(k-1), \tag{5.11}$$

where the sliding matrix G is given. We may choose, for example, $G = \sum_{i=1}^{r} g_i B_i^T$ with some suitable scalars g_i $(i = 1, \ldots, r)$. Here, denote $x(-1) = 0$.

Then, we design the following fuzzy sliding mode security controller as:

Rule j: IF $g_1(\bar{x}(k))$ is $\mathscr{M}_1^j$ and ... and $g_\wp(\bar{x}(k))$ is $\mathscr{M}_\wp^j$, THEN

$$u(k) = \frac{1}{1-\bar{\theta}} K_j \bar{x}(k) - \varpi \|\bar{x}(k-1)\| \text{sign}(\bar{s}(k)), \tag{5.12}$$

where K_j $(j = 1, \ldots, c)$ is the control gain to be designed later, and $\bar{s}(k)$ is the counterpart in (5.11) via replacing $x(k)$ and $x(k-1)$ with $\bar{x}(k)$ and $\bar{x}(k-1)$. For the j-th fuzzy rule, the firing strength is given as

$$\overrightarrow{h}_j(\bar{x}(k)) = [\underline{h}_j(\bar{x}(k)),\ \overline{h}_j(\bar{x}(k))] \tag{5.13}$$

with $\overline{h}_j(\bar{x}(k)) = \prod_{\tau=1}^{\wp} \overline{\mu}_{\mathscr{M}_\tau^j}(g_\tau(\bar{x}(k))) \geq 0$, and $\underline{h}_j(\bar{x}(k)) = \prod_{\tau=1}^{\wp} \underline{\mu}_{\mathscr{M}_\tau^j}(g_\tau(\bar{x}(k))) \geq 0$, where the membership grades $\overline{\mu}_{\mathscr{M}_\tau^j}(g_\tau(\bar{x}(k))) \in [0, 1]$ and $\underline{\mu}_{\mathscr{M}_\tau^j}(g_\tau(\bar{x}(k))) \in [0, 1]$ satisfy $\underline{\mu}_{\mathscr{M}_\tau^j}(g_\tau(\bar{x}(k))) \leq \overline{\mu}_{\mathscr{M}_\tau^j}(g_\tau(\bar{x}(k)))$.

Besides, the MFs $m_j(\bar{x}(k))$ can be represented as

$$h_j(\bar{x}(k)) = \overline{\beta}_j(\bar{x}(k))\overline{h}_j(\bar{x}(k)) + \underline{\beta}_j(\bar{x}(k))\underline{h}_j(\bar{x}(k)), \tag{5.14}$$

where weighting values $\overline{\beta}_j(\bar{x}(k))$ and $\underline{\beta}_j(\bar{x}(k))$ are chosen to satisfy $\underline{\beta}_j(\bar{x}(k))$, $\overline{\beta}_j(\bar{x}(k)) \in [0, 1]$ and $\underline{\beta}_j(\bar{x}(k)) + \overline{\beta}_j(\bar{x}(k)) = 1$.

According to the fuzzy controller (5.12) and the MFs in (5.13), we can obtain the following fuzzy sliding mode resilient controller for $k \in [k_\ell, k_{\ell+1})$,

$$u(k) = \frac{1}{1-\bar{\theta}} \sum_{j=1}^{c} h_j(\bar{x}(k)) K_j \bar{x}(k) - \varpi \|\bar{x}(k-1)\| \text{sign}(\bar{s}(k)). \tag{5.15}$$

Remark 5.3 In the fuzzy controller (5.15), MFs $h_j(\bar{x}(k))$ are dependent on the compromised triggered state $\bar{x}(k)$. Meanwhile, the number of fuzzy rule in (5.15) is not required to be the same as the ones in the fuzzy system (5.4), which increases the design flexibility, but also makes some existing handling methods to be invalid, e.g., $h_i(\bar{x}(k)) - \delta_i m_i(x(k)) \geq 0$ with $\delta_i \in (0, 1)$ in [6–9]. To remove this obstacle, the global bounds of mismatched MFs will be utilized in the subsequent analysis (see Sect. 5.5).

Substituting (5.15) into (5.4) yields the following closed-loop IT2 fuzzy system:

$$x(k+1) = \sum_{i=1}^{r}\sum_{j=1}^{c} m_i(x(k))h_j(\bar{x}(k))\Big\{\Big(A_i + \frac{1-\theta(k_\ell)}{1-\bar{\theta}}B_iK_j\Big)x(k) + \frac{1-\theta(k_\ell)}{1-\bar{\theta}}B_iK_je(k) + \frac{1-\theta(k_\ell)}{1-\bar{\theta}}B_iK_j\bar{x}(k-1) + B_iD(k)\Big\}, \tag{5.16}$$

with $D(k) \triangleq d(x(k), k) - \varpi||\bar{x}(k-1)||\text{sign}(\bar{s}(k))$.

To analyze the stability of the closed-loop system (5.16), the following definition is given.

Definition 5.1 The closed-loop IT2 fuzzy system (5.16) is said to be stochastically stable, if

$$\mathbf{E}\left\{\sum_{k=0}^{\infty} ||x(k)||^2\right\} < \infty \tag{5.17}$$

holds for $x(0) \in \mathbb{R}^n$ and $\bar{x}(-1) = 0$.

5.4 Performance Analysis Under Event-Triggered Communication and DoS Attacks

In this section, we will proved that the proposed event-triggered security controller can ensure the stochastic stability of closed-loop system and the reachability of the specified sliding surface $s(k) = 0$, meanwhile, sufficient conditions will be derived.

5.4.1 *Stochastic Stability of Closed-Loop System*

Under the fuzzy sliding mode resilient controller (5.15), the closed-loop system (5.16) can attain a desirable performance via the following conditions.

Theorem 5.1 *Given parameters* $\bar{\theta} \in [0, 1)$ *and* $\overline{\sigma} \in (0, 1)$, *if there exist matrices* $P_l > 0$, $Q_l > 0$, $\Omega > 0$, K_j *and scalars* $\xi_i > 0$ *satisfying the following conditions (for* $i,\ l = 1, \ldots, r,\ j = 1, \ldots, c$*)*

$$B_i^T P_l B_i - \xi_i I \leq 0, \tag{5.18}$$

$$\sum_{i=1}^{r}\sum_{j=1}^{c} m_i(x(k))h_j(\bar{x}(k))\Gamma_{ijl} < 0 \tag{5.19}$$

with

$$\Gamma_{ijl} = \begin{bmatrix} \Gamma_{ijl}^{11} & * & * \\ \Gamma_{ijl}^{21} & \Gamma_{ijl}^{22} & * \\ \Gamma_{ijl}^{31} & \Gamma_{ijl}^{32} & \Gamma_{ijl}^{33} \end{bmatrix},$$

$$\begin{aligned}
\Gamma_{ijl}^{11} &= -P_i + \overline{\sigma}\Omega + (1-\bar{\theta})Q_l + 2(A_i + B_iK_j)^T P_l (A_i + B_iK_j) \\
&\quad + \frac{\bar{\theta}}{1-\bar{\theta}}(B_iK_j)^T P_l B_i K_j + 4\varpi^2 \xi_i I, \\
\Gamma_{ijl}^{21} &= (1-\bar{\theta})Q_l + 2(B_iK_j)^T P_l (A_i + B_iK_j) + \frac{\bar{\theta}}{1-\bar{\theta}}(B_iK_j)^T P_l B_i K_j + \overline{\sigma}\Omega, \\
\Gamma_{ijl}^{22} &= -(1-\overline{\sigma})\Omega + (1-\bar{\theta})Q_l + 2(B_iK_j)^T P_l B_i K_j + \frac{\bar{\theta}}{1-\bar{\theta}}(B_iK_j)^T P_l B_i K_j, \\
\Gamma_{ijl}^{31} &= \frac{2\bar{\theta}}{1-\bar{\theta}}(B_iK_j)^T P_l (A_i + B_iK_j) - \frac{\bar{\theta}}{1-\bar{\theta}}(B_iK_j)^T P_l B_i K_j, \\
\Gamma_{ijl}^{32} &= \frac{2\bar{\theta}}{1-\bar{\theta}}(B_iK_j)^T P_l B_i K_j - \frac{\bar{\theta}}{1-\bar{\theta}}(B_iK_j)^T P_l B_i K_j, \\
\Gamma_{ijl}^{33} &= -Q_i + \bar{\theta} Q_l + 4m\varpi^2 \xi_i I + \frac{2\bar{\theta}^2}{(1-\bar{\theta})^2}(B_iK_j)^T P_l B_i K_j \\
&\quad + \frac{\bar{\theta}}{1-\bar{\theta}}(B_iK_j)^T P_l B_i K_j,
\end{aligned}$$

then the stochastic stability of the closed-loop system (5.16) can be guaranteed in the presence of the probability-dependent AETS (5.9) and DoS attacks.

Proof Construct an MFD Lyapunov functional as

$$V(\chi(k)) = x^T(k)P(k)x(k) + \bar{x}^T(k-1)Q(k)\bar{x}(k-1) \tag{5.20}$$

with $\chi(k) = [x^T(k), \bar{x}^T(k-1)]^T$, $P(k) = \sum_{i=1}^r m_i(x(k))P_i$, and $Q(k) = \sum_{i=1}^r m_i(x(k))Q_i$.

Taking the difference of the expression $V(\chi(k))$, one has

$$\begin{aligned}
&\mathbf{E}\{\Delta V(\chi(k))|\chi(k)\} \\
&\quad = \mathbf{E}\{x^T(k+1)P(k+1)x(k+1)|\chi(k)\} \\
&\qquad + \mathbf{E}\{\bar{x}^T(k)Q(k+1)\bar{x}(k)|\chi(k)\} - V(\chi(k)).
\end{aligned} \tag{5.21}$$

It follows from (5.21) and the closed-loop system (5.16) that

$$\begin{aligned}
&\mathbf{E}\{x^T(k+1)P(k+1)x(k+1)|\chi(k)\} \\
&\leq \sum_{i=1}^{r}\sum_{j=1}^{c} m_i(x(k))h_j(\bar{x}(k))\Big\{2\Big[(A_i + B_iK_j)x(k) + B_iK_je(k) + \frac{\bar{\theta}}{1-\bar{\theta}}\bar{x}(k-1)\Big]^T \\
&\times P(k+1)\Big[(A_i + B_iK_j)x(k) + B_iK_je(k) + \frac{\bar{\theta}}{1-\bar{\theta}}\bar{x}(k-1)\Big] \\
&+ \frac{\bar{\theta}}{1-\bar{\theta}}[x(k) + e(k) - \bar{x}(k-1)]^T(B_iK_j)^T P(k+1)B_jK_j[x(k) \\
&+ e(k) - \bar{x}(k-1)] \\
&+ 4d^T(x(k), k)B_i^T P(k+1)B_id(x(k), k) + 4\varpi^2||\bar{x}(k-1)||^2\text{sign}(\bar{s}(k))^T B_i^T \\
&\times P(k+1)B_i\text{sign}(\bar{s}(k))\Big\}. \qquad (5.22)
\end{aligned}$$

By means of the condition (5.18) and $\sum_{l=1}^{r} m_l(x(k+1)) = 1$, we can obtain $B_i^T P(k+1)B_i \leq \xi_i I$. Then, it can be deduced from (5.22) that

$$\begin{aligned}
&\mathbf{E}\{x^T(k+1)P(k+1)x(k+1)|\chi(k)\} \\
&\leq \sum_{i=1}^{r}\sum_{j=1}^{c} m_i(x(k))h_j(\bar{x}(k))\Big\{2\Big[(A_i + B_iK_j)x(k) + B_iK_je(k) + \frac{\bar{\theta}}{1-\bar{\theta}}\bar{x}(k-1)\Big]^T \\
&\times P(k+1)\Big[(A_i + B_iK_j)x(k) + B_iK_je(k) + \frac{\bar{\theta}}{1-\bar{\theta}}\bar{x}(k-1)\Big] \\
&+ \frac{\bar{\theta}}{1-\bar{\theta}}[x(k) + e(k) - \bar{x}(k-1)]^T(B_iK_j)^T P(k+1)B_jK_j[x(k) \\
&+ e(k) - \bar{x}(k-1)] \\
&+ 4\varpi^2\xi_i[x^T(k)x(k) + m\bar{x}^T(k-1)\bar{x}(k-1)]\Big\}. \qquad (5.23)
\end{aligned}$$

According to the AETS (5.9) and $\sigma(k) \in [\underline{\sigma}, \overline{\sigma}]$, the following inequality holds for $k \in [k_\ell, k_{\ell+1})$

$$\overline{\sigma}(x(k) + e(k))^T\Omega(x(k) + e(k)) - e^T(k)\Omega e(k) \geq 0. \qquad (5.24)$$

By introducing condition (5.24) into expression (5.23), one has

$$
\begin{aligned}
&\mathbf{E}\{x^T(k+1)P(k+1)x(k+1)|\chi(k)\}\\
&\le \sum_{i=1}^{r}\sum_{j=1}^{c} m_i(x(k))h_j(\bar{x}(k))\Big\{2\Big[(A_i+B_iK_j)x(k)+B_iK_j\big(e(k)+\frac{\bar{\theta}}{1-\bar{\theta}}\bar{x}(k-1)\big)\Big]^T\\
&\times P(k+1)\Big[(A_i+B_iK_j)x(k)+B_iK_j\big(e(k)+\frac{\bar{\theta}}{1-\bar{\theta}}\bar{x}(k-1)\big)\Big]\\
&+\frac{\bar{\theta}}{1-\bar{\theta}}[x(k)+e(k)-\bar{x}(k-1)]^T(B_iK_j)^TP(k+1)B_jK_j[x(k)\\
&+e(k)-\bar{x}(k-1)]+4\varpi^2\big[x^T(k)x(k)+m\bar{x}^T(k-1)\bar{x}(k-1)\big]\Big\}\\
&+\bar{\sigma}(x(k)+e(k))^T\Omega(x(k)+e(k))-e^T(k)\Omega e(k).
\end{aligned}
\tag{5.25}
$$

Furthermore, we can use the relation of $\bar{x}(k)$ in (5.8) to get

$$
\begin{aligned}
&\mathbf{E}\{\bar{x}^T(k)Q(k+1)\bar{x}(k)|\chi(k)\}\\
&=[(1-\bar{\theta})(x(k)+e(k))+\bar{\theta}\bar{x}(k-1)]^TQ(k+1)[(1-\bar{\theta})(x(k)+e(k))\\
&\quad+\bar{\theta}\bar{x}(k-1)]\\
&\quad+\bar{\theta}(1-\bar{\theta})[x(k)+e(k)-\bar{x}(k-1)]^TQ(k+1)[x(k)+e(k)-\bar{x}(k-1)].
\end{aligned}
\tag{5.26}
$$

Substituting (5.25)–(5.26) into (5.21) and combining with (5.20), we can deduce

$$
\begin{aligned}
&\mathbf{E}\{\Delta V(\chi(k))|\chi(k)\}\\
&\le \sum_{i=1}^{r}\sum_{j=1}^{c} m_i(x(k))h_j(\bar{x}(k))\Big\{2\Big[(A_i+B_iK_j)x(k)e(k)+B_iK_j\big(e(k)+\frac{\bar{\theta}}{1-\bar{\theta}}\bar{x}(k-1)\big)\Big]^T\\
&\times P(k+1)\Big[(A_i+B_iK_j)x(k)+B_iK_j\big(e(k)+\frac{\bar{\theta}}{1-\bar{\theta}}\bar{x}(k-1)\big)\Big]\\
&+\frac{\bar{\theta}}{1-\bar{\theta}}(x(k)+e(k)-\bar{x}(k-1))^T(B_iK_j)^TP(k+1)(B_jK_j)(x(k)+e(k)\\
&-\bar{x}(k-1))+4\varpi^2\xi_ix^T(k)x(k)+4m\varpi^2\xi_i\bar{x}^T(k-1)\bar{x}(k-1)\\
&+\bar{\sigma}(x(k)+e(k))^T\Omega\\
&\times(x(k)+e(k))-e^T(k)\Omega e(k)+[(1-\bar{\theta})(x(k)+e(k))+\bar{\theta}\bar{x}(k-1)]^TQ(k+1)\\
&\times[(1-\bar{\theta})(x(k)+e(k))+\bar{\theta}\bar{x}(k-1)]+\bar{\theta}(1-\bar{\theta})[x(k)+e(k)-\bar{x}(k-1)]^T\\
&\times Q(k+1)[x(k)+e(k)-\bar{x}(k-1)]-x^T(k)P_ix(k)-\bar{x}^T(k-1)Q_i\bar{x}(k-1)\Big\}\\
&=\sum_{l=1}^{r}\sum_{i=1}^{r}\sum_{j=1}^{c} m_l^+m_i(x(k))h_j(\bar{x}(k))\eta^T(k)\Gamma_{ijl}\eta(k),
\end{aligned}
\tag{5.27}
$$

with $m_l^+ \triangleq m_l(x(k+1))$ and $\eta(k)\triangleq[x^T(k), e^T(k), \bar{x}^T(k-1)]^T$.

According to the condition (5.19), one has

$$\mathbf{E}\{\Delta V(\chi(k))|\chi(k)\} \leq -\gamma||\eta(k)||^2 < 0 \tag{5.28}$$

with $\gamma \triangleq \min_{1\leq l\leq r}\{\lambda_{\min}(-\Gamma_l)\}$ and $\Gamma_l \triangleq \sum_{i=1}^{r}\sum_{j=1}^{c} m_i(x(k))h_j(\bar{x}(k))\Gamma_{ijl}$.

Taking the mathematical expectation for both sides of the expression (5.28), it can be obtained

$$\mathbf{E}\{\Delta V(\chi(k))\} \leq -\gamma\mathbf{E}\left\{||\eta(k)||^2\right\}. \tag{5.29}$$

Taking sum on both sides of the expression (5.29) from $k = 0$ to $k = T$, one has

$$\mathbf{E}\{V(\chi(T+1))\} - V(\chi(0)) \leq -\gamma\mathbf{E}\left\{\sum_{k=0}^{T}||\eta(k)||^2\right\}, \tag{5.30}$$

from which one has

$$\begin{aligned}\mathbf{E}\left\{\sum_{k=0}^{T}||\eta(k)||^2\right\} &\leq \frac{V(\chi(0)) - E\{V(\chi(T+1))\}}{\gamma} \\ &< \frac{V(\chi(0))}{\gamma}.\end{aligned} \tag{5.31}$$

As $T \to \infty$, we have $\mathbf{E}\left\{\sum_{k=0}^{\infty}||\eta(k)||^2\right\} < \frac{V(\chi(0))}{\gamma} < \infty$, that is,

$$\mathbf{E}\left\{\sum_{k=1}^{\infty}||x(k)||^2\right\} < \infty. \tag{5.32}$$

According to Definition 5.1, the closed-loop system (5.16) is stochastically stable. □

In the above Theorem 5.1, the constructed MFD Lyapunov functional (5.20) can bring more useful information on the fuzzy characteristic into the stability analysis. When all matrices P_i and Q_i $(i = 1, \ldots, r)$ are the same, the result in Theorem 5.1 will reduce to the case with the common Lyapunov functional. That is, the present MFD form (5.20) is more beneficial to obtain the conditions with less conservatism. In addition, it should be pointed out that only the upper bound $\overline{\sigma}$ is involved in sufficient conditions (5.18)–(5.19), that is, the value of $\overline{\sigma}$ may have effects on the solvability of the above stability conditions. On the other hand, in the implementation of the AETS (5.9)–(5.10), the lower bound $\underline{\sigma}$ may affect the triggering number, that is, a smaller $\underline{\sigma}$ may result in more triggering number and yield greater resources consumption. Hence, we should appropriately choose the bounds $\overline{\sigma}$ and $\underline{\sigma}$ in practice.

5.4.2 Reachability of Sliding Surface

The following conditions are given to ensure that the resilient fuzzy controller (5.15) drives the state trajectories of the controlled system (5.16) suffering from DoS attacks converge into a neighborhood of the specified sliding surface $s(k) = 0$.

Theorem 5.2 *Given parameters $\bar{\theta} \in [0, 1)$ and $\overline{\sigma} \in (0, 1)$, if there exist matrices $P_i > 0$, $Q_i > 0$, $R_i > 0$, $\Omega > 0$, K_j and scalars $\xi_i > 0$ satisfying conditions (5.18) and the following condition (for $i,\ l = 1, \ldots, r$, $j = 1, \ldots, c$)*

$$\sum_{i=1}^{r}\sum_{j=1}^{c} m_i(x(k))h_j(\bar{x}(k))\Pi_{ijl} < 0 \tag{5.33}$$

with $\Pi_{ijl} = \Gamma_{ijl} + 2\mathcal{G}_{ij}^T R_l \mathcal{G}_{ij} + \bar{\theta}(1-\bar{\theta})\mathcal{M}_{ij}^T R_l \mathcal{M}_{ij}$, $\mathcal{G}_{ij} = [(1-\bar{\theta})G(A_i + B_i K_j) + \bar{\theta}G, (1-\bar{\theta})GB_iK_j, \bar{\theta}GB_iK_j]$, and $\mathcal{M}_{ij} \triangleq [GB_iK_j, GB_iK_j, -GB_iK_j]$, then the state trajectories of the closed-loop system (5.16) under the probability-dependent AETS and DoS attack can be driven (in mean square sense) into the following sliding domain:

$$\mathcal{D} = \left\{ s(k) \mid ||s(k)|| \le \hat{\Upsilon}(k) \right\} \tag{5.34}$$

with $\hat{\Upsilon}(k) = \max\limits_{1 \le i,l \le r} \sqrt{\frac{\Upsilon(k)}{\lambda_{\min}(R_i)}}$ and $\Upsilon(k) = 4\varpi^2(1-\bar{\theta})^2||(GB_i)^T R_l G B_i||(||x(k)||^2 + m||\bar{x}(k-1)||^2)$.

Proof Construct an MFD Lyapunov functional as

$$U(\chi(k)) = V(\chi(k)) + s^T(k)R(k)s(k) \tag{5.35}$$

with $V(\chi(k))$ defined in (5.20) and $R(k) = \sum_{i=1}^{r} m_i(x(k))R_i$.

Then, it can be obtained from (5.35) that

$$\begin{aligned} &\mathbf{E}\{\Delta U(\chi(k)) \mid \chi(k)\} \\ &\quad = \mathbf{E}\{\Delta V(\chi(k)) \mid \chi(k)\} + \mathbf{E}\{s^T(k+1)R(k+1)s(k+1) \mid \chi(k)\} \\ &\quad - s^T(k)R(k)s(k). \end{aligned} \tag{5.36}$$

In the light of (5.11) and (5.16), one has

$$\begin{aligned} s(k+1) = &\sum_{i=1}^{r}\sum_{j=1}^{c} m_i(x(k))h_j(\bar{x}(k))\Big\{\Big[\Big((1-\bar{\theta})G(A_i + B_iK_j) + \bar{\theta}G\Big)x(k) \\ &+ (1-\bar{\theta})GB_iK_je(k) + \bar{\theta}GB_iK_j\bar{x}(k-1)\Big] + (\bar{\theta} - \theta(k_\ell))GB_iK_j(x(k) \\ &+ e(k) - \bar{x}(k-1)) + (1-\bar{\theta})GB_iD(k)\Big\}, \end{aligned} \tag{5.37}$$

from which one can deduce that

$$
\begin{aligned}
&\mathbf{E}\{s^T(k+1)R(k+1)s(k+1)|x(k)\} \\
&\le \sum_{i=1}^{r}\sum_{j=1}^{c} m_i(x(k))h_j(\bar{x}(k))\Big\{2\Big[\Big((1-\bar{\theta})G(A_i+B_iK_j)+\bar{\theta}G\Big)x(k) \\
&+(1-\bar{\theta})GB_iK_je(k)+\bar{\theta}GB_iK_j\bar{x}(k-1)\Big]^T R(k+1) \\
&\times\Big[\Big((1-\bar{\theta})G(A_i+B_iK_j)+\bar{\theta}G\Big)x(k)+(1-\bar{\theta})GB_iK_je(k) \\
&+\bar{\theta}GB_iK_j\bar{x}(k-1)\Big]+\bar{\theta}(1-\bar{\theta})(x(k)+e(k)-\bar{x}(k-1))^T \\
&\times(GB_iK_j)^TR(k+1)GB_iK_j(x(k)+e(k)-\bar{x}(k-1)) \\
&+2(1-\bar{\theta})^2D^T(k)(GB_i)^TR(k+1)GB_iD(k)\Big\}.
\end{aligned}
\tag{5.38}
$$

Substituting (5.38) and (5.27) into (5.36) yields

$$
\begin{aligned}
&\mathbf{E}\{\Delta U(k)|\chi(k)\} \\
&\le \sum_{l=1}^{r}\sum_{i=1}^{r}\sum_{j=1}^{c} m_l^+ m_i(x(k))h_j(\bar{x}(k))\Big\{\eta^T(k)\Gamma_{ijl}\eta(k)+2\Big[\Big((1-\bar{\theta})G(A_i+B_iK_j) \\
&+\bar{\theta}G\Big)x(k)+(1-\bar{\theta})GB_iK_je(k)+\bar{\theta}GB_iK_j\bar{x}(k-1)\Big]^T R_l \\
&\times\Big[((1-\bar{\theta})G(A_i+B_iK_j)+\bar{\theta}G)x(k)+(1-\bar{\theta})GB_iK_je(k) \\
&+\bar{\theta}GB_iK_j\bar{x}(k-1)\Big]+\bar{\theta}(1-\bar{\theta})(x(k)+e(k)-\bar{x}(k-1))^T \\
&\times(GB_iK_j)^TR_lGB_iK_j(x(k)+e(k)-\bar{x}(k-1))-\Big[\lambda_{\min}(R_i)||s(k)||^2 \\
&-4\varpi^2(1-\bar{\theta})^2||(GB_i)^TR_lGB_i||(||x(k)||^2+m||\bar{x}(k-1)||^2)\Big]\Big\} \\
&\le \sum_{l=1}^{r}\sum_{i=1}^{r}\sum_{j=1}^{c} m_l^+ m_i(x(k))h_j(\bar{x}(k))\eta^T(k)\Pi_{ijl}\eta(k) \\
&-\Big(\lambda_{\min}(R_i)||s(k)||^2-\Upsilon(k)\Big).
\end{aligned}
\tag{5.39}
$$

If the system state lies outside region $\mathcal{D}$, i.e., $||s(k)|| > \hat{\Upsilon}(k)$, the condition (5.33) can ensure

$$
\mathbf{E}\{\Delta U(k)|\chi(k)\} < 0. \tag{5.40}
$$

Taking the expectation on both sides of the expression (5.40), one has $\mathbf{E}\{\Delta U(k)\} < 0$, which means that the state trajectories are strictly decreasing outside the region $\mathcal{D}$, and can finally converge into this region $\mathcal{D}$ in mean square sense. □

In Theorems 5.1 and 5.2, the MFD conditions have been derived, however, which contain the unmatched MFs $m_i(x(k))$ and $h_j(\bar{x}(k))$ and some coupled terms such that they are difficultly solved. In addition, as stated in Remark 5.3, due to the different fuzzy rules of $m_i(x(k))$ and $h_j(\bar{x}(k))$, some existing handling methods on matching situation cannot be used. Meanwhile, because of the AETS and random DoS attacks resulting in the unpredictability of the state $\bar{x}(k)$, the specific information of the state $\bar{x}(k)$ cannot be estimated so that it is difficult to obtain the local bounds of $h_j(\bar{x}(k))$. Nevertheless, the global bounds of $h_j(\bar{x}(k))$ can be exploited to gather more information of $h_{ij}(k) \triangleq m_i(x(k))h_j(\bar{x}(k))$, which will be further discussed in the following section to obtain the solvable conditions with less conservatism.

5.5 Co-design of AETS and Controller via Global-Boundary-Information Approach

In this section, the global bounds of the mismatched MFs will be employed to transform the sufficient conditions involving the time-varying MFs $m_i(x(k))$ and $h_j(\bar{x}(k))$ into the solvable linear conditions.

Theorem 5.3 *Given parameters $\bar{\theta} \in [0, 1)$ and $\overline{\sigma} \in (0, 1)$, if there exist matrices $\mathcal{P}_i > 0$, $\mathcal{Q}_i > 0$, $\mathcal{R}_i > 0$, $\mathcal{M}_{ij\epsilon} > 0$, $\mathcal{N}_{ij\epsilon} > 0$, $\bar{\Omega} > 0$, F_j, Z, and scalars $\zeta_i > 0$ satisfying the following conditions (for i, $l = 1, \ldots, r$, $j = 1, \ldots, c$)*

$$\begin{bmatrix} -\zeta_i I & * \\ \zeta_i B_i & -\mathcal{P}_l \end{bmatrix} \leq 0, \tag{5.41}$$

$$\Lambda_{ijl} - \mathbb{M}_{ij} - \mathbb{N}_{ij} + \sum_{\iota=1}^{r}\sum_{\jmath=1}^{c}(\bar{h}_{\iota\jmath}\mathbb{M}_{\iota\jmath} - \underline{h}_{\iota\jmath}\mathbb{N}_{\iota\jmath}) < 0 \tag{5.42}$$

with

$$\Lambda_{ijl} = \begin{bmatrix} \Lambda^{11} & * & * & * & * & * \\ \Lambda^{21} & \Lambda^{22} & * & * & * & * \\ 0 & 0 & \Lambda^{33} & * & * & * \\ \Lambda^{41} & \Lambda^{42} & \Lambda^{43} & \Lambda^{44} & * & * \\ \Lambda^{51} & \Lambda^{52} & \Lambda^{53} & 0 & \Lambda^{55} & * \\ \Lambda^{61} & 0 & \Lambda^{63} & 0 & 0 & \Lambda^{66} \end{bmatrix}$$

and

$$
\begin{aligned}
\Lambda^{11} &= -Z^T - Z + \mathcal{P}_i + \overline{\sigma}\bar{\Omega} + (1-\bar{\theta})\bar{\mathcal{Q}}_l, \quad \Lambda^{21} = \overline{\sigma}\bar{\Omega} + (1-\bar{\theta})\bar{\mathcal{Q}}_l, \\
\Lambda^{22} &= -(1-\overline{\sigma})\bar{\Omega} + (1-\bar{\theta})\bar{\mathcal{Q}}_l, \quad \Lambda^{33} = -\bar{\mathcal{Q}}_i + \bar{\theta}\bar{\mathcal{Q}}_l, \\
\Lambda^{41} &= \begin{bmatrix} A_i Z + B_i F_j \\ \sqrt{\frac{\bar{\theta}}{1-\bar{\theta}}} B_i F_j \end{bmatrix}, \Lambda^{42} = \begin{bmatrix} B_i F_j \\ \sqrt{\frac{\bar{\theta}}{1-\bar{\theta}}} B_i F_j \end{bmatrix}, \quad \Lambda^{43} = \begin{bmatrix} \frac{\bar{\theta}}{1-\bar{\theta}} B_i F_j \\ -\sqrt{\frac{\bar{\theta}}{1-\bar{\theta}}} B_i F_j \end{bmatrix}, \\
\Lambda^{51} &= \begin{bmatrix} (1-\bar{\theta})G(A_i Z + B_i F_j) + \bar{\theta} G Z \\ \sqrt{\bar{\theta}(1-\bar{\theta})} G B_i F_j \end{bmatrix}, \quad \Lambda^{52} = \begin{bmatrix} (1-\bar{\theta}) G B_i F_j \\ \sqrt{\bar{\theta}(1-\bar{\theta})} G B_i F_j \end{bmatrix}, \\
\Lambda^{53} &= \begin{bmatrix} \bar{\theta} G B_i F_j \\ -\sqrt{\bar{\theta}(1-\bar{\theta})} G B_i F_j \end{bmatrix}, \Lambda^{61} = \begin{bmatrix} 2\varpi Z \\ 0 \end{bmatrix}, \Lambda^{63} = \begin{bmatrix} 0 \\ 2\sqrt{m}\varpi Z \end{bmatrix}, \\
\Lambda^{44} &= \mathrm{diag}\left\{-\frac{1}{2}\mathcal{P}_l, \ -\mathcal{P}_l\right\}, \quad \Lambda^{55} = \mathrm{diag}\left\{-\frac{1}{2}\mathcal{R}_l, \ -\mathcal{R}_l\right\}, \\
\Lambda^{66} &= \mathrm{diag}\{-\zeta_i I, \ -\zeta_i I\}, \\
\mathbb{M}_{ij} &= \mathrm{diag}\{\mathcal{M}_{ij1}, \ \mathcal{M}_{ij2}, \ \mathcal{M}_{ij3}, \ \underbrace{0, \ldots, 0}_{6}\}, \ \mathbb{N}_{ij} \\
&= \mathrm{diag}\{\mathcal{N}_{ij1}, \ \mathcal{N}_{ij2}, \ \mathcal{N}_{ij3}, \ \underbrace{0, \ldots, 0}_{6}\},
\end{aligned}
$$

then results of Theorems 5.1 and 5.2 on the stochastic stability and the reachability can be simultaneously attained. Meanwhile, the control gains K_j and the triggering matrix Ω can be calculated as $K_j = F_j Z^{-1}$ and $\Omega = Z^{-T}\bar{\Omega}Z^{-1}$.

Proof Let $P_i^{-1} \triangleq \mathcal{P}_i$, $Q_i^{-1} \triangleq \mathcal{Q}_i$, $R_i^{-1} \triangleq \mathcal{R}_i$, and $\xi_i^{-1} \triangleq \zeta_i$. By using Schur complement, condition (5.18) is equivalent to condition (5.41). In addition, due to $\Pi_{ijl} - \Gamma_{ijl} > 0$, condition (5.33) can ensure condition (5.19). Denote $h_{ij}(k) \triangleq m_i(x(k))h_j(\bar{x}(k))$ and its global upper and lower bounds as $\bar{h}_{ij}$ and $\underline{h}_{ij}$, which satisfy $\underline{h}_{ij} \le h_{ij}(k) \le \bar{h}_{ij}$ for $\forall i, j$. For matrices $M_{ij} > 0$ and $N_{ij} > 0$, one has

$$
\sum_{i=1}^{r}\sum_{j=1}^{c}(\bar{h}_{ij} - h_{ij}(k))M_{ij} \ge 0, \tag{5.43}
$$

$$
\sum_{i=1}^{r}\sum_{j=1}^{c}(h_{ij}(k) - \underline{h}_{ij})N_{ij} \ge 0, \tag{5.44}
$$

with $M_{ij} = \mathrm{diag}\{M_{ij1}, M_{ij2}, M_{ij3}\}$ and $N_{ij} = \mathrm{diag}\{N_{ij1}, N_{ij2}, N_{ij3}\}$.

By means of conditions (5.43)–(5.44) and the property of $\sum_{i=1}^{r}\sum_{j=1}^{c} h_{ij}(k) = 1$, we have from condition (5.33)

$$\begin{aligned}&\sum_{i=1}^{r}\sum_{j=1}^{c}h_{ij}(k)\Pi_{ijl}\\&\quad\le\sum_{i=1}^{r}\sum_{j=1}^{c}\Big[h_{ij}(k)\Pi_{ijl}+(\bar{h}_{ij}-h_{ij}(k))M_{ij}+(h_{ij}(k)-\underline{h}_{ij})N_{ij}\Big]\\&\quad=\sum_{i=1}^{r}\sum_{j=1}^{c}h_{ij}(k)\Big[\Pi_{ijl}-M_{ij}+N_{ij}+\sum_{\imath=1}^{r}\sum_{\jmath=1}^{c}(\bar{h}_{\imath\jmath}M_{\imath\jmath}-\underline{h}_{\imath\jmath}N_{\imath\jmath})\Big],\end{aligned}\tag{5.45}$$

for $l=1,\ldots,r$.

Thus, condition (5.33) can be guaranteed if the following inequality holds:

$$\Pi_{ijl}-M_{ij}+N_{ij}+\sum_{\imath=1}^{r}\sum_{\jmath=1}^{c}(\bar{h}_{\imath\jmath}M_{\imath\jmath}-\underline{h}_{\imath\jmath}N_{\imath\jmath})<0,\tag{5.46}$$

which is equivalent to the following condition:

$$\Psi_{ijl}-\hat{M}_{ij}-\hat{N}_{ij}+\sum_{\imath=1}^{r}\sum_{\jmath=1}^{c}(\bar{h}_{\imath\jmath}\hat{M}_{\imath\jmath}-\underline{h}_{\imath\jmath}\hat{N}_{\imath\jmath})<0,\tag{5.47}$$

with

$$\Psi_{ijl}=\begin{bmatrix}\Psi^{11}&*&*&*&*&*\\\Psi^{21}&\Psi^{22}&*&*&*&*\\0&0&\Psi^{33}&*&*&*\\\Psi^{41}&\Psi^{42}&\Psi^{43}&\Lambda^{44}&*&*\\\Psi^{51}&\Psi^{52}&\Psi^{53}&0&\Lambda^{55}&*\\\Psi^{61}&0&\Psi^{63}&0&0&\Lambda^{66}\end{bmatrix},$$

$$\begin{aligned}\Psi^{11}&=-P_i+\bar{\sigma}\Omega+(1-\bar{\theta})Q_l,\ \Psi^{21}=\bar{\sigma}\Omega+(1-\bar{\theta})Q_l,\\\Psi^{22}&=-(1-\bar{\sigma})\Omega+(1-\bar{\theta})Q_l,\ \Psi^{33}=-Q_i+\bar{\theta}Q_l,\\\Psi^{41}&=\begin{bmatrix}A_i+B_iK_j\\\sqrt{\frac{\bar{\theta}}{1-\theta}}B_iK_j\end{bmatrix},\ \Psi^{42}=\begin{bmatrix}B_iK_j\\\sqrt{\frac{\bar{\theta}}{1-\theta}}B_iK_j\end{bmatrix},\ \Psi^{43}=\begin{bmatrix}\frac{\bar{\theta}}{1-\theta}B_iK_j\\-\sqrt{\frac{\bar{\theta}}{1-\theta}}B_iK_j\end{bmatrix},\\\Psi^{51}&=\begin{bmatrix}(1-\bar{\theta})G(A_i+B_iK_j)+\bar{\theta}G\\\sqrt{\bar{\theta}(1-\bar{\theta})}GB_iK_j\end{bmatrix},\ \Psi^{52}=\begin{bmatrix}(1-\bar{\theta})GB_iK_j\\\sqrt{\bar{\theta}(1-\bar{\theta})}GB_iK_j\end{bmatrix},\\\Psi^{53}&=\begin{bmatrix}\bar{\theta}GB_iK_j\\-\sqrt{\bar{\theta}(1-\bar{\theta})}GB_iK_j\end{bmatrix},\ \Psi^{61}=\begin{bmatrix}2\varpi I\\0\end{bmatrix},\ \Psi^{63}=\begin{bmatrix}0\\2\sqrt{m}\varpi I\end{bmatrix},\\\hat{M}_{ij}&=\mathrm{diag}\{M_{ij1},\ M_{ij2},\ M_{ij3},\ \underbrace{0,\ldots,0}_{6}\},\\\hat{N}_{ij}&=\mathrm{diag}\{N_{ij1},\ N_{ij2},\ N_{ij3},\ \underbrace{0,\ldots,0}_{6}\}.\end{aligned}$$

Define $\mathcal{T} = \text{diag}\{Z, Z, Z, \underbrace{I, \ldots, I}_{6}\}$. Pre- and post-multiplying $\mathcal{T}^T$ and $\mathcal{T}$ to condition (5.47), respectively, and letting $\hat{\Omega} = Z^T \Omega Z$, $\hat{Q}_i = Z^T Q_i Z$, $\mathcal{M}_{ij\tau} = Z^T M_{ij\tau} Z$, $\mathcal{N}_{ij\tau} = Z^T N_{ij\tau} Z$ $(\tau = 1, 2, 3)$, and $F_j = K_j Z$. In addition, due to $-Z^T P_i Z \leq -Z^T - Z + \mathcal{P}_i$, it is easily seen that condition (5.42) can ensure condition (5.47). Thus, conditions (5.41)–(5.42) can ensure inequalities (5.18)–(5.19) and (5.33) in Theorems 5.1 and 5.2, simultaneously. □

Remark 5.4 By using the global bounds $\bar{h}_{ij}$ and $\underline{h}_{ij}$ of the coupled MFs $h_{ij}(k)$, some slack matrices in Theorem 5.3 are introduced to reduce the conservatism. Compared with the existing results as in [6–9], the present control design and sufficient conditions can permit different fuzzy rules in the controlled system and the controller, which extends the application of fuzzy control scheme. It is worth pointing out that under the AETS and DoS attacks, different variables $x(k)$ and $\bar{x}(k)$ are involved in system's MFs $m_i(x(k))$ and controller's MFs $h_j(\bar{x}(k))$, which makes it more difficult to obtain the upper and lower bounds $\bar{h}_{ij}$ and $\underline{h}_{ij}$. In addition, due to the unknown system's MFs $m_i(x(k))$ in IT2 fuzzy system (5.4), the global upper and lower bounds $\bar{h}_{ij}$ and $\underline{h}_{ij}$ of $h_{ij}(k)$ can only be estimated. In this chapter, a feasible approach to obtain the global bounds of $h_{ij}(k)$ is to calculate the upper and lower bounds of $m_i(x(k))$ and $h_j(\bar{x}(k))$, respectively. According to the designed controller's MFs $h_j(\bar{x}(k))$ in (5.14) and the modeling domain, we can calculate the maximum and minimum values of $h_j(\bar{x}(k))$ as its upper and lower bounds $\overline{m}_j$ and $\underline{m}_j$ satisfying the relation $\underline{m}_j \leq h_j(\bar{x}(k)) \leq \overline{m}_j$. Meanwhile, according to system's upper and lower MFs in (5.2)–(5.3) with the relation $\underline{m}_i(x(k)) \leq m_i(x(k)) \leq \overline{m}_i(x(k))$, we can choose the upper and lower bounds as $\overline{m}_i \triangleq \max\limits_{x(k)}\{\overline{m}_i(x(k))\}$ and $\underline{m}_i \triangleq \min\limits_{x(k)}\{\underline{m}_i(x(k))\}$ satisfying $\underline{m}_i \leq m_i(x(k)) \leq \overline{m}_i$. Thus, the global upper and lower bounds of $h_{ij}(k)$ can be obtained as $\bar{h}_{ij} = \overline{m}_i \bar{h}_j$ and $\underline{h}_{ij} = \underline{m}_i \underline{h}_j$ satisfying $\underline{h}_{ij} \leq h_{ij}(k) \leq \bar{h}_{ij}$.

5.6 Simulation Example

Consider an IT2 fuzzy system (5.4) with the following parameters:

$$A_1 = \begin{bmatrix} 1.2 & 0.1 & 0 \\ 0 & 0.5 & 0.1 \\ 0.2 & 0.1 & 0.9 \end{bmatrix}, A_2 = \begin{bmatrix} 0.4 & 0.1 & 0.1 \\ 0.2 & 0.5 & 0.1 \\ 0.1 & 0.1 & 0.9 \end{bmatrix}, A_3 = \begin{bmatrix} 0.8 & 0.1 & 0 \\ 0 & 1.0 & 0.1 \\ 0.1 & 0.2 & 0.8 \end{bmatrix},$$

$$B_1 = \begin{bmatrix} 0.5 & 0.8 \\ 1.0 & 0.5 \\ 0.4 & 1.0 \end{bmatrix}, B_2 = \begin{bmatrix} 1.0 & 0 \\ 0 & 1.0 \\ 0.5 & 0.6 \end{bmatrix}, B_3 = \begin{bmatrix} 0 & 0.6 \\ 0.3 & 1.0 \\ 1.0 & 0.6 \end{bmatrix},$$

$$d(x(k), k) = 0.05 \begin{bmatrix} \sqrt{2|x_1(k)x_2(k)|} \\ x_3(k) \end{bmatrix},$$

and upper and lower MFs:

$$\overline{m}_1(x(k)) = 1 - \frac{1}{1.5 + e^{-1.5(x_1(k)+0.5)}},\ \underline{m}_1(x(k)) = 1 - \frac{1}{1.5 + e^{-1.5(x_1(k)+1.5)}},$$
$$\overline{m}_3(x(k)) = \frac{1}{1.5 + e^{-1.5(x_1(k)-0.5)}},\ \underline{m}_3(x(k)) = \frac{1}{1.5 + e^{-1.5(x_1(k)-1.5)}},$$

and $\overline{m}_2(x(k)) = 1 - \underline{m}_1(x(k)) - \underline{m}_3(x(k)), \underline{m}_2(x(k)) = 1 - \overline{m}_1(x(k)) - \overline{m}_3(x(k))$.

The event-triggered fuzzy sliding mode security controller (5.12) is designed with the following MFs:

$$\overline{h}_1(\bar{x}(k)) = \frac{1}{1 + e^{-(\bar{x}_1(k)+0.5)}},\ \underline{h}_1(\bar{x}(k)) = \frac{1}{1 + e^{-(\bar{x}_1(k)+1.5)}}$$

and the weighting coefficient $\overline{\beta}_1(\bar{x}(k)) = 0.6$. Then, the MFs can be obtained as $h_1(\bar{x}(k)) = 0.6\overline{h}_1(\bar{x}(k)) + 0.4\underline{h}_1(\bar{x}(k))$ and $h_2(\bar{x}(k)) = 1 - h_1(\bar{x}(k))$.

Choose $g_1 = 0.2,\ g_2 = 0.5$, and $g_3 = 0.3$ for the sliding matrix G in (5.11). Given $\overline{\sigma} = 0.4$ and $\bar{\theta} = 0.4$, by solving conditions (5.41)–(5.42) in Theorem 5.3, we obtain the event-triggering matrix and control gains as

$$\Omega = \begin{bmatrix} 251.5828 & 81.2797 & 100.0489 \\ 81.2797 & 152.6297 & 125.7167 \\ 100.0489 & 125.7167 & 187.5534 \end{bmatrix},$$
$$K_1 = \begin{bmatrix} -0.0319 & -0.0198 & -0.0696 \\ -0.0872 & -0.1071 & -0.0753 \end{bmatrix},$$
$$K_2 = \begin{bmatrix} -0.0319 & -0.0196 & -0.0698 \\ -0.0872 & -0.1073 & -0.0752 \end{bmatrix}.$$

Suppose that the weighting parameters of system's MFs are given as $\overline{\alpha}_1(x(k)) = (\sin(2x_1(k)) + 1)/2,\ \underline{\alpha}_1(x(k)) = 1 - \overline{\alpha}_1(x(k)),\ \overline{\alpha}_2(x(k)) = (\cos(2x_1(k)) + 1)/2$, and $\underline{\alpha}_2(x(k)) = 1 - \overline{\alpha}_2(x(k))$. Given triggering parameters $\underline{\sigma} = 0.1$, $\varrho = 0.2$, the simulation results under $x(0) = [-1.2,\ 1.5,\ -0.6]^T$ are shown in Fig. 5.2a–d, in which all vertical lines represent the triggering instants and the grey vertical lines denote that the current triggered state is blocked by DoS attacks. The state trajectories in Fig. 5.2a show that under the proposed resilient control strategy composed of the probability-dependent AETS (5.9) and the controller (5.15), the closed-loop system is stable. It can also be observed from Fig. 5.2b that the control signal $u(k)$ is piecewise under the AETS and DoS attacks, and Fig. 5.2c shows the reachability of the specified sliding surface. In addition, it can be seen from Fig. 5.2d that the triggering parameter $\sigma(k)$ will be adaptively adjusted and finally attain a stable value with the convergence of the system state.

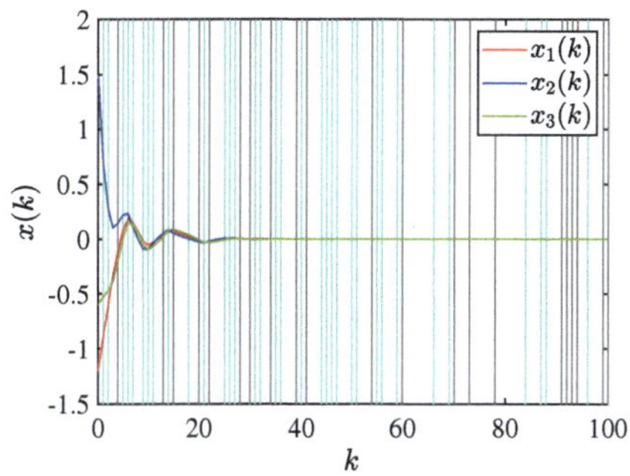

(a) System state $x(k)$

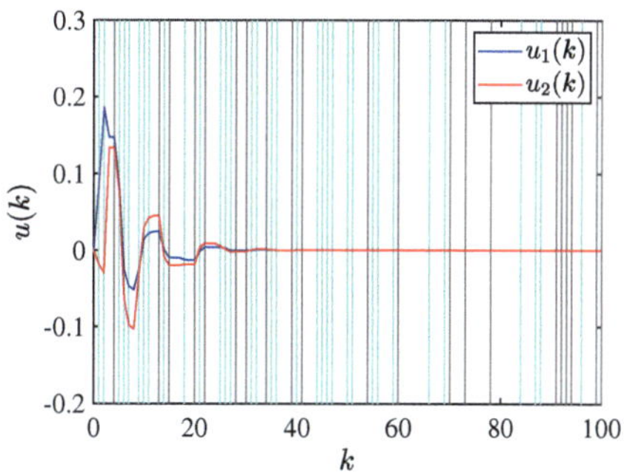

(b) Control input $u(k)$

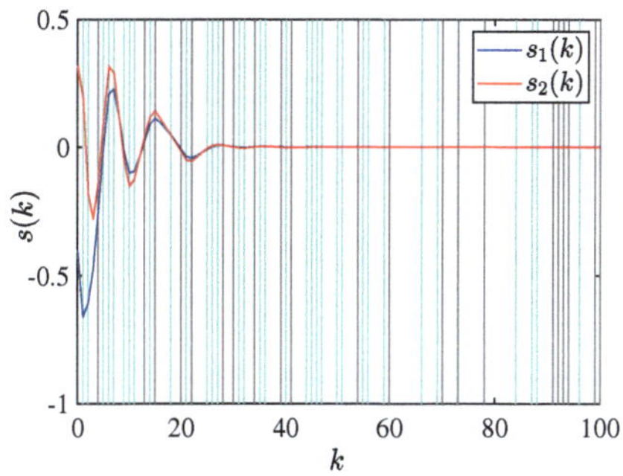

(c) Sliding variable $s(k)$

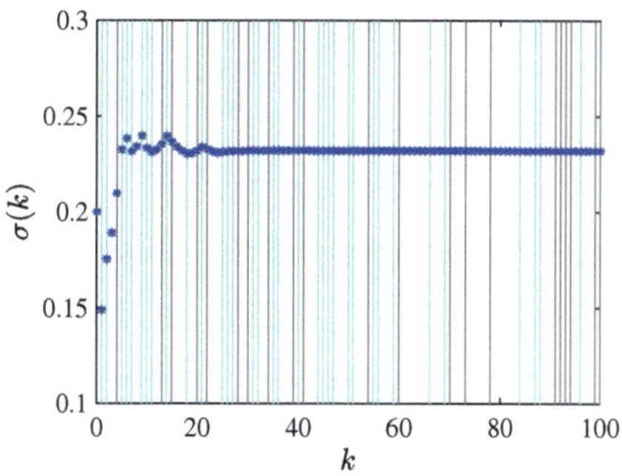

(d) Adaptive parameter $\sigma(k)$

Fig. 5.2 Simulation results under AETS (5.9)

Under the above AETS, for different attacks with $\bar{\theta} = \{0, 0.1, 0.3, 0.5\}$, and the same parameters $\underline{\sigma}$, $\overline{\sigma}$, and ϱ, the effect of attack on the adaptive parameter $\sigma(k)$ and the stability of the controlled system is further compared. It is shown in Fig. 5.3 that the state trajectories will converge as time k increases, meanwhile, it is observed from Fig. 5.4 that the convergence rate of the system state is slower as the attack probability $\bar{\theta}$ increases. Besides, it is also seen from Fig. 5.5 that under the proposed probability-dependent AETS, the triggering parameter $\sigma(k)$ will adaptively decrease as the probability $\bar{\theta}$ increases, that is, the proposed AETS can adaptively adjust the triggering threshold for mitigating the effect due to attacks and ensuring the desirable performance.

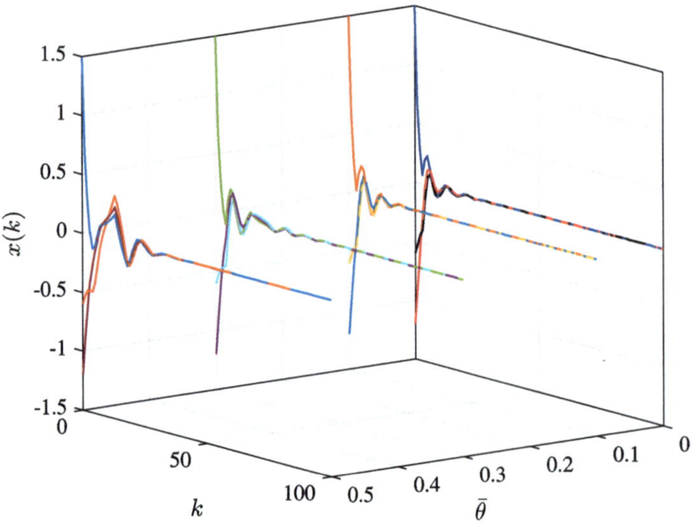

Fig. 5.3 State trajectories $x(k)$ with different probability $\bar{\theta}$

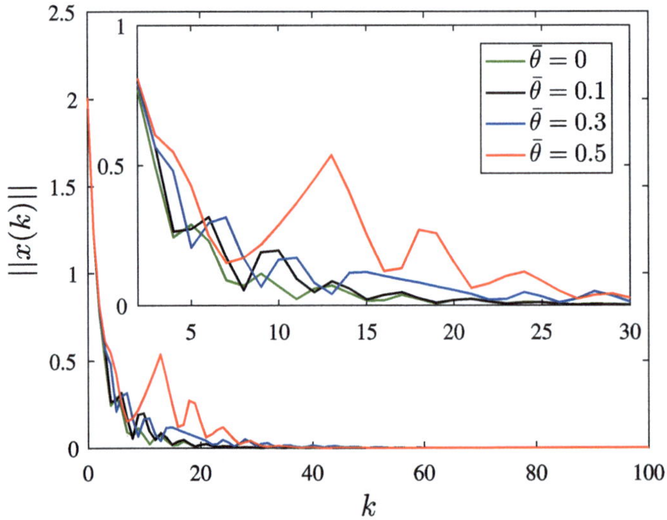

Fig. 5.4 The variation of norm $\|x(k)\|$ with different probability $\bar{\theta}$

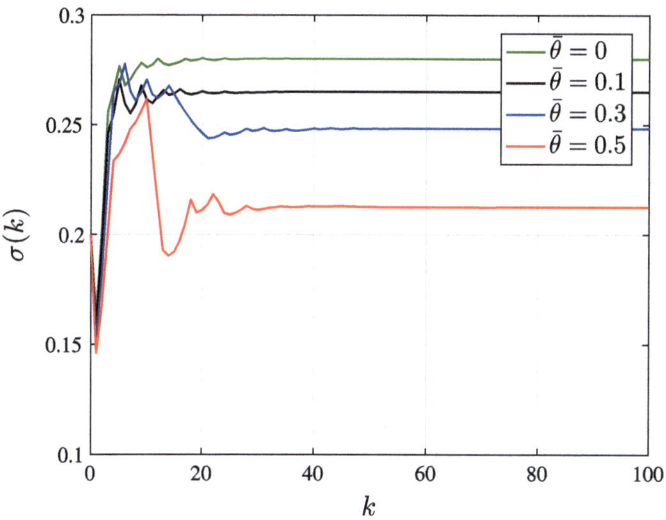

Fig. 5.5 Adaptive parameters $\sigma(k)$ with different probability $\bar{\theta}$

5.7 Conclusion

This chapter has addressed the event-triggered sliding mode resilient control problem for IT2 fuzzy systems under the unreliable communication network. A probability-dependent AETS has been proposed to cope with the randomly occurring DoS attacks. According to the variation trend of the system state and the attack probability, an adaptive rule has been formulated to dynamically update the triggering threshold. Then, the one-step memorized state-based sliding function has been constructed, and the corresponding fuzzy sliding mode controller has been designed. By means of the global bounds of mismatched MFs, the co-design conditions of fuzzy sliding mode controller and the AETS have been obtained for reaching the desired performance of the closed-loop IT2 T-S fuzzy systems.

References

1. Li, J., Liu, Y., Yu, J., et al.: Resilient sliding mode control for a class of cyber-physical systems with multiple transmission channels under denial-of-service attacks. J. Franklin Inst. **359**(11), 5302–5321 (2022)
2. Lin, H., Lam, J., Wang, Z.: Secure state estimation for systems under mixed cyber-attacks: security and performance analysis. Inf. Sci. **546**, 943–960 (2021)
3. Ding, D., Wang, Z., Han, Q.L., et al.: Security control for discrete-time stochastic nonlinear systems subject to deception attacks. IEEE Trans. Syst. Man Cybern. Syst. **48**(5), 779–789 (2018)
4. Gao, X., Deng, F., Zhang, H., et al.: Adaptive neural state estimation of Markov jump systems under scheduling protocols and probabilistic deception attacks. IEEE Trans. Cybern. **53**(3), 1830–1842 (2023)

5. Hu, S., Cheng, Z., Yue, D., et al.: Bandwidth allocation-based switched dynamic triggering control against DoS attacks. IEEE Trans. Syst. Man Cybern. Syst. **51**(10), 6050–6061 (2021)
6. Liu, J., Yin, T., Cao, J., et al.: Security control for T-S fuzzy systems with adaptive event-triggered mechanism and multiple cyber-attacks. IEEE Trans. Syst. Man Cybern. Syst. **51**(10), 6544–6554 (2021)
7. Ran, G., Li, C., Rathinasamy, S., et al.: Adaptive event-triggered asynchronous control for interval type-2 fuzzy Markov jump systems with cyber-attacks. IEEE Trans. Control Netw. Syst. **9**(1), 88–99 (2022)
8. Li, X., Ye, D.: Asynchronous event-triggered control for networked interval type-2 fuzzy systems under DoS attacks. IEEE Trans. Fuzzy Syst. **29**(2), 262–274 (2021)
9. Zhao, N., Shi, P., Xing, W., et al.: Resilient adaptive event-triggered fuzzy tracking control and filtering for nonlinear networked systems under denial of service attacks. IEEE Trans. Fuzzy Syst. **30**(8), 3191–3201 (2022)
10. Peng, C., Yang, M., Zhang, J., et al.: Network-based H_∞ control for T-S fuzzy systems with an adaptive event-triggered communication scheme. Fuzzy Sets Syst. **329**, 61–76 (2017)
11. Wang, Y., Xia, Y., Ahn, C.K., et al.: Exponential stabilization of Takagi-Sugeno fuzzy systems with aperiodic sampling: An aperiodic adaptive event-triggered method. IEEE Trans. Syst. Man Cybern. Syst. **49**(2), 444–454 (2019)

Chapter 6
Self-triggered Control via Local-Boundary-Information-Dependent Approach

6.1 Introduction

For the ETSs as in the previous chapters, an additional monitor device is required to continually check the triggering condition and determine the next triggering instant. This may increase the device implementation cost, even is impractical in many complex cases. Different from event-triggered schemes, the self-triggered scheme does not require the additional monitor device to monitor the triggering condition, and it can calculate the next triggering instant according to the latest triggering information.

On the other hand, for IT2 fuzzy system under triggering-based schemes (including event-triggered and self-triggered schemes), the fuzzy controller and its MFs can only be updated at triggering instant, which causes the asynchronism of the MFs in the fuzzy system and controller and also hinders the implementation of MFD method. Some results as in [1–3] and Chaps. 3 and 5 only utilize the global boundary information of MFs under the time-triggering case to obtain stability conditions such that these above methods may bring greater conservatism due to ignoring many local information in operation points. Moreover, the above works cannot be applied to gain the local boundary information under triggering-based schemes, since it is difficult to identify the scope of the controller's MFs in local operation domain. To be specific, the self-triggered scheme makes the range of state variables be different in the local operation domain, and the local boundary information of MFs in the controller cannot be directly gained via the existing approaches (see Remark 6.4 later). Therefore, we shall encounter the following challenging problems for designing a self-triggering IT2 fuzzy controller by using the local boundary information of MFs: (1) How to determine the next triggering instant just depending on the current triggering information and MFs? (2) How to extract the local boundary information of MFs for obtaining MFD stability conditions?

This chapter investigates the self-triggered control scheme for IT2 fuzzy systems subject to external disturbance. The transmission of the system state is determined via a self-triggered scheme, under which the next triggering instant can be cal-

Y. Yang et al., *Analysis and Synthesis of Interval Type-2 Fuzzy Systems Under Communication Protocols*, Studies in Systems, Decision and Control 636,
https://doi.org/10.1007/978-981-95-2355-9_6

culated by using the current latest triggering information, instead of continuously monitoring the prespecified triggering condition. Then, a self-triggered IT2 fuzzy controller is designed, whose fuzzy rules and MFs can be flexibly chosen independent of the fuzzy model. Under the constraint of imperfectly premise matching, the relationship between the triggered state and system state is established to handle the asynchronous MFs, and gain their boundaries in the local operation region. Meanwhile, the local-boundary-information-dependent approach is employed to facilitate the relaxed feasible criteria. Finally, one numerical example and one-link manipulator with motor are provided to verify the designed self-triggered control scheme, meanwhile, the comparisons with the conventional event-triggered scheme and the global-boundary-information-based control strategy are also made, respectively.

6.2 System Model

Consider the following IT2 fuzzy systems with the i-th rule:

Rule i: IF $f_1(x(t))$ is $\mathscr{F}_1^i$ and $\cdots$ and $f_\hbar(x(t))$ is $\mathscr{F}_\hbar^i$, THEN

$$\dot{x}(t) = A_i x(t) + B_i u(t) + D_i d(t), \tag{6.1}$$

where $x(t) \in \mathbb{R}^n$ and $u(t) \in \mathbb{R}^m$ denote system state and control input, respectively. $d(t)$ is the external disturbance satisfying $||d(t)|| \leq \varpi$ with ϖ a known positive constant. For the i-th IF-THEN rule, the premise variable and the IT2 fuzzy set are signified as $f_\iota(x(k))$ and $\mathscr{F}_\iota^i$ with $\iota \in \{1, \cdots, \hbar\}$ and $i \in \{1, \cdots, r\}$. Matrices A_i, B_i, and D_i are known. The firing strength of MFs with respect to the i-th rule satisfies

$$\overrightarrow{m}_i(x(t)) = \left[\underline{m}_i(x(t)), \overline{m}_i(x(t))\right],$$

with

$$\overline{m}_i(x(t)) = \prod_{\iota=1}^{\hbar} \overline{\mu}_{\mathscr{F}_\iota^i}(f_\iota(x(t))) > 0, \tag{6.2}$$

$$\underline{m}_i(x(t)) = \prod_{\iota=1}^{\hbar} \underline{\mu}_{\mathscr{F}_\iota^i}(f_\iota(x(t))) > 0, \tag{6.3}$$

where $\overline{\mu}_{\mathscr{F}_\iota^i}(f_\iota(x(t)))$ and $\underline{\mu}_{\mathscr{F}_\iota^i}(f_\iota(x(t))) \in [0, 1]$ are the upper and lower grades of the MFs and satisfy $\underline{\mu}_{\mathscr{F}_\iota^i}(f_\iota(x(t))) \leq \overline{\mu}_{\mathscr{F}_\iota^i}(f_\iota(x(t)))$. $\overline{m}_i(x(t))$ and $\underline{m}_i(x(t))$ represent the upper and lower MFs, respectively.

Based on the singleton fuzzification, product inference and the center-average defuzzification, the fuzzy system (6.1) can be inferred as

$$\dot{x}(t) = \sum_{i=1}^{r} m_i(x(t))\left[A_i x(t) + B_i u(t) + D_i d(t)\right], \tag{6.4}$$

where the MFs $m_i(x(t))$ satisfy $\sum_{i=1}^{r} m_i(x(t)) = 1$ with the following form

$$m_i(x(t)) = \underline{m}_i(x(t))\underline{\alpha}_i(x(t)) + \overline{m}_i(x(t))\overline{\alpha}_i(x(t)).$$

and the weighting coefficients $\overline{\alpha}_i(x(t))$ and $\underline{\alpha}_i(x(t))$ can be represented as:

$$\underline{\alpha}_i(x(t)), \overline{\alpha}_i(x(t)) \in [0, 1], \ \underline{\alpha}_i(x(t)) + \overline{\alpha}_i(x(t)) = 1.$$

Notice that the known upper and lower MFs $\overline{m}_i(x(t))$ and $\underline{m}_i(x(t))$ in (6.2)–(6.3) are related to the system state $x(t)$, which implies that they may not be available for the controller when the state $x(t)$ is not transmitted in real time. This will bring difficulties to the design of fuzzy controller.

6.3 Self-triggered Control Design

As shown in Fig. 6.1, the transmission of the system state $x(t)$ is determined by the self-triggered scheme that will be designed later. Under this circumstance, the controller can only receive the triggered state (denoting as $x(t_k)$), where the triggering sequence is denoted as $\{t_k\}_{k=0}^{\infty}$. That is, the controller will be updated only when the system state $x(t_k)$ is released at the triggering instants. During the triggering interval, the previous value will be held via the ZOH strategy.

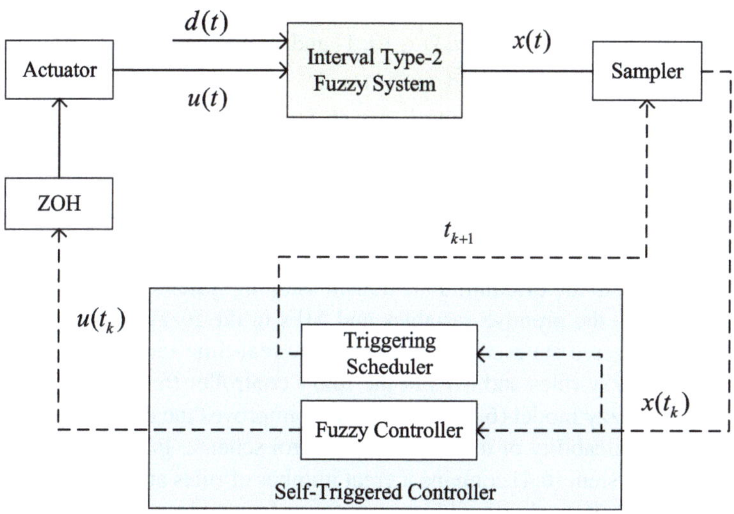

Fig. 6.1 Self-triggered control system

6.3.1 Interval Type-2 Fuzzy Controller

In the following, a c-rule IT2 fuzzy controller is designed as:

Rule j: IF $g_1(x(t_k))$ is $\mathscr{M}_1^j$ and $\cdots$ and $g_\wp(x(t_k))$ is $\mathscr{M}_\wp^j$, THEN

$$u(t) = K_j x(t_k), \quad \forall t \in [t_k, t_{k+1}), \tag{6.5}$$

with the control gains K_j to be designed later. $\mathscr{M}_\tau^j$ is the fuzzy set with regard to the premise variable $g_\tau(x(t_k))$ in the j-th fuzzy rule, where the numbers of fuzzy rules and premise variables are denoted by $j \in \{1, \ldots, c\}$ and $\tau \in \{1, \ldots, \wp\}$. The corresponding firing strength of j-th rule satisfies

$$\vec{h}_j(x(t_k)) = [\underline{h}_j(x(t_k)), \overline{h}_j(x(t_k))],$$

with $\overline{h}_j(x(t_k)) = \prod_{\tau=1}^{\wp} \overline{\mu}_{\mathscr{M}_\tau^j}(g_\tau(x(t_k)))$, and $\underline{h}_j(x(t_k)) = \prod_{\tau=1}^{\wp} \underline{\mu}_{\mathscr{M}_\tau^j}(g_\tau(x(t_k)))$ satisfying $0 \le \underline{h}_j(x(t_k)) \le \overline{h}_j(x(t_k)) \le 1$. Meanwhile, the lower and upper grade of the membership denoting as $\underline{\mu}_{\mathscr{M}_\tau^j}(g_\tau(x(t_k)))$ and $\overline{\mu}_{\mathscr{M}_\tau^j}(g_\tau(x(t_k)))$ meet $0 \le \underline{\mu}_{\mathscr{M}_\tau^j}(g_\tau(x(t_k))) \le \overline{\mu}_{\mathscr{M}_\tau^j}(g_\tau(x(t_k))) \le 1$.

Then, MFs $h_j(x(t_k))$ are represented as

$$h_j(x(t_k)) = \frac{\underline{\beta}_j(x(t_k))\underline{h}_j(x(t_k)) + \overline{\beta}_j(x(t_k))\overline{h}_j(x(t_k))}{\sum_{j=1}^{c}(\underline{\beta}_j(x(t_k))\underline{h}_j(x(t_k)) + \overline{\beta}_j(x(t_k))\overline{h}_j(x(t_k)))}$$

where $\sum_{j=1}^{c} h_j(x(t_k)) = 1$, $\underline{\beta}(x(t_k)) \in [0, 1]$ and $\overline{\beta}_j(x(t_k)) \in [0, 1]$ are two chosen functions satisfying $\underline{\beta}_j(x(t_k)) + \overline{\beta}_j(x(t_k)) = 1$.

Furthermore, the IT2 fuzzy controller in (6.5) can be obtainedS as

$$u(t) = \sum_{j=1}^{c} h_j(x(t_k)) K_j x(t_k), \quad \forall t \in [t_k, t_{k+1}). \tag{6.6}$$

Remark 6.1 Due to the discontinuous transmission of system states under the self-triggered scheme, the premise variables and MFs in the fuzzy controller (6.6) are based on the triggered states $x(t_k)$ rather than the real-time states $x(t)$. In addition, the number of fuzzy rules and MFs in the fuzzy controller (6.6) are different from the ones in the fuzzy model (6.1), which further improves the design flexibility and enhances the applicability of the IT2 fuzzy control scheme. In particular, when the modeled fuzzy system (6.4) contains a great number of rules and complex MFs, the fuzzy controller is allowed to flexibly choose fewer fuzzy rules and simpler MFs, e.g., triangle, trapezoid, and so forth, which can reduce the implementation complexity of the IT2 fuzzy controller, resulting in a lower implementation cost.

Substituting the self-triggered fuzzy controller (6.6) into the controlled IT2 fuzzy system (6.4) yields the following closed-loop system as

$$\dot{x}(t) = \sum_{i=1}^{r}\sum_{j=1}^{c} m_i(x(t))h_j(x(t_k))\Big[A_i x(t) + B_i K_j x(t_k) + D_i d(t)\Big], \quad t \in [t_k, t_{k+1}). \tag{6.7}$$

In the following, the main objectives are to propose a self-triggered strategy for determining the next triggering instant only using the current latest triggering information, and give the feasible conditions for guaranteeing the uniform ultimate boundedness of the closed-loop system (6.7).

6.3.2 Self-triggered Scheme

In this subsection, the self-triggered scheme will be proposed such that the next triggering instant can be calculated beforehand. Moreover, the difference between the current system state $x(t)$ and the latest triggered value $x(t_k)$ can be ensured to be bounded during the triggering interval $[t_k, t_{k+1})$.

To this end, the following theorem is developed to propose an effective self-triggered scheme.

Theorem 6.1 *Given positive constants δ_1, δ_2, and ζ, the following self-triggered scheme is designed to obtain the next triggering instant $t_{k+1} = t_k + T_k$ with*

$$T_k = \frac{1}{\bar{A}} ln \left[\frac{\bar{A}\sqrt{\delta_1 + \delta_2 e^{-\zeta t_k}}}{\varphi(t_k)} + 1 \right], \tag{6.8}$$

with

$$\bar{A} = \max_{1\le i\le r}\{||A_i||\}, \; \bar{B}(t_k) = \max_{1\le i\le r}\{||A_i + B_i \sum_{j=1}^{c} h_j(x(t_k))K_j||\},$$
$$\bar{D} = \max_{1\le i\le r}\{||D_i||\}, \; \varphi(t_k) = \bar{B}(t_k)||x(t_k)|| + \bar{D}\varpi.$$

Under the self-triggered scheme (6.8), the difference $e(t) = x(t_k) - x(t)$ satisfies

$$||e(t)|| \le \sqrt{\delta_1 + \delta_2 e^{-\zeta t_k}}, \; t \in [t_k, t_{k+1}), \tag{6.9}$$

with $t_0 = 0$.

Proof. Define the set $\Omega = \{t \ge t_k \mid e(t) = 0\}$. Then, along the closed-loop system (6.7), we have for $t \in [t_k, t_{k+1}) \setminus \Omega$

$$\begin{aligned}
\frac{d}{dt}\|e(t)\| &\le \left\|\sum_{i=1}^{r}\sum_{j=1}^{c} m_i(x(t))h_j(x(t_k))\Big[A_i e(t) + (A_i + B_i K_j)x(t_k) + D_i d(t)\Big]\right\| \\
&\le \sum_{i=1}^{r} m_i(x(t))\Big\{\|A_i\|\|e(t)\| + \|(A_i + B_i\sum_{j=1}^{c} h_j(x(t_k))K_j)\|\|x(t_k)\| \\
&+ \|D_i\|\|d(t)\|\Big\} \\
&\le \bar{A}||e(t)|| + \bar{B}(t_k)||x(t_k)|| + \bar{D}\varpi \\
&= \bar{A}||e(t)|| + \varphi(t_k).
\end{aligned} \tag{6.10}$$

According to the Comparison Lemma in [4], we can obtain from (6.10)

$$\begin{aligned}
||e(t)|| &\le \frac{\varphi(t_k)}{\bar{A}}\left(e^{\bar{A}(t-t_k)} - 1\right) \le \frac{\varphi(t_k)}{\bar{A}}\left(e^{\bar{A}(t_{k+1}-t_k)} - 1\right) \\
&= \frac{\varphi(t_k)}{\bar{A}}\left(e^{\bar{A}T_k} - 1\right).
\end{aligned} \tag{6.11}$$

Thus, under the self-triggered scheme (6.8), we have from (6.11)

$$||e(t)|| \le \sqrt{\delta_1 + \delta_2 e^{-\zeta t_k}}, \tag{6.12}$$

for $\forall t \in [t_k,\ t_{k+1})$. □

Remark 6.2 The self-triggered scheme (6.8) can calculate next triggering instant t_{k+1} according to the current triggering information, i.e., t_k, $x(t_k)$, and the controller's MFs $h_j(x(t_k))$, without requiring an extra equipment to persistently judge pre-defined triggering condition. Moreover, it is noted from (6.8) and (6.9) that the parameters δ_1, δ_2, and ζ will affect the triggering instant and the bound of the difference $e(t)$. The parameter δ_1 can be taken as the absolute threshold, while ζ represents the decay rate of threshold δ_2. Furthermore, for the triggering interval T_k, it can be guaranteed to be absolutely greater than zero, and the upper bound $\sqrt{\delta_1 + \delta_2 e^{-\zeta t_k}}$ of the difference $e(t)$ will decrease with time.

Apparently, the traditional ET scheme with the following triggering condition (6.13) can also ensure the similar error boundedness (6.9) for $\forall t \in [t_q, t_{q+1})$:

$$t_{q+1} = \inf\Big\{t \;\Big|\; t > t_q, ||x(t) - x(t_q)|| \ge \delta(t_q)\Big\}, \tag{6.13}$$

with $\delta(t_q) = \sqrt{\delta_1 + \delta_2 e^{-\zeta t_q}}$ and $t_0 = 0$. In (6.13), the triggering sequence is defined as $\{t_q\}_{q=0}^{\infty}$ for avoiding confusion with the self-triggering case $\{t_k\}_{k=0}^{\infty}$. In order to compare the ET scheme (6.13) and the proposed self-triggered scheme (6.8), the following corollary is provided to show the inter-interval under the ET condition (6.13).

Corollary 6.1 *Under the ET condition (6.13), the triggering interval* $\Delta t_q = t_{q+1} - t_q$ *satisfies*

$$\Delta t_q \geq \frac{1}{\bar{A}} ln \left[\frac{\bar{A}\sqrt{\delta_1 + \delta_2 e^{-\zeta t_q}}}{\varphi(t_q)} + 1 \right], \tag{6.14}$$

where $\varphi(t_q)$ *is defined as (6.8) by replacing* t_k *with* t_q.

Proof. Similar to expressions (6.10) and (6.11), we have for $t \in [t_q, t_{q+1})$

$$||x(t_q) - x(t)|| \leq \frac{\varphi(t_q)}{\bar{A}} \left(e^{\bar{A}(t-t_q)} - 1 \right), \tag{6.15}$$

which means

$$||x(t_q) - x(t_{q+1}^-)|| \leq \frac{\varphi(t_q)}{\bar{A}} \left(e^{\bar{A}(t_{q+1}^- - t_q)} - 1 \right), \tag{6.16}$$

where t_{q+1}^- is the left limit of t_{q+1}.

Note the fact that $||x(t_q) - x(t_{q+1}^-)|| \leq \sqrt{\delta_1 + \delta_2 e^{-\zeta t_q}}$ and $||x(t_q) - x(t_{q+1}^+)|| \geq \sqrt{\delta_1 + \delta_2 e^{-\zeta t_q}}$, where t_{q+1}^+ denotes the right limit of t_{q+1}. According to the continuity of system state $x(t)$, one has $||x(t_q) - x(t_{q+1}^-)|| = ||x(t_q) - x(t_{q+1}^+)|| = ||x(t_q) - x(t_{q+1})|| = \sqrt{\delta_1 + \delta_2 e^{-\zeta t_q}}$.

Thus, one has from (6.16)

$$\sqrt{\delta_1 + \delta_2 e^{-\zeta t_q}} \leq \frac{\varphi(t_q)}{\bar{A}} \left(e^{\bar{A}(t_{q+1} - t_q)} - 1 \right), \tag{6.17}$$

which can further deduce that

$$e^{\bar{A}(t_{q+1} - t_q)} \geq \frac{\bar{A}\sqrt{\delta_1 + \delta_2 e^{-\zeta t_q}}}{\varphi(t_q)} + 1.$$

Taking the natural logarithm for both sides of the above expression, we can obtain

$$t_{q+1} - t_q \geq \frac{1}{\bar{A}} \ln \left[\frac{\bar{A}\sqrt{\delta_1 + \delta_2 e^{-\zeta t_q}}}{\varphi(t_q)} + 1 \right], \tag{6.18}$$

that is, the condition (6.14) holds. □

Remark 6.3 From the above analysis, it can be observed that the self-triggered scheme (6.8) can not only ensure the same error boundedness as the ETS (6.13), but also remove the additional hardware device. On the other hand, comparing the expressions (6.8) and (6.14), it is known that the number of triggering under the self-triggered scheme (6.8) could be more than the one under the ETS (6.13), which will be further clarified in the simulation later.

6.4 Membership-Function-Dependent Stability Analysis Under Self-triggered Scheme

In the following, we shall analyze the boundedness of the closed-loop system (6.7) under the self-triggered scheme (6.8).

Theorem 6.2 *Given any constants* $\delta_1 > 0$, $\underline{h}_{ijl} \in [0, 1]$, *and* $\overline{h}_{ijl} \in [0, 1]$, $i = 1, \ldots, r$, $j = 1, \ldots, c$, $l \in \mathbb{L} = \{1, \cdots, L\}$, *if there exist matrices* $P > 0$, $Y_{ijl} \geq 0$ *and* K_j *satisfying*

$$\Phi_{ij} - \hat{Y}_{ijl} \leq 0, \qquad \forall i, j, l, \tag{6.19}$$

$$\sum_{i=1}^{r}\sum_{j=1}^{c}\left(\underline{h}_{ijl}\Phi_{ij} + \tilde{h}_{ijl}\hat{Y}_{ijl}\right) < 0, \quad \forall l, \tag{6.20}$$

with

$$\hat{Y}_{ijl} = \text{diag}\{Y_{ijl}, 0, 0\}, \ \tilde{h}_{ijl} = \overline{h}_{ijl} - \underline{h}_{ijl},$$

$$\Phi_{ij} = \begin{bmatrix} He(PA_i + PB_iK_j) & PB_iK_j & PD_i \\ * & -I & 0 \\ * & * & -I \end{bmatrix},$$

then, the closed-loop system (6.7) is uniformly ultimately bounded with the following convergence domain

$$\mathcal{B} \triangleq \left\{x(t)\middle| ||x(t)|| \leq \sqrt{\frac{\delta_1 + \varpi^2}{\epsilon}}\right\}, \tag{6.21}$$

with $\epsilon = \min_{l \in \mathbb{L}}\{\lambda_{\min}(-\Lambda_l)\}$, $\Lambda_l \triangleq \sum_{i=1}^{r}\sum_{j=1}^{c}\left(\underline{h}_{ijl}\Theta_{ij} + \tilde{h}_{ijl}Y_{ijl}\right)$, *and* $\Theta_{ij} = (A_i + B_iK_j)^TP + P(A_i + B_iK_j) + PB_iK_j(B_iK_j)^TP + PD_iD_i^TP$.

Proof. Choose Lyapunov function as

$$V(t) = x^T(t)Px(t). \tag{6.22}$$

According to the closed-loop system (6.7), one has from (6.22) that

$$\begin{aligned}\dot{V}(t) &= \sum_{i=1}^{r}\sum_{j=1}^{c} m_i(x(t))h_j(x(t_k))\Big\{\Big[A_i x(t) + B_i K_j x(t_k) + D_i d(t)\Big]^T P x(t) \\ &+ x^T(t)P\Big[A_i x(t) + B_i K_j x(t_k) + D_i d(t)\Big]\Big\} \\ &= \sum_{i=1}^{r}\sum_{j=1}^{c} m_i(x(t))h_j(x(t_k))\Big\{\Big[(A_i + B_i K_j)x(t) + B_i K_j e(t) + D_i d(t)\Big]^T P x(t) \\ &+ x^T(t)P\Big[(A_i + B_i K_j)x(t) + B_i K_j e(t) + D_i d(t)\Big]\Big\} \\ &= \sum_{i=1}^{r}\sum_{j=1}^{c} m_i(x(t))h_j(x(t_k))\Big\{x^T(t)\Big[(A_i + B_i K_j)^T P + P(A_i + B_i K_j)\Big]x(t) \\ &+ 2x^T(t)P\Big[B_i K_j e(t) + D_i d(t)\Big]\Big\}.\end{aligned} \tag{6.23}$$

Furthermore, for the last term in (6.23), we have

$$2x^T(t)PB_iK_je(t) \le x^T(t)PB_iK_j(B_iK_j)^T Px(t) + ||e(t)||^2, \tag{6.24}$$

$$2x^T(t)PD_id(t) \le x^T(t)PD_iD_i^T Px(t) + ||d(t)||^2. \tag{6.25}$$

Submitting (6.24)–(6.25) into the expression (6.23) yields

$$\begin{aligned}\dot{V}(t) &\le \sum_{i=1}^{r}\sum_{j=1}^{c} m_i(x(t))h_j(x(t_k))x^T(t)\Big[(A_i + B_i K_j)^T P + P(A_i + B_i K_j) \\ &+ PB_iK_j(B_iK_j)^T P + PD_iD_i^T P\Big]x(t) + ||e(t)||^2 + ||d(t)||^2 \\ &\le \sum_{i=1}^{r}\sum_{j=1}^{c} m_i(x(t))h_j(x(t_k))x^T(t)\Big[(A_i + B_i K_j)^T P + P(A_i + B_i K_j) \\ &+ PB_iK_j(B_iK_j)^T P + PD_iD_i^T P\Big]x(t) + \delta(t_k)^2 + \varpi^2 \\ &\le x^T(t)\Theta x(t) + \delta(t_k)^2 + \varpi^2,\end{aligned} \tag{6.26}$$

with $\Theta \triangleq \sum_{i=1}^{r}\sum_{j=1}^{c} m_i(x(t))h_j(x(t_k))\Theta_{ij}$ and $\delta(t_k) \triangleq \sqrt{\delta_1 + \delta_2 e^{\xi t_k}}$.

For the expression (6.26), if $\Theta_{ij} < 0$ holds for any $i \in \{1, \ldots, r\}$ and $j \in \{1, \ldots, c\}$, the closed-loop system (6.7) can be guaranteed to be bounded, which is generally called as MFI approach. It should be pointed out that $\Theta_{ij} < 0$ is independent of MFs, $m_i(x(t))$ and $h_j(x(t_k))$, so that the resultant stability conditions are conservative. One feasible method is to take the MFs into consideration for obtaining the relaxed MFD stability conditions. However, the existing results in [1–3] only considered the global boundary of MFs, which is still conservative to a certain extent due to ignoring the local characteristic of MFs. Thus, in the following, we shall utilize the local boundary information of MFs to obtain MFD stability conditions.

Firstly, we divide the overall operation domain Γ into L connected regional sub-spaces, i.e., $\Gamma = \bigcup_{l=1}^{L} \Gamma_l$. Denote $h_{ij}(x(t), x(t_k)) \triangleq m_i(x(t))h_j(x(t_k))$. Within l-th sub-space Γ_l, the MFs $h_{ijl}(x(t), x(t_k))$ can be approximated via taking $\underline{h}_{ijl}$ and $\overline{h}_{ijl}$ as their bounds, where $\underline{h}_{ijl}$ and $\overline{h}_{ijl}$ are given constants in Theorem 6.2 satisfying $0 \leq \underline{h}_{ijl} \leq h_{ijl}(x(t), x(t_k)) \leq \overline{h}_{ijl} \leq 1$.

Therefore, within l-th sub-space, we can obtain

$$\begin{aligned}
&\sum_{i=1}^{r}\sum_{j=1}^{c} h_{ijl}(x(t), x(t_k))\Theta_{ij} \\
&= \sum_{i=1}^{r}\sum_{j=1}^{c} (h_{ijl}(x(t), x(t_k)) - \underline{h}_{ijl} + \underline{h}_{ijl})\Theta_{ij} \\
&= \sum_{i=1}^{r}\sum_{j=1}^{c} \left(\underline{h}_{ijl}\Theta_{ij} + (h_{ijl}(x(t), x(t_k)) - \underline{h}_{ijl})\Theta_{ij}\right).
\end{aligned} \tag{6.27}$$

Besides, the slack matrix $Y_{ijl} \geq 0$ is introduced for satisfying $Y_{ijl} - \Theta_{ij} \geq 0$ for $\forall i, j, l$, which can be ensured by the condition (6.19) according to Schur complement lemma.

Furthermore, it follows from (6.27) that

$$\begin{aligned}
&\sum_{i=1}^{r}\sum_{j=1}^{c} h_{ijl}(x(t), x(t_k))\Theta_{ij} \\
&\leq \sum_{i=1}^{r}\sum_{j=1}^{c} \left(\underline{h}_{ijl}\Theta_{ij} + (h_{ijl}(x(t), x(t_k)) - \underline{h}_{ijl})Y_{ijl}\right) \\
&\leq \sum_{i=1}^{r}\sum_{j=1}^{c} \left(\underline{h}_{ijl}\Theta_{ij} + \tilde{h}_{ijl}Y_{ijl}\right).
\end{aligned} \tag{6.28}$$

It is easily shown via Schur complement that the condition (6.20) can ensure

$$\Lambda_l \triangleq \sum_{i=1}^{r}\sum_{j=1}^{c} \left(\underline{h}_{ijl}\Theta_{ij} + \tilde{h}_{ijl}Y_{ijl}\right) < 0. \tag{6.29}$$

It should be pointed out that the conditions (6.19) and (6.20) in Theorem 6.2 hold for all sub-space Γ_l, $l = 1, \cdots, L$. Thus, within the overall operation domain $\Gamma = \bigcup_{l=1}^{L} \Gamma_l$, conditions (6.28) and (6.29) can ensure for the matrix Θ in (6.26):

$$\Theta \leq \sum_{l=1}^{L} \xi_l \Lambda_l = \sum_{l=1}^{L} \sum_{i=1}^{r} \sum_{j=1}^{c} \xi_l \left(\underline{h}_{ijl} \Theta_{ij} + \tilde{h}_{ijl} Y_{ijl} \right) < 0, \tag{6.30}$$

where $\xi_l \in \{0, 1\}$, and $\xi_l = 1$ represents that the state $x(t)$ is in the l-th sub-space, otherwise, $\xi_l = 0$.

It can be obtained from (6.26) and (6.30) that

$$\begin{aligned} \dot{V}(t) &\leq x^T(t) \left(\sum_{l=1}^{L} \xi_l \Lambda_l \right) x(t) + \delta(t_k)^2 + \varpi^2 \\ &\leq -\epsilon ||x(t)||^2 + \delta(t_k)^2 + \varpi^2 \\ &= -\epsilon \left(||x(t)||^2 - \frac{\psi(t_k)}{\epsilon} \right), \end{aligned} \tag{6.31}$$

with $\epsilon = \min_{l \in \mathbb{L}} \{\lambda_{\min}(-\Lambda_l)\}$ and $\psi(t_k) = \delta(t_k)^2 + \varpi^2$.

When $t_k \to \infty$, the term $\delta_2 e^{-\zeta t_k} \to 0$ such that $\psi(t_k) \to \delta_1 + \varpi^2$. Therefore, we can conclude from (6.31) that the resultant system (6.7) under the self-triggered controller (6.6) is uniformly ultimately bounded with the convergence domain $\mathcal{B}$. □

For the self-triggered scheme (6.8) in Theorem 6.1 and the MFD stability conditions (6.19)–(6.20) in Theorem 6.2, the following result is further given to obtain the control gains K_j.

Theorem 6.3 *Given any constants $\gamma > 0$, $\delta_1 > 0$, $\underline{h}_{ijl} \in [0, 1]$, and $\overline{h}_{ijl} \in [0, 1]$, $i \in \mathbb{M}$, $j \in \mathbb{G}$, $l \in \mathbb{L} = \{1, \cdots, L\}$, if there exist matrices $Q > 0$, $Z_{ijl} \geq 0$, and L_j satisfying*

$$\hat{\Phi}_{ij} - \hat{Z}_{ijl} \leq 0, \quad \forall i, j, l, \tag{6.32}$$

$$\sum_{i=1}^{r} \sum_{j=1}^{c} (\underline{h}_{ijl} \hat{\Phi}_{ij} + \tilde{h}_{ijl} \hat{Z}_{ijl}) < 0, \quad \forall l, \tag{6.33}$$

with

$$\hat{Z}_{ijl} = \mathrm{diag}\{Z_{ijl}, 0, 0\}, \ \tilde{h}_{ijl} = \overline{h}_{ijl} - \underline{h}_{ijl},$$

$$\hat{\Phi}_{ij} = \begin{bmatrix} He(A_i Q + B_i L_j) & B_i L_j & D_i \\ * & -2\gamma Q + \gamma^2 I & 0 \\ * & * & -I \end{bmatrix},$$

then, the closed-loop system (6.7) is uniformly ultimately bounded with the convergence domain defined in (6.21). Meanwhile, the control gain can be calculated as $K_j = L_j Q^{-1}$.

Proof. Pre- and post-multiplying the expressions (6.19) and (6.20) in Theorem 6.2 by $\text{diag}\{P^{-1}, P^{-1}, I\}$, and letting $Q = P^{-1}, L_j = K_j Q, Z_{ijl} = QY_{ijl}Q$, the conditions (6.19) and (6.20) are equivalent to

$$\bar{\Phi}_{ij} - \hat{Z}_{ijl} \le 0, \quad \forall i, j, l, \tag{6.34}$$

$$\sum_{i=1}^{r}\sum_{j=1}^{c}(\underline{h}_{ijl}\bar{\Phi}_{ij} + \tilde{h}_{ijl}\hat{Z}_{ijl}) < 0, \quad \forall l, \tag{6.35}$$

with

$$\bar{\Phi}_{ij} = \begin{bmatrix} \text{He}(A_i Q + B_i L_j) & B_i L_j & D_i \\ * & -Q^2 & 0 \\ * & * & -I \end{bmatrix}.$$

Due to $(Q - \gamma I)^T(Q - \gamma I) \ge 0$ with any positive constant γ, one has

$$-Q^2 \le -2\gamma Q + \gamma^2 I. \tag{6.36}$$

By means of (6.36), it is easily shown that the conditions (6.32) and (6.33) imply that the expressions (6.34) and (6.35) hold. Thus, we obtain that if the conditions (6.32)–(6.33) hold, the closed-loop system (6.7) is uniformly ultimately bounded. □

Notice that the local upper and lower bounds, $\overline{h}_{ijl}$ and $\underline{h}_{ijl}$, are used in the stability conditions (6.32)–(6.33) for obtaining the control gains K_j. Hence, parameters $\overline{h}_{ijl}$ and $\underline{h}_{ijl}$ should be determined by means of MFs $m_i(x(t))$ and $h_j(x(t_k))$ in the fuzzy system (6.4) and the designed controller (6.6). However, the implementation of the self-triggered scheme (6.8) causes MFs $m_i(x(t))$ and $h_j(x(t_k))$ to be asynchronous, depending on the current state $x(t)$ and the latest triggered value $x(t_k)$, respectively. If states $x(t)$ and $x(t_k)$ are treated as two independent variables to derive the corresponding bounds $\underline{h}_{ijl}$ and $\overline{h}_{ijl}$ in sub-space Γ_l, it may introduce more conservatism due to it ignores the relationship between $x(t)$ and $x(t_k)$. To address this limitation, we intend to establish the relationship between $x(t)$ and $x(t_k)$ to extract the local bounds $\overline{h}_{ijl}$ and $\underline{h}_{ijl}$ in each sub-space Γ_l.

Since the self-triggered scheme (6.8) can ensure $||e(t)|| = ||x(t_k) - x(t)|| \le \sqrt{\delta_1 + \delta_2 e^{-\zeta t_k}}$, it can be further obtained that each component $x_c(t)$ of state $x(t)$ and its triggered value $x_c(t_k)$ satisfy $x_c(t_k) \in [x_c(t) - \sqrt{\delta_1 + \delta_2 e^{-\zeta t_k}}, x_c(t) + \sqrt{\delta_1 + \delta_2 e^{-\zeta t_k}}] \subseteq [x_c(t) - \sqrt{\delta_1 + \delta_2}, x_c(t) + \sqrt{\delta_1 + \delta_2}]$, for $c = 1, 2, \cdots, n$. This implies that $x_c(t_k)$ can be estimated via the following relation

$$x_c(t_k) \in [x_c(t) - \bar{\delta}, x_c(t) + \bar{\delta}], \tag{6.37}$$

with $\bar{\delta} = \sqrt{\delta_1 + \delta_2}$.

Define $\overline{h}_{ijl} \triangleq \overline{m}_{il}\overline{h}_{jl}$ and $\underline{h}_{ijl} \triangleq \underline{m}_{il}\underline{h}_{jl}$, where $\overline{m}_{il}$ and $\underline{m}_{il}$ are the upper and lower bounds of $m_i(x(t))$ in sub-space Γ_l, and $\overline{h}_{jl}$ and $\underline{h}_{jl}$ are upper and lower bounds of

$h_j(x(t_k))$ in sub-space Γ_l. Suppose that the MFs of fuzzy system are dependent on the state $x_c(t)$, i.e., $m_i(x_c(t))$, then the upper and lower bounds, $\overline{m}_{il}$ and $\underline{m}_{il}$, can be obtained by calculating the maximum of $\overline{m}_i(x_c(t))$ and the minimum of $\underline{m}_i(x_c(t))$ during the l-th sub-space $x_c(t) \in [a, b]$. For the controller's MFs $h_j(x_c(t_k))$, according to the relation (6.37), the scope of $h_j(x_c(t_k))$ with respect to sub-space Γ_l is equivalent to the scope of $h_j(x_c(t))$ for $x_c(t) \in [a - \bar{\delta}, b + \bar{\delta}]$, and $\overline{h}_{jl}$ and $\underline{h}_{jl}$ can be obtained via the the maximum and minimum of $h_j(x_c(t))$ for $x_c(t) \in [a - \bar{\delta}, b + \bar{\delta}]$. Thus, the bounds $\overline{h}_{ijl}$ and $\underline{h}_{ijl}$ are obtained in the sub-space Γ_l.

Remark 6.4 The results in [1–3] and Chapters 2∼5 only utilized the global boundary information of MFs. Generally, the global bounds of event-triggered controller's MFs $h_j(x(t_k))$ can be replaced by the global maximum and minimum of time-driven controller's MFs $h_j(x(t))$, which may introduce conservatism. On the contrary, the present chapter establishes the connection (6.37) between the current state $x(t)$ and the latest triggered value $x(t_k)$ for obtaining the local information of MFs. Once the parameters δ_1 and δ_2 are given, the varying range of triggered state $x(t_k)$ can be estimated via the threshold $\bar{\delta}$ and the state $x(t)$. That is, the range of the MFs $h_j(x(t_k))$ in the controller can be transformed into the one merely related to the system state $x(t)$. Then, based on the local boundary information $\overline{h}_{ijl}$ and $\underline{h}_{ijl}$ in the sub-space Γ_l of operation domain, the MFD analysis approach can be carried out to obtain stability conditions (6.32)–(6.33) in Theorem 6.3. Therefore, the proposed approach is more beneficial for reducing conservatism and obtaining relaxed conditions.

6.5 Simulation Example

Consider a 3-rule IT2 fuzzy system (6.4) with the following parameters

$$
\begin{aligned}
A_1 &= \begin{bmatrix} -0.1 & 0.4 \\ 0.2 & 0.1 \end{bmatrix}, A_2 = \begin{bmatrix} -0.3 & 0.7 \\ 0.1 & 0.5 \end{bmatrix}, A_3 = \begin{bmatrix} 0.5 & 0.2 \\ 0.8 & -0.5 \end{bmatrix}, \\
B_1 &= \begin{bmatrix} 1.0 \\ 0.8 \end{bmatrix}, B_2 = \begin{bmatrix} 2.0 \\ 0.8 \end{bmatrix}, B_3 = \begin{bmatrix} 2.0 \\ 0.5 \end{bmatrix}, \\
D_1 &= \begin{bmatrix} 0.1 \\ 0.2 \end{bmatrix}, D_2 = \begin{bmatrix} 0.3 \\ 0.2 \end{bmatrix}, D_3 = \begin{bmatrix} 0.1 \\ 0.3 \end{bmatrix}, \\
d(t) &= 0.1 \sin(0.5t),
\end{aligned}
$$

and the upper and lower MFs

$$
\begin{aligned}
\overline{m}_1(x_1(t)) &= 1 - \frac{1}{1 + e^{(-x_1(t)-2.5)}}, \quad \underline{m}_1(x_1(t)) = 1 - \frac{1}{1 + e^{(-x_1(t)-3.5)}}, \\
\overline{m}_3(x_1(t)) &= \frac{1}{1 + e^{(-x_1(t)+2.5)}}, \quad \underline{m}_3(x_1(t)) = \frac{1}{1 + e^{(-x_1(t)+3.5)}},
\end{aligned}
$$

and $\overline{m}_2(x_1(t)) = 1 - \underline{m}_1(x_1(t)) - \underline{m}_3(x_1(t))$, $\underline{m}_2(x_1(t)) = 1 - \overline{m}_1(x_1(t)) - \overline{m}_3(x_1(t))$.

Aiming at the above IT2 T-S fuzzy system, we design a 2-rule fuzzy controller (6.6) with the following upper and lower MFs:

$$\overline{h}_1(x_1(t)) = 1 - \frac{1}{1 + e^{(-x_1(t)-0.5)}}, \quad \underline{h}_1(x_1(t)) = 1 - \frac{1}{1 + e^{(-x_1(t)-1.5)}},$$

and $\overline{h}_2(x_1(t)) = 1 - \underline{h}_2(x_1(t))$, $\underline{h}_2(x_1(t)) = 1 - \overline{h}_2(x_1(t))$.

Under the self-triggered control scheme (6.8), the fuzzy controller can only use the triggered state $x(t_k)$, and the weighting coefficients of the upper and lower MFs are specified as $\underline{\beta}_j(x_1(t_k)) = \overline{\beta}_j(x_1(t_k)) = 0.5$ for $j = 1, 2$. Here, the triggering parameters are selected as $\delta_1 = 0.006$, $\delta_2 = 0.034$, and $\zeta = 0.5$. According to the expression $\bar{\delta} = \sqrt{\delta_1 + \delta_2} = \sqrt{0.006 + 0.034} = 0.2$ and the relation (6.37), the range of the triggered state $x_1(t_k)$ can be represented as $x_1(t_k) \in [x_1(t) - 0.2, x_1(t) + 0.2]$. Thus, the upper and lower bounds of MFs $h_j(x_1(t_k))$ $(i = 1, 2)$ can be obtained via the maximum and minimum of $h_j(x_1(t))$ during the interval $[x_1(t) - 0.2, x_1(t) + 0.2]$.

The operation domain $x_1(t) \in [-5, 5]$ is divided into 10 equal-size connected sub-spaces with $\underline{x}_{1l} = -6 + l$ and $\overline{x}_{1l} = -5 + l$, and the sub-space Γ_l is denoted as $x_1(t) \in [\underline{x}_{1l}, \overline{x}_{1l}]$ for $l = 1, \cdots, 10$. The control gains can be calculated via the conditions (6.32)–(6.33) in Theorem 6.3

$$K_1 = \begin{bmatrix} -0.0543 & -1.2987 \end{bmatrix}, \quad K_2 = \begin{bmatrix} -0.4981 & -1.2189 \end{bmatrix}.$$

In the simulation, we choose $\underline{\alpha}_1(x_1(t)) = \frac{\sin(2x_1(t))+1}{2}$, $\overline{\alpha}_1(x_1(t)) = 1 - \underline{\alpha}_1(x_1(t))$, $\underline{\alpha}_3(x_1(t)) = \frac{\cos(2x_1(t))+1}{2}$, $\overline{\alpha}_3(x_1(t)) = 1 - \underline{\alpha}_3(x_1(t))$ to obtain the MFs $m_1(x_1(t))$, $m_3(x_1(t))$, and $m_2(x_1(t)) = 1 - m_1(x_1(t)) - m_3(x_1(t))$.

The simulation results with initial states $x(0) = [-4, 3]^T$ are depicted in Fig. 6.2a–d. It can be seen from the subfigure of Fig. 6.2b that the control input $u(t)$ is held as a constant during each triggering interval, which can avoid the continuous transmission and save the communication resources. The triggering instants and the inter-interval are depicted Fig. 6.2c, which reveals the release instants of system state $x(t)$. The difference $e(t)$ and the triggering threshold are shown in Fig. 6.2d.

Meanwhile, we make the simulation (Fig. 6.3a–d) on the conventional ETS (6.13) with the same parameters δ_1, δ_2, and ζ. By comparing Fig. 6.2e and e, it can be seen that the ETS (6.13) has fewer triggering number than the self-triggered scheme (6.8). Nevertheless, they both can ensure the desirable dynamic performance, as shown in Figs. 6.2a and 6.3(a). Moreover, it is also observed from Figs. 6.2(d) and 6.3(d) that the difference $e(t)$ can be ensured to be less than the threshold under both two triggering schemes.

Besides, we also make the comparison with the global boundary information approach, that is, the upper and lower boundaries, $\overline{h}_{ij}$ and $\underline{h}_{ij}$, in the overall operation domain (i.e., $L = 1$), is used in the stability conditions (6.32)–(6.33) in Theorem 6.3. Unfortunately, no feasible solution can be found for the above conditions with global

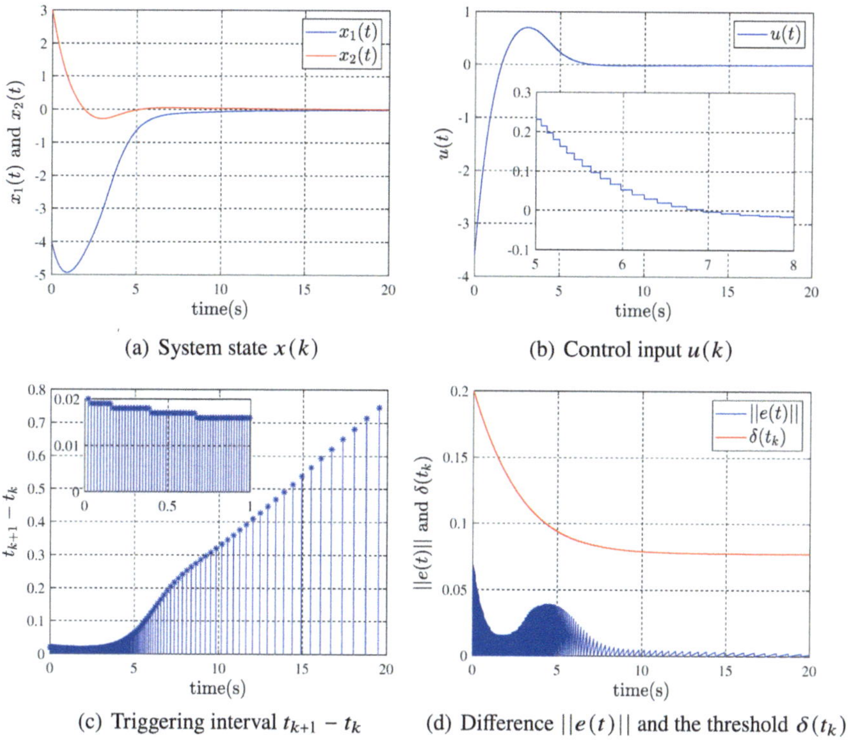

Fig. 6.2 Simulation results under the self-triggered scheme (6.8)

boundary information, which also means that the local boundary information may expand the solution space and reduce the conservatism of stability conditions.

In addition, as the statement in Remark 6.1, the controller can adopt MFs with simpler structure. Here, we re-choose the upper and lower MFs of controller as the following trapezoid functions to design the 2-rule IT2 fuzzy controller (6.6)

$$\overline{h}_1(x_1(t)) = \begin{cases} 1 & x_1(t) \in [-5, -2.2] \\ \dfrac{-x_1(t) + 2.8}{5} & x_1(t) \in [-2.2,\ 2.8] \\ 0 & x_1(t) \in [2.8,\ 5], \end{cases} \tag{6.38}$$

$$\underline{h}_1(x_1(t)) = \begin{cases} 1 & x_1(t) \in [-5, -2.8] \\ \dfrac{-x_1(t) + 2.2}{5} & x_1(t) \in [-2.8,\ 2.2] \\ 0 & x_1(t) \in [2.2,\ 5], \end{cases} \tag{6.39}$$

and $\overline{h}_2(x_1(t)) = 1 - \underline{h}_1(x_1(t))$, $\underline{h}_2(x_1(t)) = 1 - \overline{h}_1(x_1(t))$.

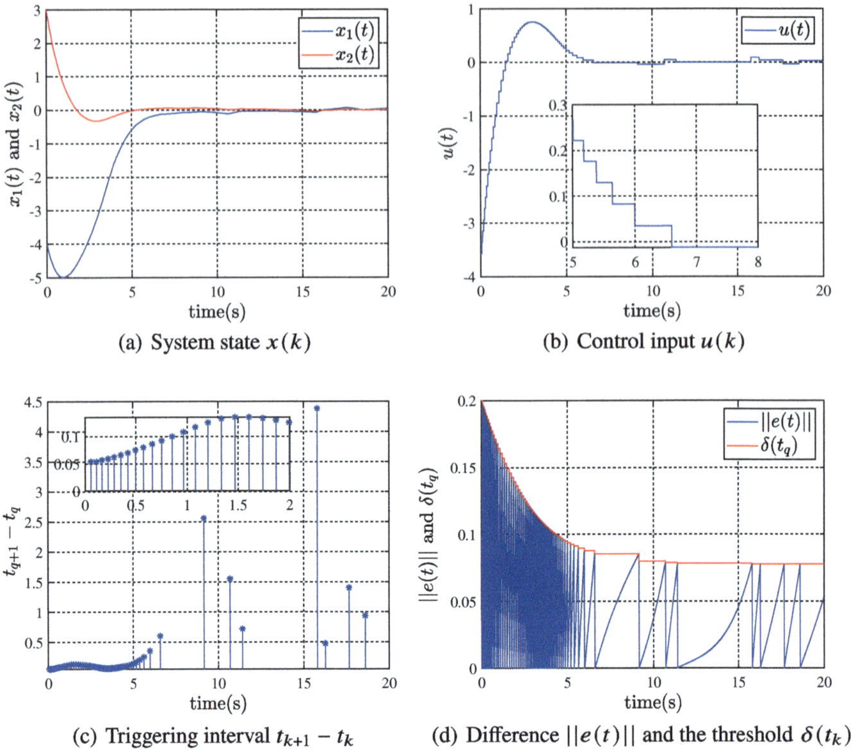

Fig. 6.3 Simulation results under the ETS (6.13)

Taking the same self-triggering parameters and 10 sub-spaces as the above case, the control gains can be obtained via solving the conditions (6.32)–(6.33) in Theorem 6.3:

$$K_1 = \begin{bmatrix} -0.1053 & -1.2392 \end{bmatrix}, \quad K_2 = \begin{bmatrix} -0.5146 & -1.1463 \end{bmatrix}.$$

The simulation results are depicted in Fig. 6.4a–d. By comparing Fig. 6.2a–c with Fig. 6.4a–c, it is seen that similar control performance can be attained for different MFs of the controller.

On the other hand, it needs to be pointed out that due to the lower bounds of the MFs in (6.38) and (6.39) to be 0, so the global lower bounds $\underline{h}_{ij}$ of $h_{ij}(x(t), x(t_k))$ are also 0. If using the global boundary information, the condition (6.33) in Theorem 6.3 will reduce to an infeasible inequality, $\sum_{i=1}^{r} \sum_{j=1}^{c} \tilde{h}_{ij} \hat{Z}_{ij} < 0$ with $\hat{Z}_{ij} = \text{diag}\{Z_{ij}, 0, 0\}$, for $Z_{ij} > 0$. This further demonstates that the local boundary information approach may have a wider range of applications.

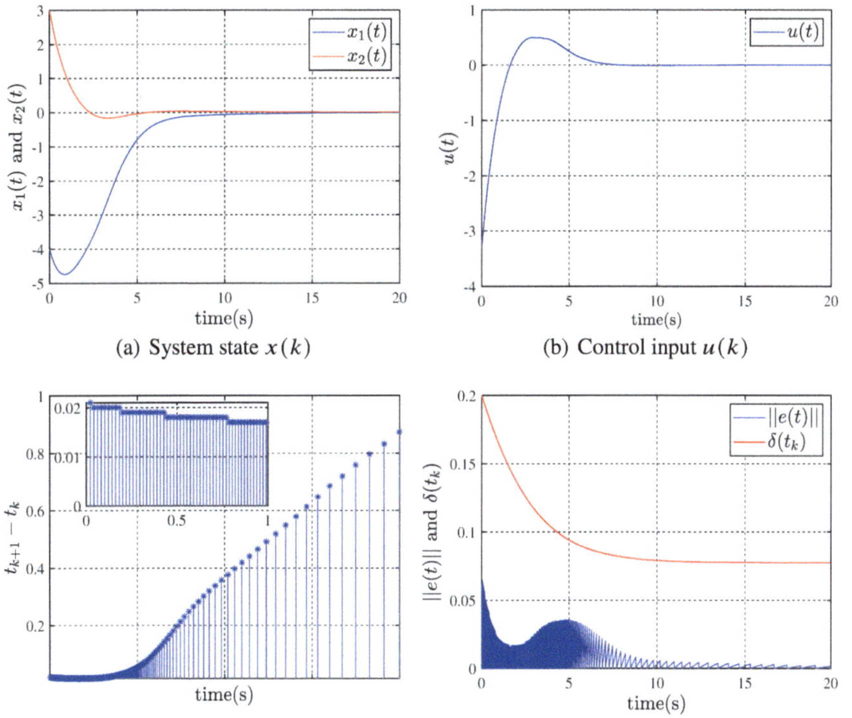

(a) System state $x(k)$

(b) Control input $u(k)$

(c) Triggering interval $t_{k+1} - t_k$

(d) Difference $||e(t)||$ and the threshold $\delta(t_k)$

Fig. 6.4 Simulation results under the self-triggered scheme with trapezoid MFs (6.38)–(6.39)

6.6 Conclusion

The self-triggered fuzzy control issue of IT2 fuzzy system has been investigated in this chapter. A self-triggered strategy has been proposed according to the current triggering information and the controller's MFs. By choosing different fuzzy rules and MFs, we have designed the IT2 fuzzy controller to ensure the uniform ultimate boundedness of the closed-loop system. Under the self-triggered strategy, the relation between the triggered state and the system state has been established to extract the local boundary information of MFs and derive the MFD stability conditions. Finally, the simulation comparisons have been made with the existing ETS and the global-boundary-information-based method, which verifies the superiority of the local-boundary-information-dependent self-triggered control scheme.

References

1. Pan, Y., Yang, G.H.: Event-triggered fuzzy control for nonlinear networked control systems. Fuzzy Sets Syst. **329**, 91–107 (2017)
2. Li, X., Ye, D.: Asynchronous event-triggered control for networked interval type-2 fuzzy systems against DoS attacks. IEEE Trans. Fuzzy Syst. **29**(2), 262–274 (2021)
3. Zhang, Z., Su, S.F., Niu, Y.: Dynamic event-triggered control for interval type-2 fuzzy systems under fading channel. IEEE Trans. Cybern. **51**(11), 5342–5351 (2021)
4. Khalil, H.K.: Nonlinear systems, 3rd edn. Prentice-Hall, Englewood Cliffs, NJ, USA (2002)
5. Xie, X., Zhou, Q., Yue, D., Li, H.: Relaxed control design of discrete-time Takagi-Sugeno fuzzy systems: an event-triggered real-time scheduling approach. IEEE Trans. Syst. Man Cybern. Syst. **48**(12), 2251–2262 (2018)

Part II
Channel-Scheduling Protocols

Chapter 7
Sliding Mode Control Design Under Multi-node Stochastic Scheduling Protocol

7.1 Introduction

In Chaps. 2–6, the ETS and self-triggered scheme have been utilized to regulate the transmission process of system state or measurement output, which can reduce the signal transmission frequency and improve the utilization of limited communication resources. Aiming at the access constraints over the communication channels, channel-scheduling protocols, e.g., RR protocol, WTOD protocol, and SS protocol, also provide a feasible approach by reasonably choosing a part of sensor/actuator nodes. The SS protocol has been widely utilized to describe the carrier-sense multiple access with collision avoidance (CSMA/CA) protocols in some wireless communication systems, e.g. Ethernet, IEEE 802.11-based wireless local area networks, and IEEE 802.15.4-based wireless sensor networks. For preventing from the data collisions, the discrete-time Markov chain has been proposed in [1] to govern the communication protocol, which randomly determines which sensor/actuator node to access the shared network according to a discrete-time Markov chain [2–5]. However, it should be pointed out that the existing works on SS protocol permit only one node to access the communication channels at any instant, which is obviously an extreme case of all nodes accessing the network. Actually, some network conditions may allow more nodes to achieve the information exchange [6]. Furthermore, it has also been shown in [7, 8] that more nodes accessing the communication channels may be beneficial to improve the system performance. Hence, an interesting issue is how to propose a feasible SMC strategy to achieve the desirable dynamic performance of the IT2 fuzzy system under the case that more than one nodes are permitted to access the communication channels.

This chapter investigates the SMC issue for the discrete-time IT2 T-S fuzzy systems under the multi-node SS (MN-SS) protocol, in which a part of the sensor/actuator nodes are chosen to transmit signals through the communication channels at each moment. The access status of sensors and actuators is described as two independent Markov chains, which are further mapped into a new variable obeying

Y. Yang et al., *Analysis and Synthesis of Interval Type-2 Fuzzy Systems Under Communication Protocols*, Studies in Systems, Decision and Control 636,
https://doi.org/10.1007/978-981-95-2355-9_7

the Markov chain to facilitate the control design and analysis. By constructing the new MFs dependent on the received system states, a scheduling signal-dependent fuzzy SMC law is designed. The relations of MFs between the fuzzy model and the SMC law are explicitly established, and the MFD sufficient conditions are derived to ensure the stochastic stability of the controlled system and the reachability of the specified sliding surface. Besides, an optimization-based solving algorithm is proposed to achieve less energy consumption of the control input via searching the sliding matrix.

7.2 Problem Description

7.2.1 System Model

Consider the discrete-time IT2 T-S fuzzy systems with the following rules:

Rule i: IF $f_1(x(k))$ is $\mathscr{F}_1^i$ and $\cdots$ and $f_\hbar(x(k))$ is $\mathscr{F}_\hbar^i$, THEN

$$x(k+1) = A_i x(k) + B_i(u(k) + d(x(k), k)) \tag{7.1}$$

where $f_\iota(x(k))$ $(\iota = 1, \ldots, \hbar)$ is the premise variable, $\mathscr{F}_\iota^i$ $(i = 1, \ldots, r)$ is the IT2 fuzzy set, $x(k) \in \mathbb{R}^n$ is the system state, and $u(k) \in \mathbb{R}^m$ is the control input. $d(x(k), k)$ is the external disturbance satisfying $||d(x(k), k)|| \leq \varpi ||x(k)||$ with ϖ a known constant. A_i and B_i are known matrices. The firing strength is defined for the i-th rule into the following interval set:

$$\vec{m}_i(x(k)) = [\underline{m}_i(x(k)), \overline{m}_i(x(k))],$$

with

$$\overline{m}_i(x(k)) = \prod_{\iota=1}^{\hbar} \overline{\mu}_{\mathscr{F}_\iota^i}(f_\iota(x(k))) > 0, \tag{7.2}$$

$$\underline{m}_i(x(k)) = \prod_{\iota=1}^{\hbar} \underline{\mu}_{\mathscr{F}_\iota^i}(f_\iota(x(k))) > 0, \tag{7.3}$$

where the upper and lower grade functions $\overline{\mu}_{\mathscr{F}_\iota^i}(f_\iota(x(k)))$ and $\underline{\mu}_{\mathscr{F}_\iota^i}(f_\iota(x(k))) \in [0, 1]$ satisfy $\underline{\mu}_{\mathscr{F}_\iota^i}(f_\iota(x(k))) \leq \overline{\mu}_{\mathscr{F}_\iota^i}(f_\iota(x(k)))$. $\overline{m}_i(x(k))$ and $\underline{m}_i(x(k))$ represent the upper and lower MFs, respectively. Then, the fuzzy system (7.1) can be inferred as

$$x(k+1) = \sum_{i=1}^{r} m_i(x(k)) \left[A_i x(k) + B_i(u(k) + d(x(k), k))\right] \tag{7.4}$$

where $m_i(x(k))$ is membership function satisfying

$$m_i(x(k)) = \underline{m}_i(x(k))\underline{\alpha}_i(x(k)) + \overline{m}_i(x(k))\overline{\alpha}_i(x(k)),$$

and $\sum_{i=1}^{r} m_i(x(k)) = 1$ with weighting coefficients $\overline{\alpha}_i(x(k))$ and $\underline{\alpha}_i(x(k))$ satisfying $\underline{\alpha}_i(x(k)), \overline{\alpha}_i(x(k)) \in [0, 1]$ and $\underline{\alpha}_i(x(k)) + \overline{\alpha}_i(x(k)) = 1$.

It should be noticed that the MFs $m_i(x(k))$ in the IT2 fuzzy system (7.4) are characterized by the upper and lower bounds (i.e., $\overline{m}_i(x(k))$ and $\underline{m}_i(x(k))$), and the coefficients $\overline{\alpha}_i(x(k))$ and $\underline{\alpha}_i(x(k))$ are related to the uncertain parameters of the modeled nonlinear system. Thus, the MFs $m_i(x(k))$ are uncertain and cannot be directly used in the design of the controller. Furthermore, only the upper and lower MFs $\overline{m}_i(x(k))$ in (7.2) and $\underline{m}_i(x(k))$ in (7.3) are known, and dependent on the system states $x(k)$. This is the feature of the IT2 fuzzy systems. In this chapter, the SS protocol will be introduced to schedule the transmission nodes on the S/C channels and the C/A channels, under which only a part of system states can be transmitted from sensors to the controller. This implies that the upper and lower MFs $\overline{m}_i(x(k))$ and $\underline{m}_i(x(k))$ dependent on the system state $x(k)$ cannot be directly utilized for synthesizing the controller as in some existing works [9, 10], which will bring some difficulties for the subsequent analysis and design.

7.2.2 Multi-node Stochastic Scheduling Protocol

As illustrated in Fig. 7.1, the IT2 fuzzy system (7.4) comprises n sensor nodes and m actuator nodes interconnected via a shared network. Due to limitations in network bandwidth and data rate, the communication channels are capable of exchanging only a subset of signals over the S/C and C/A channels. To alleviate these communication constraints and enhance signaling efficiency, the MN-SS protocol is implemented to regulate the transmission process of system state and control input signal.

Suppose that the S/C channel only accommodates c_1 $(1 \le c_1 \le n)$ sensor nodes to transmit the state signals. Define the binary-valued function $\psi_i(k) : \mathbb{Z}^+ \to \{0, 1\}$ representing the access status of the i-th sensor node at instant k. When the sensor node i is permitted to access the channel, denoting $\psi_i(k) = 1$, the controller will receive the state information $x_i(k)$, otherwise, $\psi_i(k) = 0$ means that no signal from sensor i is sent to controller. That is, the available state signal from the sensor node i for the controller at instant k is denoted as

$$\bar{x}_i(k) = \begin{cases} x_i(k), & \text{if } \psi_i(k) = 1, \\ 0, & \text{otherwise} . \end{cases} \tag{7.5}$$

Define $\bar{x}(k) \triangleq [\bar{x}_1(k), \ldots, \bar{x}_n(k)]^T$. The system state received by the controller can be represented as

$$\bar{x}(k) = \Psi_{\rho(k)} x(k), \tag{7.6}$$

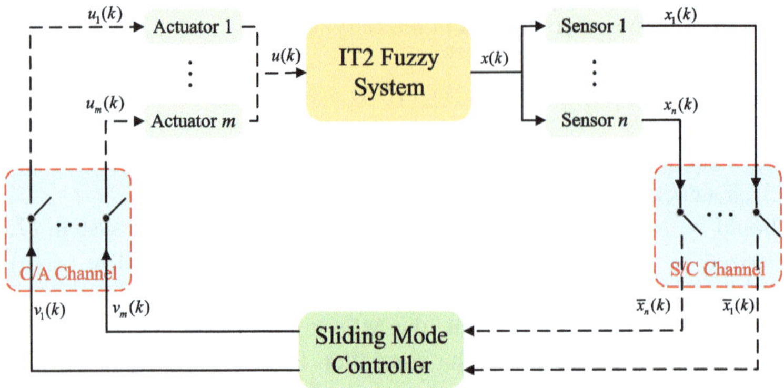

Fig. 7.1 System structure with communication networks

with $\Psi_{\rho(k)} = \text{diag}\{\psi_1(k), \psi_2(k), \ldots, \psi_n(k)\}$, and $\rho(k) \in \mathbb{S}_1 \triangleq \{1, 2, \ldots, N_1\}$ with $N_1 = C_n^{c_1}$ represents the access status of the sensor nodes at instant k. It is assumed that $\rho(k)$ obeys a discrete-time Markov chain with the following conditional probability

$$\mathbf{Pr}\{\rho(k+1) = t|\rho(k) = l\} = \lambda_{lt}, \quad l, t \in \mathbb{S}_1, \tag{7.7}$$

with $\lambda_{lt} \in [0, 1]$ and $\sum_{t=1}^{N_1} \lambda_{lt} = 1$. Define the transition probability matrix (TPM) as $\Lambda = [\lambda_{lt}]_{N_1 \times N_1}$.

Likewise, it is assumed that the C/A channel can accommodate c_2 $(1 \le c_2 \le m)$ actuator nodes to access the network and obtain the control signal simultaneously. Let $\phi_i(k) : \mathbb{Z}^+ \to \{0, 1\}$ represent the access status of the i-th actuator node at instant k. Then, the received control signal by the actuator can be represented as

$$u(k) = \Phi_{\sigma(k)} v(k), \tag{7.8}$$

with $\Phi_{\sigma(k)} = \text{diag}\{\phi_1(k), \phi_2(k), \ldots, \phi_m(k)\}$. $\sigma(k) \in \mathbb{S}_2 \triangleq \{1, 2, \ldots, N_2\}$ with $N_2 = C_m^{c_2}$ represents the access status of the actuator nodes at instant k, and obeys the following conditional probability

$$\mathbf{Pr}\{\sigma(k+1) = \jmath|\sigma(k) = \imath\} = \gamma_{\imath\jmath}, \quad \imath, \jmath \in \mathbb{S}_2. \tag{7.9}$$

with $\gamma_{\imath\jmath} \in [0, 1]$ and $\sum_{\jmath=1}^{N_2} \gamma_{\imath\jmath} = 1$. Define the TPM as $\Omega = [\gamma_{\imath\jmath}]_{N_2 \times N_2}$.

Remark 7.1 When the number of selected nodes is fixed to one, i.e., $c_1 = 1$ and $c_2 = 1$, the proposed MN-SS protocol is reduced to the conventional single-node SS protocol case, which has been widely investigated in [2–5]. So, the SS protocol is a special case of the MN-SS protocol. The superiority of the MN-SS protocol for improving the control performance will be demonstrated in the simulation example later. Besides, in the existing results [2–5], the single-node SS protocol is employed in only one channel to schedule the transmission node, where an implicit assumption is that another channel without the SS protocol can allow the transmission of all signals. This requirement is difficult to meet in practice. In this chapter, we introduce the MN-SS protocol to effectively orchestrate the transmission nodes at both sides of the controller, which is more consistent with the characteristics of network transmission.

Notice that there are two Markov chains in (7.7) and (7.9) to govern the access of the sensor/actuator nodes. In order to facilitate the control design and the stability analysis, we shall introduce a new stochastic variable $\theta(k)$ to define a one-to-one mapping with Markov chains $\rho(k)$ and $\sigma(k)$ by using the following mapping technique as in [11]:

$$\theta(k) = R(\rho(k), \sigma(k)) = \rho(k) + N_1(\sigma(k) - 1) \tag{7.10}$$

with $\theta(k) \in \mathbb{S} \triangleq \{1, \cdots, N\}$ and $N = N_1 \times N_2$.

Correspondingly, for the given $\theta(k) \in \mathbb{S}$, an unique pair $(\rho(k), \sigma(k))$ can be determined as

$$\begin{cases} \rho(k) \triangleq \eta_1(\theta(k)) = mod(\theta(k) - 1, N_1) + 1, \\ \sigma(k) \triangleq \eta_2(\theta(k)) = \lfloor \frac{\theta(k)-1}{N_1} \rfloor + 1. \end{cases} \tag{7.11}$$

That is, the variable $\theta(k)$ and the pair $(\rho(k), \sigma(k))$ are one-to-one mapping.

Meanwhile, according to the transition probabilities (7.7) and (7.9) of Markov chains $\rho(k)$ and $\sigma(k)$, respectively, we can obtain the conditional probability of the variable $\theta(k)$ shown below:

$$\begin{aligned} \pi_{pq} &= \mathbf{Pr}\{\theta(k+1) = q | \theta(k) = p\} \\ &= \mathbf{Pr}\{\rho(k+1) = \eta_1(q), \sigma(k+1) = \eta_2(q) | \rho(k) = \eta_1(p), \sigma(k) = \eta_2(p)\} \\ &= \mathbf{Pr}\{\rho(k+1) = \eta_1(q) | \rho(k) = \eta_1(p)\} \mathbf{Pr}\{\sigma(k+1) = \eta_2(q) | \sigma(k) = \eta_2(p)\} \\ &= \lambda_{\eta_1(p)\eta_1(q)} \gamma_{\eta_2(p)\eta_2(q)}, \end{aligned} \tag{7.12}$$

where $\lambda_{\eta_1(p)\eta_1(q)}$ and $\gamma_{\eta_2(p)\eta_2(q)}$ are given in (7.7) and (7.9), respectively. It can be seen from (7.12) that the stochastic variable $\theta(k)$ also obeys a Markov chain with the TPM $\Pi = [\pi_{pq}]_{N \times N}$.

Definition 7.1 The IT2 T-S fuzzy system (7.4) under the MN-SS protocols (7.7) and (7.9) is said to be stochastically stable if

$$\mathbf{E}\left\{ \sum_{k=0}^{\infty} ||x(k)||^2 \middle| x(0), \theta(0) \right\} < \infty \tag{7.13}$$

holds for any $x(0) \in \mathbb{R}^n$ and $\theta(0) \in \mathbb{S}$

In this chapter, we shall propose a SMC strategy to guarantee the stochastic stability of the IT2 fuzzy system (7.4) under the MN-SS protocols in (7.7) and (7.9).

7.3 Scheduling Signal-Dependent Sliding Mode Control Design

Firstly, choose the following sliding function

$$s(k) = Gx(k), \tag{7.14}$$

where $G \in \mathbb{R}^{m\times n}$ is the sliding matrix to be designed later.

Note that the upper and lower MFs, i.e., $\overline{m}_i(x(k))$ and $\underline{m}_i(x(k))$, in the fuzzy model (7.4) are dependent on the system state $x(k)$. However, as shown in Fig. 7.1, the MN-SS protocol (7.6) is introduced into the S/C channel to schedule the sensor nodes. This means that the controller can only obtain the information from the selected sensor nodes, i.e., the state $\bar{x}(k)$. Thus, we modify the upper and lower MFs in (7.2) and (7.3) as $\overline{m}_j(\bar{x}(k))$ and $\underline{m}_j(\bar{x}(k))$ to construct the new MFs $h_j(\bar{x}(k)) = \underline{\beta}_j(\bar{x}(k))\underline{m}_j(\bar{x}(k)) + \overline{\beta}_j(\bar{x}(k))\overline{m}_j(\bar{x}(k))$ satisfying $\sum_{j=1}^{r} h_j(\bar{x}(k)) = 1$ with the coefficients $\underline{\beta}_j(\bar{x}(k))$ and $\overline{\beta}_j(\bar{x}(k))$ for $j = 1, \ldots, r$, which is only related to the received state $\bar{x}(k)$.

Then, we design the scheduling signal-dependent fuzzy SMC law as

$$v(k) = \sum_{j=1}^{r} h_j(\bar{x}(k))K_{j,\theta(k)}\bar{x}(k) - \varpi||\bar{x}(k)||\text{sign}(\bar{s}(k)), \tag{7.15}$$

where $K_{j,\theta(k)}$ are the control gains to be designed later, and $\bar{s}(k)$ is the sliding variable $s(k)$ in (7.14) with $x(k)$ replaced by $\bar{x}(k)$.

Remark 7.2 Employing the mapping technique (7.10), the token pair $(\rho(k), \sigma(k))$ is mapped into a new scheduling signal $\theta(k)$, which can simultaneously reflect the access status of the sensor/actuator nodes. Thus, the control gain $K_{j,\theta(k)}$ in the SMC law (7.15) is only dependent on the new scheduling signal $\theta(k)$, which also avoids the co-existence of $\rho(k)$ and $\sigma(k)$.

Substituting the fuzzy sliding mode controller (7.15) into the IT2 T-S fuzzy systems (7.4), we obtain the following closed-loop system as

$$\begin{aligned} x(k+1) = &\sum_{i=1}^{r}\sum_{j=1}^{r} m_i(x(k))h_j(\bar{x}(k))\Big[(A_i + B_i\Phi_{\eta_2(\theta(k))}K_{j,\theta(k)}\Psi_{\eta_1(\theta(k))})x(k) \\ &+ B_i\Big(d(x(k),k) - \hat{D}_{\theta(k)}(\bar{x}(k))\Big)\Big] \end{aligned} \tag{7.16}$$

with $\hat{D}_{\theta(k)}(\bar{x}(k)) \triangleq \varpi||\bar{x}(k)||\Phi_{\eta_2(\theta(k))}\text{sign}(\bar{s}(k))$.

7.4 Reachability and Stability Analysis Under MN-SS Protocol

In this section, the stochastic stability of the closed-loop systems and the reachability of the specified sliding surface $s(k) = 0$ will be analyzed, and the corresponding sufficient conditions will be derived.

7.4.1 Stochastic Stability of Closed-Loop System

In this subsection, we shall prove that the stochastic stability of the closed-loop system (7.16) under the scheduling signal-dependent fuzzy SMC scheme with the MN-SS protocols in (7.7) and (7.9) can be ensured via the following MFD conditions.

Theorem 7.1 *For the IT2 T-S fuzzy system (7.4) with the MN-SS protocols (7.7) and (7.9), if there exist matrices $P_p > 0$, K_{jp}, and scalars $\varepsilon_{ip} > 0$, and $\epsilon_{ip} > 0$ satisfying the following conditions (for $p \in \mathbb{S}$, and $i, j = 1, \ldots, r$)*

$$B_i^T \mathcal{P}_p B_i \leq \varepsilon_{ip} I, \tag{7.17}$$

$$\Phi_{\eta_2(i)} B_i^T \mathcal{P}_p B_i \Phi_{\eta_2(p)} \leq \epsilon_{ip} I, \tag{7.18}$$

$$\sum_{i=1}^{r} \sum_{j=1}^{r} m_i(x(k)) h_j(\bar{x}(k)) \Theta_{ijp} < 0 \tag{7.19}$$

with

$$\mathcal{P}_p = \sum_{q=1}^{N} \pi_{pq} P_q,$$
$$\Theta_{ijp} = -P_p + 3[A_{ijp}^T \mathcal{P}_p A_{ijp} + \varpi^2(\varepsilon_{ip} I + m\epsilon_{ip} \Psi_{\eta_1(p)}^T \Psi_{\eta_1(p)})],$$

then, the closed-loop system (7.16) can be guaranteed to be stochastically stable.

Proof Choose the token-dependent Lyapunov function

$$V(x(k), \theta(k)) = x^T(k) P_{\theta(k)} x(k). \tag{7.20}$$

Taking the difference of the Lyapunov function $V_1(x(k), \theta(k))$, one obtains

$$\begin{aligned}
&\mathbf{E}\{\Delta V(x(k), \theta(k))|x(k), \theta(k)\} \\
&\quad = \mathbf{E}\{V(x(k+1), \theta(k+1))|x(k), \theta(k)\} - V(x(k), \theta(k)) \\
&\quad = \mathbf{E}\{x(k+1)^T P_{\theta(k+1)} x(k+1)|x(k), \theta(k)\} \\
&\qquad - x^T(k) P_{\theta(k)} x(k).
\end{aligned} \tag{7.21}$$

Let $\theta(k) = p$ and $\theta(k+1) = q$. According to the closed-loop system (7.16) and expression (7.21), one can obtain

$$\begin{aligned}
&\mathbf{E}\{x(k+1)^T P_q x(k+1)|x(k), p\} \\
&\quad \leq \sum_{i=1}^{r}\sum_{j=1}^{r} m_i(x(k)) h_j(\bar{x}(k)) \Big[A_{ijp} x(k) + B_i\Big(d(x(k), k) - \hat{D}_p(\bar{x}(k))\Big)\Big]^T \\
&\quad \times \mathcal{P}_p \Big[A_{ijp} x(k) + B_i\Big(d(x(k), k) - \hat{D}_p(\bar{x}(k))\Big)\Big] \\
&\quad \leq \sum_{i=1}^{r}\sum_{j=1}^{r} m_i(x(k)) h_j(\bar{x}(k)) \Big\{3x^T(k) A_{ijp}^T \mathcal{P}_p A_{ijp} x(k) \\
&\qquad + 3d^T(k, x(k)) B_i^T \mathcal{P}_p B_i d(x(k), k) + 3\hat{D}_p^T(\bar{x}(k)) B_i^T \mathcal{P}_p B_i \hat{D}_p(\bar{x}(k))\Big\}
\end{aligned} \tag{7.22}$$

with $A_{ij\theta(k)} \triangleq A_i + B_i \Phi_{\eta_2(\theta(k))} K_{j,\theta(k)} \Psi_{\eta_1(\theta(k))}$.

It can be deduced from the conditions (7.17)–(7.18) that

$$3d^T(x(k), k) B_i^T \mathcal{P}_p B_i d(x(k), k) \leq 3\bar{d}^2 \varepsilon_{ip} x^T(k) x(k), \tag{7.23}$$

$$3\hat{D}_p^T(\bar{x}(k)) B_i^T \mathcal{P}_p B_i \hat{D}_p(\bar{x}(k)) \leq 3m\bar{d}^2 \epsilon_{ip} \bar{x}^T(k) \bar{x}(k). \tag{7.24}$$

Note that $\bar{x}(k) = \Psi_{\eta_1(p)} x(k)$. Thus, by combining (7.22) and (7.23)–(7.24), we have

$$\begin{aligned}
&\mathbf{E}\{x(k+1)^T P_q x(k+1)|x(k), p\} \\
&\leq \sum_{i=1}^{r}\sum_{j=1}^{r} m_i(x(k)) h_j(\bar{x}(k)) \Big\{3x^T(k)[A_{ijp}^T \mathcal{P}_p A_{ijp} \\
&\quad + \varpi^2 \varepsilon_{ip} I + m\varpi^2 \epsilon_{ip} \Psi_{\eta_1(p)}^T \Psi_{\eta_1(p)}] x(k)\Big\}.
\end{aligned} \tag{7.25}$$

Substituting the expression (7.25) into (7.21) yields

$$\mathbf{E}\{\Delta V(x(k), p)|x(k), p\} \leq \sum_{i=1}^{r}\sum_{j=1}^{r} m_i(x(k)) h_j(\bar{x}(k)) x^T(k) \Theta_{ijp} x(k). \tag{7.26}$$

If the condition (7.19) holds, we can have from (7.26)

$$\mathbf{E}\{\Delta V(x(k), p)|x(k), j\} < 0. \tag{7.27}$$

Hence, there exists a sufficiently small positive scalar ϱ such that

$$\mathbf{E}\{\Delta V(x(k), p)|x(k), p\} \leq -\varrho||x(k)||^2 < 0. \tag{7.28}$$

Taking the expectation with respect to the both sides of the expression (7.28), we obtain

$$\mathbf{E}\{\Delta V(x(k), p)\} \leq -\varrho\mathbf{E}\{||x(k)||^2\}. \tag{7.29}$$

Taking sum on the both sides of expression (7.29) with respect to k from 0 to T with $T > 0$, it can be obtained that

$$\mathbf{E}\{V(x(T+1), \theta(T+1))\} - V(x(0), \theta(0)) \leq -\varrho\mathbf{E}\left\{\sum_{k=0}^{T} ||x(k)||^2\right\}, \tag{7.30}$$

from which we can further obtain

$$\begin{aligned}\mathbf{E}\left\{\sum_{k=0}^{T} ||x(k)||^2\right\} &\leq \frac{\mathbf{E}\{V(x(T+1), \theta(T+1))\} - V(x(0), \theta(0))}{-\varrho} \\ &< \frac{\mathbf{E}\{V(x(0), \theta(0))\}}{\varrho}.\end{aligned} \tag{7.31}$$

As $T \to \infty$, the inequality (7.31) indicates that the closed-loop system (7.16) can achieve the stochastic stability according to Definition 7.1. □

Remark 7.3 Since the MN-SS protocols, governed by the Markov chain, are utilized to schedule both the sensor and actuator nodes, it is more reasonable to incorporate the stochastic characteristics of two Markov chains into the Lyapunov function. An effective method is to construct a quadratic Lyapunov function involving two Markov chains $\rho(k)$ and $\sigma(k)$. To achieve this, we design the token-dependent quadratic Lyapunov function (7.20) by introducing the auxiliary variable $\theta(k)$. This method not only captures the influence of both scheduling signals simultaneously but also enhances the stability analysis process.

7.4.2 Reachability of Sliding Surface

In the following, we shall prove that the state trajectories of the IT2 fuzzy system (7.4) can be driven into a small neighbourhood of the specified sliding surface $s(k) = 0$ in mean square sense.

Theorem 7.2 *For the IT2 T-S fuzzy system (7.4) with the MN-SS protocols (7.7) and (7.9), if there exist matrices $P_p > 0$, $R_p > 0$, G, K_{jp}, and scalars $\varepsilon_{ip} > 0$, $\epsilon_{ip} > 0$, $\xi_p > 0$, $\zeta_p > 0$ satisfying conditions (7.17)–(7.18) and the following conditions (for $p \in \mathbb{S}$, and $i, j = 1, \ldots, r$)*

$$B_i^T G^T \mathcal{R}_p G B_i \leq \xi_p I, \tag{7.32}$$

$$\Phi_{\eta_2(p)} B_i^T G^T \mathcal{R}_p G B_i \Phi_{\eta_2(p)} \leq \zeta_p I, \tag{7.33}$$

$$\sum_{i=1}^{r}\sum_{j=1}^{r} m_i(x(k)) h_j(\bar{x}(k)) \hat{\Theta}_{ijp} < 0 \tag{7.34}$$

with

$$\mathcal{R}_p = \sum_{q=1}^{N} \pi_{pq} R_q, \quad \hat{\Theta}_{ijp} = \Theta_{ijp} + 3A_{ijp}^T G^T \mathcal{R}_p G A_{ijp},$$

then, the state trajectories will be driven into the following region in mean square sense

$$\mathcal{D} = \left\{ s(k) \mid ||s(k)|| \leq \max_{1 \leq p \leq N} \sqrt{\frac{\mathcal{N}_p}{\lambda_{\min}(R_p)}} \right\} \tag{7.35}$$

with $\mathcal{N}_p \triangleq 3(\bar{d}^2 \xi_p + \bar{d}^2 \zeta_p m ||\Psi_{\eta_1(p)}||^2)||x(k)||^2$.

Proof It follows from expressions (7.14) and (7.15) that

$$\begin{aligned} s(k+1) = & \sum_{i=1}^{r}\sum_{j=1}^{r} m_i(x(k)) h_j(\bar{x}(k)) G\Big[A_{ij\theta(k)} x(k) \\ & + B_i\Big(d(k, x(k)) - \hat{D}_{\theta(k)}(\bar{x}(k))\Big)\Big]. \end{aligned} \tag{7.36}$$

Choose the following token-dependent Lyapunov function

$$U(x(k), \theta(k)) = V(x(k), \theta(k)) + s(k)^T R_{\theta(k)} s(k), \tag{7.37}$$

with $V_1(x(k), \theta(k))$ defined in (7.20).

It can be obtained from (7.37) that

$$\begin{aligned}&\mathbf{E}\{\Delta U(x(k),\theta(k))|x(k),\theta(k)\}\\&=\mathbf{E}\{\Delta V(x(k),\theta(k))|x(k),\theta(k)\}\\&\quad+\mathbf{E}\{s^T(k+1)R_{\theta(k+1)}s(k+1)|s(k),\theta(k)\}-s^T(k)R_{\theta(k)}s(k).\end{aligned} \tag{7.38}$$

For $\theta(k)=p$ and $\theta(k+1)=q$, one can obtain from the expression (7.36) that

$$\begin{aligned}&\mathbf{E}\{s^T(k+1)R_q s(k+1)|s(k),p\}\\&\le\sum_{i=1}^{r}\sum_{j=1}^{r}m_i(x(k))h_j(\bar{x}(k))\Big[A_{ijp}x(k)+B_i\Big(d(x(k),k)-\hat{D}_p(\bar{x}(k))\Big)\Big]^T G^T\\&\times\mathcal{R}_p G\Big[A_{ijp}x(k)+B_i\Big(d(x(k),k)-\hat{D}_p(\bar{x}(k))\Big)\Big]\\&\le\sum_{i=1}^{r}\sum_{j=1}^{r}m_i(x(k))h_j(\bar{x}(k))\Big\{3x^T(k)A_{ijp}^T G^T\mathcal{R}_p G A_{ijp}x(k)\\&+3d^T(x(k),k)B_i^T G^T\mathcal{R}_p G B_i d(x(k),k)\\&+3\hat{D}_p^T(\bar{x}(k))B_i^T G^T\mathcal{R}_p G B_i\hat{D}_p(\bar{x}(k))\Big\}.\end{aligned} \tag{7.39}$$

By means of the conditions (7.32) and (7.33), we have from (7.39)

$$\begin{aligned}&\mathbf{E}\{s^T(k+1)R_q s(k+1)|s(k),p\}\\&\le\sum_{i=1}^{r}\sum_{j=1}^{r}m_i(x(k))h_j(\bar{x}(k))\Big\{3x^T(k)[A_{ijp}^T G^T\mathcal{R}_p G A_{ijp}\\&+\varpi^2\xi_p I+m\varpi^2\zeta_p\Psi_{\eta_1(p)}^T\Psi_{\eta_1(p)}]x(k)\Big\}.\end{aligned} \tag{7.40}$$

Taking (7.26) and (7.40) into consideration, we can further obtain from (7.38)

$$\begin{aligned}&\mathbf{E}\{\Delta U(x(k),p)|x(k),p\}\\&\le x^T(k)\sum_{i=1}^{r}\sum_{j=1}^{r}m_i(x(k))h_j(\bar{x}(k))\hat{\Theta}_{ijp}x(k)\\&-\Big[\lambda_{\min}(R_p)||s(k)||^2-3(\varpi^2\xi_p+\varpi^2\zeta_p m||\Psi_{\eta_1(p)}||^2)||x(k)||^2\Big].\end{aligned} \tag{7.41}$$

Thus, under the condition (7.34), when the state trajectories stay outside of the region $\mathcal{D}$, i.e.,

$$||s(k)||>\sqrt{\frac{\mathcal{N}_p}{\lambda_{\min}(R_p)}}, \tag{7.42}$$

one has from (7.41) that $\mathbf{E}\{\Delta U(x(k), p)|x(k), p\} < 0$. Taking the expectation of both sides of the above inequality, we have $\mathbf{E}\{\Delta U(x(k), p)\} < 0$, which implies that the state trajectories are strictly decreasing outside the region $\mathcal{D}$, and can finally converge into this region $\mathcal{D}$ in mean square sense. □

It is noted that $m_i(x(k))$ and $h_j(\bar{x}(k))$ are coupled in the conditions (7.19) and (7.34) of Theorems 7.1 and 7.2, which is difficultly solved. In what follows, we will devote to conquer the obstacle via establishing the relation between $m_i(x(k))$ and $h_j(\bar{x}(k))$, under which more relaxed conditions will be obtained for guaranteeing the feasibility of the conditions in Theorems 7.1 and 7.2.

Inspired by the idea in [12], we assume that the MFs $m_i(x(k))$ and $h_j(\bar{x}(k))$ satisfy the following relations (for $j = 1, \cdots r$)

$$|h_j(\bar{x}(k)) - m_j(x(k))| \le \vartheta_j, \tag{7.43}$$

$$h_j(\bar{x}(k)) = \delta_j(k) m_j(x(k)). \tag{7.44}$$

It follows from (7.43)–(7.44) that

$$\begin{aligned} 0 < \delta_1^j \le 1 - \frac{\vartheta_j}{\underline{m}_j(x(k))} \le 1 - \frac{\vartheta_j}{m_j(x(k))} \le \delta_j(k) \\ \le 1 + \frac{\vartheta_j}{m_j(x(k))} \le 1 + \frac{\vartheta_j}{\underline{m}_j(x(k))} \le \delta_2^j < \infty, \end{aligned} \tag{7.45}$$

where δ_1^j and δ_2^j are the lower and upper bounds of $\delta_j(k)$.

Define $\delta_1 \triangleq \min_j\{\delta_1^j\}$ and $\delta_2 \triangleq \max_j\{\delta_2^j\}$. We can further obtain

$$\frac{\delta_1}{\delta_2} \le \frac{\delta_1^i}{\delta_2^j} = \frac{\min_i\{\delta_i(k)\}}{\max_j\{\delta_j(k)\}} \le \frac{\delta_i(k)}{\delta_j(k)} \le \frac{\max_i\{\delta_i(k)\}}{\min_j\{\delta_j(k)\}} = \frac{\delta_2^i}{\delta_1^j} \le \frac{\delta_2}{\delta_1}. \tag{7.46}$$

Thus, we have

$$\kappa_1 \le \frac{\delta_i(k)}{\delta_j(k)} \le \kappa_2 \tag{7.47}$$

with $\kappa_1 \triangleq \frac{\delta_1}{\delta_2}$ and $\kappa_2 \triangleq \frac{\delta_2}{\delta_1}$.

7.5 MFD Stability Conditions and Performance Optimization

In this section, we derive the MFD-based sufficient conditions for guaranteeing the stochastic stability of the closed-loop systems and the reachability of the sliding surface. Additionally, an optimization-solving algorithm is proposed to enhance the control performance.

7.5.1 Sufficient Conditions Based on Membership-Function-Dependent Approach

Now, by using the relations (7.43)–(7.44) and (7.47), we further give the following theorem that can guarantee the conditions in Theorems 7.1 and 7.2, simultaneously.

Theorem 7.3 *For given positive scalars κ_1 and κ_2, if there exist matrices $P_p > 0$, $\mathbf{P}_p > 0$, $R_p > 0$, $\mathbf{R}_p > 0$, G, K_{jp} and positive scalars ε_{ip}, ϵ_{ip}, ξ_p, ζ_p satisfying the following conditions (for $p \in \mathbb{S}$, $i, j = 1, \ldots, r$)*

$$\begin{bmatrix} -\varepsilon_{ip} I & * \\ \mathcal{B}_{ip} & -\hat{P} \end{bmatrix} \leq 0, \quad \begin{bmatrix} -\epsilon_{ip} I & * \\ \bar{\mathcal{B}}_{ip} & -\hat{P} \end{bmatrix} \leq 0, \tag{7.48}$$

$$\begin{bmatrix} -\xi_i I & * \\ \mathcal{F}_{ip} & -\hat{R} \end{bmatrix} \leq 0, \quad \begin{bmatrix} -\zeta_i I & * \\ \bar{\mathcal{F}}_{ip} & -\hat{R} \end{bmatrix} \leq 0, \tag{7.49}$$

and

$$\hat{\Gamma}_{iip} < 0, \tag{7.50}$$

$$\hat{\Gamma}_{ijp} + \kappa_1 \hat{\Gamma}_{jip} < 0, \tag{7.51}$$

$$\hat{\Gamma}_{ijp} + \kappa_2 \hat{\Gamma}_{jip} < 0, \quad 1 \leq i < j \leq r \tag{7.52}$$

with

$$\hat{\Gamma}_{ijp} = \begin{bmatrix} \hat{\Gamma}_{ijp}^{11} & * & * \\ \hat{\Gamma}_{ijp}^{31} & -\frac{1}{3}\hat{P} & * \\ \hat{\Gamma}_{ijp}^{41} & 0 & -\frac{1}{3}\hat{R} \end{bmatrix},$$

and

$$\begin{aligned}
\hat{P} &= \mathrm{diag}\{\mathbf{P}_1, \mathbf{P}_2, \cdots, \mathbf{P}_N\}, \quad \hat{R} = \mathrm{diag}\{\mathbf{R}_1, \mathbf{R}_2, \cdots, \mathbf{R}_N\}, \\
\mathcal{B}_{ip} &= [\sqrt{\pi_{p1}}, \sqrt{\pi_{p2}}, \cdots, \sqrt{\pi_{pN}}]^T \otimes B_i, \\
\bar{\mathcal{B}}_{ip} &= [\sqrt{\pi_{p1}}, \sqrt{\pi_{p2}}, \cdots, \sqrt{\pi_{pN}}]^T \otimes B_i \Phi_{\eta_2(p)}, \\
\mathcal{F}_{ip} &= [\sqrt{\pi_{p1}}, \sqrt{\pi_{p2}}, \cdots, \sqrt{\pi_{pN}}]^T \otimes G B_i, \\
\bar{\mathcal{F}}_{ip} &= [\sqrt{\pi_{p1}}, \sqrt{\pi_{p2}}, \cdots, \sqrt{\pi_{pN}}]^T \otimes G B_i \Phi_{\eta_2(p)}, \\
\hat{\Gamma}_{ijp}^{11} &= -P_p + 3\varpi^2(\varepsilon_{ip} I + m\epsilon_{ip} \Psi_{\eta_1(p)}^T \Psi_{\eta_1(p)}), \\
\hat{\Gamma}_{ijp}^{31} &= [\sqrt{\pi_{p1}}, \sqrt{\pi_{p2}}, \cdots, \sqrt{\pi_{pN}}]^T \otimes (A_i + B_i \Phi_{\eta_2(p)} K_{jp} \Psi_{\eta_1(p)}), \\
\hat{\Gamma}_{ijp}^{41} &= [\sqrt{\pi_{p1}}, \sqrt{\pi_{p2}}, \cdots, \sqrt{\pi_{pN}}]^T \otimes G(A_i + B_i \Phi_{\eta_2(p)} K_{jp} \Psi_{\eta_1(p)}),
\end{aligned}$$

with $\mathbf{P}_p = P_p^{-1}$, $\mathbf{R}_p = R_p^{-1}$, the SMC law (7.15) can guarantee that the stochastic stability of the closed-loop system (7.16) and the reacahability of the sliding region $\mathcal{D}$ defined in (7.35).

Proof Let $\mathbf{P}_p = P_p^{-1}$ and $\mathbf{R}_p = R_p^{-1}$. By using the Schur complement, it is easy to illustrate that the condition (7.48) can guarantee the conditions (7.17) and (7.18) in Theorem 7.1, and the condition (7.49) can ensure the conditions (7.32) and (7.33) in Theorem 7.2. In the following, we shall prove that if the conditions (7.50)–(7.52) hold, the conditions (7.19) and (7.34) in Theorems 7.1 and 7.2 will hold. Firstly, due to $3A_{ijp}^T G^T \mathcal{R}_p G A_{ijp} > 0$, it is easy to show that the condition (7.19) in Theorem 7.1 can be ensured by the inequality (7.34).

By means of the relation (7.44), the left side of the inequality (7.34) can be further represented as

$$\begin{aligned}
&\sum_{i=1}^{r}\sum_{j=1}^{r} m_i(x(k))h_j(\bar{x}(k))\hat{\Theta}_{ijp} \\
&= \sum_{i=1}^{r}\sum_{j=1}^{r} \delta_j(k)m_i(x(k))m_j(x(k))\hat{\Theta}_{ijp} \\
&= \sum_{i=1}^{r} \delta_i(k)m_i^2(x(k))\hat{\Theta}_{iip} \\
&+ \sum_{i=1}^{r-1}\sum_{j>i} \delta_j(k)m_i(x(k))m_j(x(k))\Big[\hat{\Theta}_{ijp} + \frac{\delta_i(k)}{\delta_j(k)}\hat{\Theta}_{jip}\Big].
\end{aligned} \tag{7.53}$$

Besides, in the light of the relation (7.47), there exist $\varsigma_1 \geq 0$ and $\varsigma_2 \geq 0$ satisfying $\varsigma_1 + \varsigma_2 = 1$ such that

$$\frac{\delta_i(k)}{\delta_j(k)} = \varsigma_1\kappa_1 + \varsigma_2\kappa_2. \tag{7.54}$$

It is easy to obtain from (7.54) that

$$\hat{\Theta}_{ijp} + \frac{\delta_i(k)}{\delta_j(k)}\hat{\Theta}_{jip} = \sum_{\ell=1}^{2} \varsigma_\ell(\hat{\Theta}_{pqi} + \kappa_\ell\hat{\Theta}_{qpi}). \tag{7.55}$$

In addition, the conditions (7.50)–(7.52) are equivalent to the following inequalities (for $1 \leq i < j \leq r$)

$$\hat{\Theta}_{iip} < 0, \tag{7.56}$$

$$\hat{\Theta}_{ijp} + \kappa_1\hat{\Theta}_{jip} < 0, \tag{7.57}$$

$$\hat{\Theta}_{ijp} + \kappa_2\hat{\Theta}_{jip} < 0. \tag{7.58}$$

It can be easily observed from (7.53) and (7.55) that the conditions (7.56)–(7.58) can ensure the condition (7.34).

Thus, the conditions (7.48)–(7.52) in Theorem 7.3 can guarantee the conditions (7.17)–(7.19) and (7.32)–(7.34) in Theorems 7.1 and 7.2, simultaneously. □

Remark 7.4 It can be seen that the relations (7.44) and (7.47) are used to analyze the stability in the expressions (7.53) and (7.55), by which the information of MFs is considered in the stability conditions for reducing the conservatism. Besides, different from the type-1 T-S fuzzy system with known MFs, the MFs in IT2 T-S fuzzy system are unknown, except the upper and lower MFs. Thus, we can observe from the expression (7.45) that only the known lower bound $\underline{m}_j(x(k))$ is used to obtain the relation (7.47), which is the key distinction from the existing work [12] for the type-1 T-S fuzzy system case.

7.5.2 SA-Based Optimization Solving Algorithm

It should be pointed out that the sliding matrix G plays an important role in the SMC design, which will affect the dynamic performance of the controlled system. In addition, a smaller gain K_{jp} implies less energy consumption of the control input, which is desirable in practical engineering. However, there exists the coupling term, i.e., $GB_i\Phi_{\eta_2(p)}K_{jp}\Psi_{\eta_1(p)}$, in the conditions (7.48)–(7.52) in Theorem 7.3. If the matrix G is chosen a priori, the coupling term will be reduced to a linear one on K_{jp}. Therefore, a natural idea is to search a suitable sliding matrix G and obtain the optimized gain K_{jp}. To this end, we construct the following optimization problem:

$$\min_{G}\sum_{j=1}^{r}\sum_{p=1}^{N}||K_{jp}||, \qquad (7.59)$$
$$\text{s.t. conditions } (7.48)-(7.52).$$

As well known, simulated annealing (SA) algorithm is a classical random search algorithm, which simulates the annealing technique applied in metallurgy. SA has shown the distinct ability of finding the global minimum of certain objective functions under specific conditions. In this chapter, SA algorithm is utilized to solve the minimization problem (7.59) via searching a suitable sliding matrix G.

Firstly, in order to carry out the SA algorithm, we represent the matrix G as the candidate solution:

$$w=[G_{11},\cdots,G_{1n},G_{21},\cdots,G_{mn}], \qquad (7.60)$$

and define the objective function as $E=\sum_{j=1}^{r}\sum_{p=1}^{N}||K_{jp}||$.

And then, the following SA-based solving algorithm is given to solve the minimization problem (7.59):

Algorithm 7.1 SA-based solving algorithm

1: (Set parameters) Set the initial temperature T_0, the final temperature $T_f(< T_0)$, the cooling factor c_0, $(0 < c_0 < 1)$, the iteration number L at each temperature, and the tolerant error τ_0. Specify the search space $S = \{G_{ij} \in [\underline{G}_{ij}, \overline{G}_{ij}], i = 1, \cdots, m, j = 1, \cdots, n\}$.

2: (Initialization) Let the current temperature $T = T_0$, generate randomly an initial solution w in the specified search space and let the counter $l = 0$, then solve the conditions (7.48)–(7.52) and calculate the initial fitness value E.

3: (Generate new solution) Via perturbing the current solution w, generate a candidate solution w', then solve the conditions (7.48)–(7.52) and calculate the fitness value E'.

4: (Acceptance criterion) Calculate the error $\Delta E = E' - E$. If $\Delta E \leq 0$, the solution w' is updated as the new solution. Otherwise, w' is accepted as the new solution with the probability $\bar{p}$, which is determined by the following Metropolis criterion:

$$\bar{p} = \begin{cases} 1, & \text{if } \Delta E \leq 0, \\ \exp(-\Delta E / T), & \text{otherwise.} \end{cases} \tag{7.61}$$

That is, for $\Delta E > 0$, randomly generate a variable $\chi (\in [0, 1])$, if $\chi < \bar{p}$, w' will be accepted as the new solution.

5: (Iteration) Let $l = l + 1$. If the iteration number l at the current temperature attains the maximum value L, then decrease the current temperature. Otherwise, return to **Generate new solution**.

6: (Cooling operation) Decrease the current temperature $T = c_0 T$, and reset the counter $l = 0$.

7: (Termination condition) If the termination condition is met, that is, the temperature T is less than the finial temperature T_f, or the descent speed of the fitness value E is less than the tolerant threshold τ_0, stop the algorithm and return the optimized solution w. Otherwise, return to **Generate new solution**.

8: (Control design) By means of the above obtained solution w in **Termination condition**, we can obtain the sliding matrix G and the control gains K_{jp} to design the SMC law (7.15).

Remark 7.5 According to expression (7.61), it can be seen that the SA algorithm will accept suboptimal solutions with a certain probability, which can ensure a comprehensive coverage of the predefined search space. At the same time, as the temperature decreases, the probability of accepting poor solutions will decrease; specifically, when the temperature approaches zero, the Metropolis criterion no longer accepts deteriorated solutions. Therefore, this algorithm will converge to the optimal solution of the minimization problem. Moreover, it is worthy to notice that the proposed SA-based solution algorithm has two significant advantages. Firstly, by adopting a random search strategy, it can effectively solve the coupling problem between the control gain K_{jp} and the sliding matrix F without introducing additional conservatism. Secondly, by selecting an appropriate fitness function for the SA algorithm, a desirable control performance can be achieved, such as reducing resource consumption.

7.6 Simulation Example

Consider a 3-order IT2 fuzzy system (7.4) with 2 rules and the following parameters

$$A_1 = \begin{bmatrix} 1.2 & 0.1 & 0.2 \\ -0.2 & 0.5 & 0 \\ 0.3 & 0.1 & 0.2 \end{bmatrix}, \ A_2 = \begin{bmatrix} 0.8 & 0 & 0.5 \\ 0.3 & 0.6 & -0.1 \\ 0.1 & -0.1 & 0.3 \end{bmatrix},$$

$$B_1 = \begin{bmatrix} 0.5 & 1 \\ 0 & 1 \\ 1 & 1 \end{bmatrix}, \ B_2 = \begin{bmatrix} 0.5 & 1 \\ 0.2 & 0.8 \\ 1 & 0.6 \end{bmatrix},$$

$$d(x(k), k) = \begin{bmatrix} 0.04\sqrt{|x_1(k)x_2(k)|} \\ 0.03x_3(k) \end{bmatrix},$$

and the upper and lower MFs

$$\overline{m}_1(x(k)) = \frac{1}{2 + e^{-(x_1(k)+6)}}, \quad \underline{m}_1(x(k)) = \frac{1}{2 + e^{-(x_1(k)+2)}}$$

$$\overline{m}_2(x(k)) = 1 - \underline{m}_1(x(k)), \quad \underline{m}_2(x(k)) = 1 - \overline{m}_1(x(k))$$

Suppose that two among three sensor nodes are stochastically selected to access the S/C channel and transmit the state signals at each instant. Thus, there are three transmission cases, i.e., $\Psi_1 = \mathrm{diag}\{1, 1, 0\}$, $\Psi_2 = \mathrm{diag}\{1, 0, 1\}$, $\Psi_3 = \mathrm{diag}\{0, 1, 1\}$. Suppose that the TPM is given as

$$\Lambda = \begin{bmatrix} 0.4 & 0.4 & 0.2 \\ 0.3 & 0.3 & 0.4 \\ 0.2 & 0.2 & 0.6 \end{bmatrix}.$$

Besides, in the C/A channel, suppose that only one node among two controller nodes will be randomly selected at each instant, i.e., $\Phi_1 = \mathrm{diag}\{1, 0\}$, $\Phi_2 = \mathrm{diag}\{0, 1\}$, and the TPM is given as

$$\Omega = \begin{bmatrix} 0.6 & 0.4 \\ 0.5 & 0.5 \end{bmatrix}.$$

According to the expression (7.10), the relation between the pair $(\rho(k), \sigma(k))$ and the variable $\theta(k)$ is given in Table 7.1.

Table 7.1 The relation of $(\rho(k), \sigma(k))$ and $\theta(k)$

$(\rho(k), \sigma(k))$	$\theta(k)$	$(\rho(k), \sigma(k))$	$\theta(k)$
(1, 1)	1	(1, 2)	4
(2, 1)	2	(2, 2)	5
(3, 1)	3	(3, 2)	6

Moreover, the TPM of the new Markov chain $\theta(k)$ can be calculated as follows:

$$\Pi = \begin{bmatrix} 0.24 & 0.24 & 0.12 & 0.16 & 0.16 & 0.08 \\ 0.18 & 0.18 & 0.24 & 0.12 & 0.12 & 0.16 \\ 0.12 & 0.12 & 0.36 & 0.08 & 0.08 & 0.24 \\ 0.20 & 0.20 & 0.10 & 0.20 & 0.20 & 0.10 \\ 0.15 & 0.15 & 0.20 & 0.15 & 0.15 & 0.20 \\ 0.10 & 0.10 & 0.30 & 0.10 & 0.10 & 0.30 \end{bmatrix}.$$

Due to the introduction of MN-SS protocol, the controller only can receive the state $\bar{x}(k)$. That is, the upper and lower MFs $\overline{m}(x(k))$ and $\underline{m}(x(k))$ in the fuzzy systems cannot be directly employed to design the controller. Thus, by only using the received signal $\bar{x}(k)$, the new MFs (7.15) are constructed as $h_1(\bar{x}(k)) = \underline{\beta}_1(\bar{x}(k))\underline{m}_1(\bar{x}(k)) + \overline{\beta}_1(\bar{x}(k))\overline{m}_1(\bar{x}(k))$ and $h_2(\bar{x}(k)) = 1 - h_1(\bar{x}(k))$. Meanwhile, choose the coefficients $\underline{\beta}_1(\bar{x}(k)) = 0.6$, $\overline{\beta}_1(\bar{x}(k)) = 0.4$, and parameters $\kappa_1 = 0.7519$, $\kappa_2 = 1.3299$ in (7.47).

Now, we perform the proposed SA-based solving algorithm to search the sliding matrix G for obtaining the optimized control gain K_{jp}. The matrix G is transformed into a vector as $w = [G_{11}, G_{12}, G_{13}, G_{21}, G_{22}, G_{23}]$. Select the initial and final temperature as $T_0 = 100$ and $T_f = 0.1$, the cooling factor $c_0 = 0.9$, the iteration number $L = 10$, and the tolerant error $\tau_0 = 10^{-8}$. The search space of each element in w is specified within the interval $[\underline{G}, \overline{G}]$ with $\underline{G} = -2$ and $\overline{G} = 2$. At each iteration, by perturbing each element of the current solution w, generate the candidate solution $G'_{ij} = G_{ij} + S(rand(\overline{G} - \underline{G}) + \underline{G})$ where $rand \in [0, 1]$ is a random number and choose the step factor as $S = 0.3$. By performing the **Algorithm** 7.1, the evaluation of the fitness value is shown in Fig. 7.2.

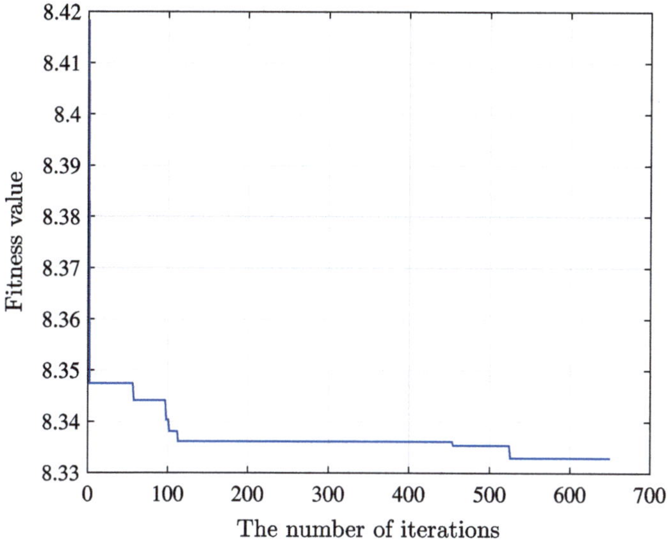

Fig. 7.2 The iteration of fitness value

Meanwhile, the optimized sliding matrix and the control gains are obtained as

$$G = \begin{bmatrix} 0.6706 & 0.4758 & -0.4671 \\ 1.3574 & -0.8939 & -0.9981 \end{bmatrix},$$

$$K_{11} = \begin{bmatrix} -0.9294 & -0.2385 & 0 \\ 0 & 0 & 0 \end{bmatrix}, K_{12} = \begin{bmatrix} -0.8027 & 0 & -0.2759 \\ 0 & 0 & 0 \end{bmatrix},$$

$$K_{13} = \begin{bmatrix} 0 & 0.2827 & -0.4919 \\ 0 & 0 & 0 \end{bmatrix}, K_{14} = \begin{bmatrix} 0 & 0 & 0 \\ -0.6658 & -0.2774 & 0 \end{bmatrix},$$

$$K_{15} = \begin{bmatrix} 0 & 0 & 0 \\ -0.5166 & 0 & -0.1624 \end{bmatrix}, K_{16} = \begin{bmatrix} 0 & 0 & 0 \\ 0 & 0.0297 & -0.2861 \end{bmatrix},$$

$$K_{21} = \begin{bmatrix} -0.7617 & -0.1274 & 0 \\ 0 & 0 & 0 \end{bmatrix}, K_{22} = \begin{bmatrix} -0.9773 & 0 & -0.5937 \\ 0 & 0 & 0 \end{bmatrix},$$

$$K_{23} = \begin{bmatrix} 0 & 0.1081 & -0.7086 \\ 0 & 0 & 0 \end{bmatrix}, K_{24} = \begin{bmatrix} 0 & 0 & 0 \\ -0.6810 & -0.2161 & 0 \end{bmatrix},$$

$$K_{25} = \begin{bmatrix} 0 & 0 & 0 \\ -0.6140 & 0 & -0.1910 \end{bmatrix}, K_{26} = \begin{bmatrix} 0 & 0 & 0 \\ 0 & -0.0450 & -0.4143 \end{bmatrix}.$$

In the simulation, suppose $\underline{\alpha}_1(x(k)) = \sin^2(x_1(k))$ and $\overline{\alpha}_1(x(k)) = 1 - \underline{\alpha}_1(x(k))$ for the MFs $m_1(x(k))$ and $m_2(x(k))$ in the IT2 T-S fuzzy system. Under the initial condition $x(0) = [-2.0,\ 1.1,\ 0.1]^T$, the simulation results under 100 independent trials are shown in Fig. 7.3a–d, form which it can be concluded that the proposed SMC law can ensure the stability of the closed-loop fuzzy system even though only

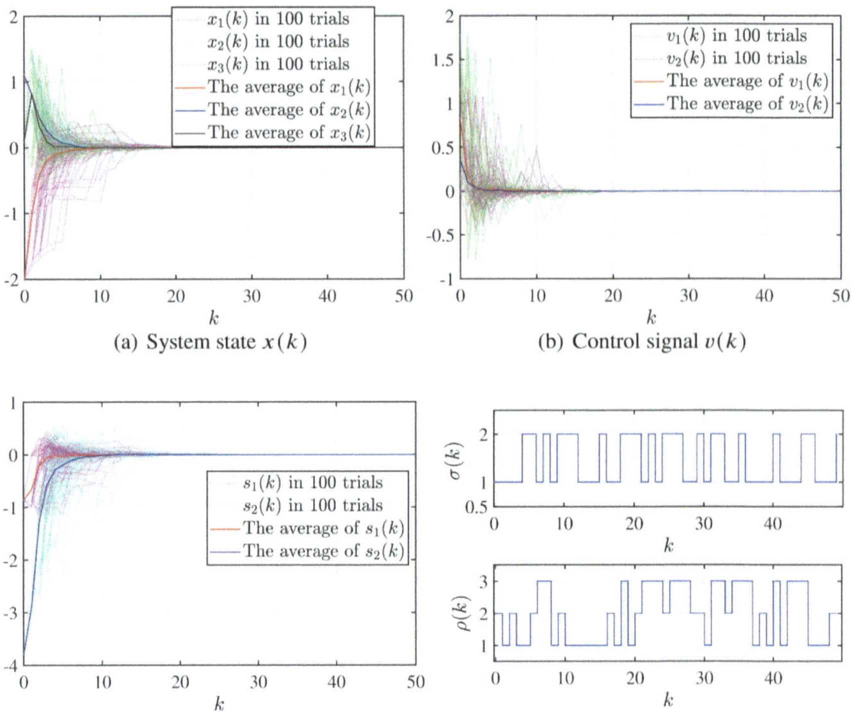

(a) System state $x(k)$

(b) Control signal $v(k)$

(c) Sliding variable $s(k)$

(d) A possible sequence of $\sigma(k)$ and $\rho(k)$

Fig. 7.3 Simulation results under the MN-SS protocol

a part of information are exchanged over the communication channels. A possible sequence of the scheduling signals $\sigma(k)$ and $\rho(k)$ under the MN-SS protocol is shown in Fig. 7.3d, in which $\sigma(k) = 1$ and 2 represent the transmission of signals $v_1(k)$ and $v_2(k)$ at the current instant, respectively. Meanwhile, $\rho(k) = 1$ means that the system states $x_1(k)$ and $x_2(k)$ are simultaneously transmitted at the current instant, and $\rho(k) = 2$ and $\rho(k) = 3$ denote the transmission of system states $\{x_1(k), x_3(k)\}$ and $\{x_2(k), x_3(k)\}$, respectively.

As comparison, we further consider the case that the network channels allow only one sensor node to exchange its information at each moment, under which the MN-SS protocol reduces to the single-node SS protocol case. For the proposed SMC scheme (7.15) with (7.14), we solve the optimization problem with the same parameters in the SA-based solving algorithm and obtain the optimized sliding matrix and control gains as follows:

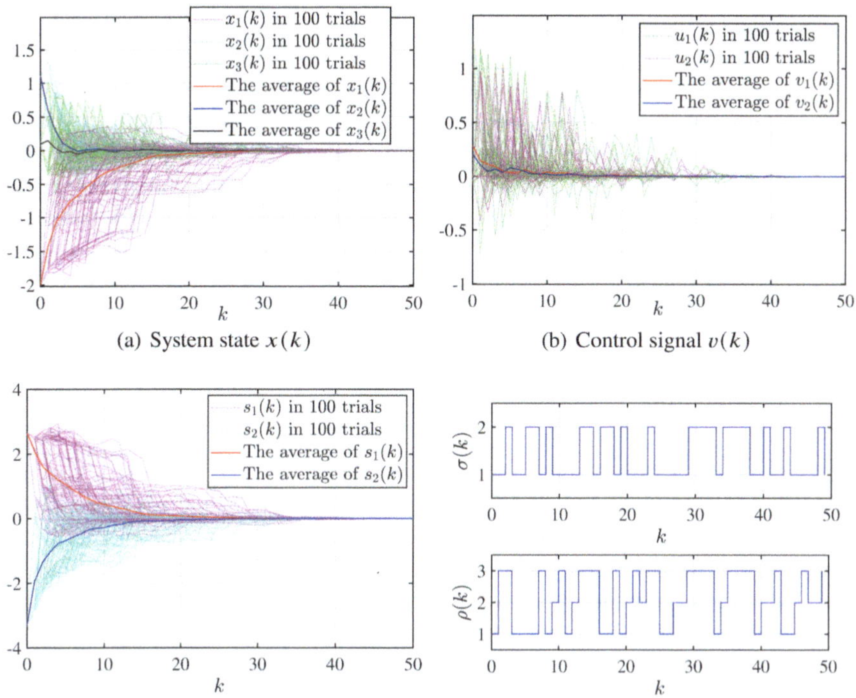

(c) Sliding variable $s(k)$

(d) A possible sequence of $\sigma(k)$ and $\rho(k)$

Fig. 7.4 Simulation results under the conventional single-node SS protocol

$$G = \begin{bmatrix} -1.6215 & -0.5928 & 0.6396 \\ 0.9028 & -1.5477 & 1.9071 \end{bmatrix},$$

$$K_{11} = \begin{bmatrix} -0.5991 & 0 & 0 \\ 0 & 0 & 0 \end{bmatrix}, K_{12} = \begin{bmatrix} 0 & 0.0815 & 0 \\ 0 & 0 & 0 \end{bmatrix},$$

$$K_{13} = \begin{bmatrix} 0 & 0 & -0.6825 \\ 0 & 0 & 0 \end{bmatrix}, K_{14} = \begin{bmatrix} 0 & 0 & 0 \\ -0.4200 & 0 & 0 \end{bmatrix},$$

$$K_{15} = \begin{bmatrix} 0 & 0 & 0 \\ 0 & -0.0052 & 0 \end{bmatrix}, K_{16} = \begin{bmatrix} 0 & 0 & 0 \\ 0 & 0 & -0.3832 \end{bmatrix},$$

$$K_{21} = \begin{bmatrix} -0.7318 & 0 & 0 \\ 0 & 0 & 0 \end{bmatrix}, K_{22} = \begin{bmatrix} 0 & 0.0548 & 0 \\ 0 & 0 & 0 \end{bmatrix},$$

$$K_{23} = \begin{bmatrix} 0 & 0 & -0.7433 \\ 0 & 0 & 0 \end{bmatrix}, K_{24} = \begin{bmatrix} 0 & 0 & 0 \\ -0.5941 & 0 & 0 \end{bmatrix},$$

$$K_{25} = \begin{bmatrix} 0 & 0 & 0 \\ 0 & -0.0287 & 0 \end{bmatrix}, K_{26} = \begin{bmatrix} 0 & 0 & 0 \\ 0 & 0 & -0.3650 \end{bmatrix}.$$

Then, under the same initial condition $x(0) = [-2,\ 1.1,\ 0.1]^T$, the simulation results are shown in Fig. 7.4a–d, from which it can be observed that the state trajectories in Figs. 7.4a, c converge slower than the ones in Fig. 7.3a and c under the MN-SS protocol. This is also consistent with the advantage of the proposed MN-SS protocol. Besides, the scheduling signals $\rho(k)$ and $\sigma(k)$ shown in Fig. 7.4d represent the selected sensor and actuator nodes, respectively. By the above comparison on two cases, it can be concluded that the proposed control scheme under the MN-SS protocol can achieve better control performance by effectively scheduling the sensor/actuator nodes.

7.7 Conclusion

This chapter has addressed the SMC problem for IT2 T-S fuzzy systems, where two MN-SS protocols have been employed simultaneously to schedule the transmission nodes of the S/C and C/A channels. A scheduling signal-dependent fuzzy SMC law has been proposed by employing the received state signals, such that the stochastic stability of the closed-loop systems and the reachability of the sliding surface can be ensured. By establishing the relation of MFs between the fuzzy sliding mode controller and the IT2 T-S fuzzy systems, the information of the MFs has been integrated to obtain the sufficient conditions with less conservatism. Besides, an optimization problem has been established, and the SA algorithm has been utilized to search the sliding matrix for obtaining the optimized control design parameters with less resources consumption.

References

1. Donkers, M., Heemels, W., Bernardini, D., Bemporad, A., Shneer, V.: Stability analysis of stochastic networked control systems. Automatica **48**(5), 917–925 (2012)
2. Cao, Z., Niu, Y., Karimi, H.R.: Dynamic output feedback sliding mode control for Markovian jump systems under stochastic communication protocol and its application. Int. J. Robust Nonlinear Control **30**(17), 7307–7325 (2020)
3. Ding, D., Wang, Z., Han, Q.L.: Neural-network-based output-feedback control with stochastic communication protocols. Automatica **106**, 221–229 (2019)
4. Zhang, J., Peng, C., Xie, X., Yue, D.: Output feedback stabilization of networked control systems under a stochastic scheduling protocol. IEEE Trans. Cybern. **50**(6), 2851–2860 (2020)
5. Zhang, Z., Niu, Y., Cao, Z., Song, J.: Security sliding mode control of interval type-2 fuzzy systems subject to cyber attacks: the stochastic communication protocol case. IEEE Trans. Fuzzy Syst. **29**(2), 240–251 (2021)
6. Tobagi, F.: Multiaccess protocols in packet communication systems. IEEE Trans. Commun. **28**(4), 468–488 (1980)
7. Guo, G.: Linear systems with medium-access constraint and Markov actuator assignment. IEEE Trans. Circuits Syst. I Regul. Pap. **57**(11), 2999–3010 (2010)
8. Guo, G., Lu, Z., Han, Q.L.: Control with Markov sensors/actuators assignment. IEEE Trans. Autom. Control **57**(7), 1799–1804 (2012)

9. Lam, H.K., Li, H., Deters, C., Secco, E.L., Wurdemann, H.A., Althoefer, K.: Control design for interval type-2 fuzzy systems under imperfect premise matching. IEEE Trans. Ind. Electron. **61**(2), 956–968 (2014)
10. Lam, H.K., Seneviratne, L.D.: Stability analysis of interval type-2 fuzzy-model-based control systems. IEEE Trans. Syst. Man Cybern. B Cybern. **38**(3), 617–628 (2008)
11. Xiong, J., Lam, J.: Stabilization of discrete-time Markovian jump linear systems via time-delayed controllers. Automatica **42**(5), 747–753 (2006)
12. Zhang, J., Peng, C.: Event-triggered H_∞ filtering for networked Takagi-Sugeno fuzzy systems with asynchronous constraints. IET Signal Process. **9**(5), 403–411 (2015)

Chapter 8
Co-design of Fuzzy Sliding Mode Control and Stochastic Scheduling Protocol

8.1 Introduction

In the existing works involving in the SS protocol as in Chap. 7, the transition probability matrix (TPM) of Markov chain was implicitly assumed to be known/predefined. That is, the existing works only focus on the design of controller under the *given/known* SS protocols, without considering the possible cooperative effect of the controller and scheduling protocols on the dynamic performance of NCSs. Since the objective of scheduling protocols is to regulate the transmission of communication networks, it should also be possible and necessary to design the scheduling protocol according to the practical requirement. In fact, Sehgal [1] has already carried out an interesting research on the dynamic buffer allocation for stochastic communication in networks-in-package via constructing a Markov jump model-based communication scheme. Moreover, a few relevant works have also been reported on the hybrid design scheme on the controller and TPM for Markov jumping systems under the conventional point-to-point circumstance [2, 3]. These pioneer works motivate us to think that the coordinative design of scheduling protocol and controller can also bring benefits to the system performance.

This chapter focuses on the co-design issue of SS protocol and sliding mode controller for the IT2 T-S fuzzy systems over the communication network. To alleviate the data congestion in the shared communication network, the transmission of sensor nodes is governed by the SS protocol obeying a discrete-time Markov chain, which will be co-designed along with sliding mode controller. To deal with the mismatched MFs caused by the special structure of IT2 fuzzy model and the insertion of SS protocol, the relation between the MFs in the fuzzy model and the ones in the controller will be established. Moveover, based on the MFD technique, the sufficient conditions are derived to ensure the stochastic stability of the closed-loop system and the reachability of the sliding surface. In addition, a PSO-based optimization algorithm is proposed to search token-dependent parameters and sliding matrix with the optimization objective of reducing the sliding domain.

Y. Yang et al., *Analysis and Synthesis of Interval Type-2 Fuzzy Systems Under Communication Protocols*, Studies in Systems, Decision and Control 636,
https://doi.org/10.1007/978-981-95-2355-9_8

8.2 Problem Description

8.2.1 System Model

Consider the following IT2 T-S fuzzy systems with the i-th rule:

Rule i: IF $f_1(x(k))$ is $\mathscr{F}_1^i$ and ... and $f_\hbar(x(k))$ is $\mathscr{F}_\hbar^i$, THEN

$$x(k+1) = A_i x(k) + B_i(u(k) + d(x(k), k)), \tag{8.1}$$

where $x(k) \in \mathbb{R}^n$ and $u(k) \in \mathbb{R}^m$ are the system state and the control input, respectively. $f_\iota(x(k))$ $(\iota = 1, \ldots, \hbar)$ is the premise variable and $\mathscr{F}_\iota^i$ $(i = 1, \ldots, r)$ is the IT2 fuzzy set of i-th rule. A_i and B_i are known matrices. $d(x(k), k)$ is the external disturbance satisfying $\|d(x(k), k)\| \leq \varpi ||x(k)||$ with ϖ a known constant.

The firing strength of the i-th fuzzy rule is represented as

$$\vec{m}_i(x(k)) = \left[\underline{m}_i(x(k)), \overline{m}_i(x(k))\right],$$

with upper and lower MFs $\underline{m}_i(x(k)) = \prod_{\iota=1}^{\hbar} \underline{\mu}_{\mathscr{F}_\iota^i}(f_\iota(x(k))) > 0$, $\overline{m}_i(x(k)) = \prod_{\iota=1}^{\hbar} \overline{\mu}_{\mathscr{F}_\iota^i}(f_\iota(x(k))) > 0$ and $0 \leq \underline{m}_i(x(k)) \leq \overline{m}_i(x(k))$; $\overline{\mu}_{\mathscr{F}_\iota^i}(f_\iota(x(k)))$ and $\underline{\mu}_{\mathscr{F}_\iota^i}(f_\iota(x(k)))$ denote the lower and upper grades of the membership function, respectively.

Then, the IT2 T-S fuzzy system (8.1) can be inferred as

$$x(k+1) = \sum_{i=1}^{r} m_i(x(k))[A_i x(k) + B_i(u(k) + d(x(k), k))], \tag{8.2}$$

where $m_i(x(k))$ is the membership function of the i-th subsystem satisfying $\sum_{i=1}^{r} m_i(x(k)) = 1$, and $m_i(x(k)) = \underline{\alpha}_i(x(k))\underline{m}_i(x(k)) + \overline{\alpha}_i(x(k))\overline{m}_i(x(k))$, $\underline{\alpha}_i(x(k)), \overline{\alpha}_i(x(k)) \in [0, 1]$, $\underline{\alpha}_i(x(k)) + \overline{\alpha}_i(x(k)) = 1$ with $\overline{\alpha}_i(x(k))$ and $\underline{\alpha}_i(x(k))$ as weighting coefficients.

8.2.2 Stochastic Scheduling Protocol

As shown in Fig. 8.1, the state signal $x(k)$ is transmitted from sensors to the controller over the shared communication network. To reduce the transmission burden and relieve the risk of data collision, the SS protocol is introduced into the S/C channel to orchestrate sensor nodes, by which only one sensor at each instant will be stochastically selected to access the communication channels.

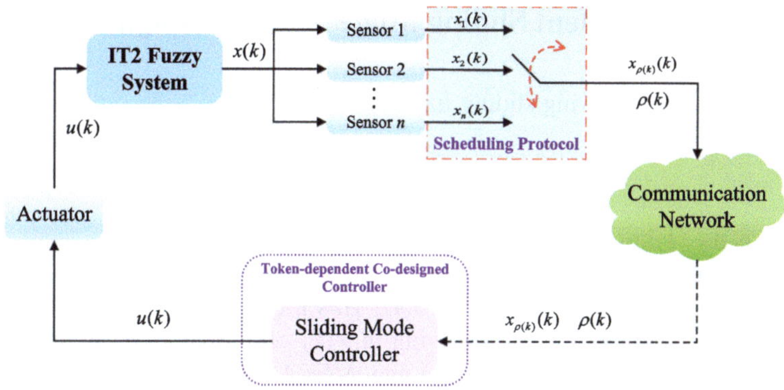

Fig. 8.1 Control system under SS protocol

Let the token $\rho(k) \in \mathbb{M} \triangleq \{1, 2, \ldots, n\}$ represent the selected sensor node at instant k, which obeys a discrete-time Markov chain with the following conditional probability

$$\pi_{pq} = \mathbf{Pr}\{\rho(k+1) = q | \rho(k) = p\}, \quad p, q \in \mathbb{M}, \tag{8.3}$$

with $0 \leq \pi_{pq} \leq 1$ and $\sum_{q=1}^{n} \pi_{pq} = 1$. Define $\Pi = [\pi_{pq}]_{n \times n}$ as the TPM.

Denote $x(k) = [x_1(k), \ldots, x_n(k)]^T$. According to the selection manner of the SS protocol under the zero-input strategy, the available state information for the controller can be denoted as

$$\bar{x}(k) = \Phi_{\rho(k)} x(k), \tag{8.4}$$

in which $\bar{x}(k) = [\bar{x}_1(k), \ldots, \bar{x}_n(k)]^T$ and $\Phi_{\rho(k)} = \text{diag}\{\delta(\rho(k) - 1), \ldots, \delta(\rho(k) - n)\}$ with $\delta(\cdot)$ being the Kronecker delta function, and $\Phi_{\rho(k)} \in \{\Phi_1, \ldots, \Phi_n\}$. In order to achieve the desirable dynamic performance of IT2 fuzzy system (8.2) under the limited available state information, we shall give a co-design scheme of the TPM governing the transmission of sensor nodes and the feasible fuzzy SMC law.

Definition 8.1 The IT2 fuzzy system (8.2) under the SS protocol (8.4) is said to be stochastically stable if

$$\mathbf{E}\left\{\sum_{k=1}^{\infty} ||x(k)||^2 \mid x(0), \rho(0)\right\} < \infty \tag{8.5}$$

holds for any $x(0) \in \mathbb{R}^n$ and $\rho(0) \in \mathbb{M}$.

8.3 Token-Dependent Sliding Mode Control Design

Firstly, choose the following sliding function

$$s(k) = Gx(k), \tag{8.6}$$

where G is the sliding matrix to be designed later.

In addition, due to the insertion of the SS protocol, the controller can only use the received state signal $\bar{x}(k)$ in (8.4). This implies that the upper and lower bounds $\underline{m}_i(x(k))$ and $\overline{m}_i(x(k))$ involving in $x(k)$ cannot be directly employed to design the fuzzy controller. Hence, the new MFs dependent on the available state $\bar{x}(k)$ are constructed as $h_j(\bar{x}(k)) = \underline{\beta}_j(\bar{x}(k))\underline{m}_j(\bar{x}(k)) + \overline{\beta}_j(\bar{x}(k))\overline{m}_j(\bar{x}(k))$ with $\sum_{j=1}^{r} h_j(\bar{x}(k)) = 1$; and $\underline{\beta}_j(\bar{x}(k))$ and $\overline{\beta}_j(\bar{x}(k))$ are two coefficients.

Then, we design the following scheduling signal-dependent fuzzy sliding mode controller as

$$u(k) = \sum_{j=1}^{r} h_j(\bar{x}(k))K_{j,\rho(k)}\bar{x}(k) - \varpi||\bar{x}(k)||\text{sign}(\bar{s}(k)), \tag{8.7}$$

where $\bar{s}(k)$ is the sliding variable by replacing $x(k)$ in (8.6) with $\bar{x}(k)$, and the control gain $K_{r,\rho(k)}$ will be designed in conjunction with the TPM Π of the SS protocol.

Substituting the sliding mode controller (8.7) into the IT2 T-S fuzzy system (8.2), we obtain the following closed-loop system

$$\begin{aligned} x(k+1) = \sum_{i=1}^{r}\sum_{j=1}^{r} m_i(x(k))h_j(\bar{x}(k))\Big[&(A_i + B_iK_{j,\rho(k)}\Phi_{\rho(k)})x(k) \\ &+ B_i(d(x(k),k) - \varpi||\bar{x}(k)||\text{sign}(\bar{s}(k)))\Big]. \end{aligned} \tag{8.8}$$

To tackle the mismatched MFs $m_i(x(k))$ and $h_j(\bar{x}(k))$ in (8.8) and facilitate the subsequent stability analysis, the following relations between the MFs $m_i(x(k))$ and $h_j(\bar{x}(k))$, $(i = 1, \ldots, r)$ are established:

$$|h_i(\bar{x}(k)) - m_i(x(k))| \leq \eta_i, \tag{8.9}$$

$$h_i(\bar{x}(k)) = \varsigma_i(k)m_i(x(k)), \tag{8.10}$$

where η_i is a constant and can be calculated or estimated according to the controller's MFs $h_i(\bar{x}(k))$ and the known upper and lower MFs (i.e., $\overline{m}_i(x(k))$, $\underline{m}_i(x(k))$) within the modeling domain.

From the above relations (8.9)–(8.10) and $\underline{m}_i(x(k)) \leq m_i(x(k))$, one has

$$
\begin{aligned}
0 < \underline{\varsigma}_i &\triangleq 1 - \frac{\eta_i}{\underline{m}_i(x(k))} \leq 1 - \frac{\eta_i}{m_i(x(k))} \leq \varsigma_i(k) \\
&\leq 1 + \frac{\eta_i}{m_i(x(k))} \leq 1 + \frac{\eta_i}{\underline{m}_i(x(k))} \triangleq \overline{\varsigma}_i < \infty,
\end{aligned} \tag{8.11}
$$

where $\min_{x(k)}\{\underline{m}_i(x(k))\}$ is the minimum value of $\underline{m}_i(x(k))$ in the modeling domain, and two constants $\underline{\varsigma}_i$ and $\overline{\varsigma}_i$ can be calculated as the lower and upper bounds of the variable $\varsigma_i(k)$ for $i = 1, \ldots, r$.

Define $\overline{\varsigma} \triangleq \max_{i \in \mathbb{H}}\{\overline{\varsigma}_i\}$ and $\underline{\varsigma} = \min_{i \in \mathbb{H}}\{\underline{\varsigma}_i\}$ for $\forall i \in \mathbb{H} \triangleq \{i = 1, \ldots, r\}$. We can obtain

$$
\frac{\varsigma_i(k)}{\varsigma_j(k)} \in [\phi_1, \phi_2], \tag{8.12}
$$

with $\phi_1 = \underline{\varsigma}/\overline{\varsigma}$ and $\phi_2 = \overline{\varsigma}/\underline{\varsigma}$.

Remark 8.1 To ensure $0 < \varsigma_i(k) < \infty$ in (8.11), the minimum value of the $m_i(x(k))$ should be greater than zero. Due to $\underline{m}_i(x(k)) \leq m_i(x(k))$, the $m_i(x(k)) > 0$ can be guaranteed by $\underline{m}_i(x(k)) > 0$. Here, the condition $\underline{m}_i(x(k)) > 0$ can be attained by properly modeling the fuzzy system for the nonlinear systems, as shown in simulation example later.

8.4 Performance Analysis Under Stochastic Scheduling Protocol

In what follows, we shall analyze the stochastic stability of the closed-loop system (8.8) and the reachability of the sliding surface $s(k) = 0$, by which we will provide the co-design scheme to determine the control gain $K_{j,\rho(k)}$ and TPM Π.

8.4.1 Stochastic Stability of Closed-Loop System

Theorem 8.1 *Consider the IT2 fuzzy system (8.2) with given positive scalars τ_1, τ_2, ϕ_1, and ϕ_2, if there exist scalars $\zeta_i > 0$, $\pi_{pq} \in [0, 1]$ with $\sum_{q=1}^{n} \pi_{pq} = 1$, matrices $P_p > 0$, G, $K_{j,p}$ satisfying the following conditions (for $p, q = 1, \ldots, n, i, j = 1, \ldots, r$)*

$$
B_i^T \mathcal{P}_p B_i \leq \zeta_i I, \tag{8.13}
$$

$$
\Lambda_{ijp} < 0, \quad (i = j), \tag{8.14}
$$

$$
\Lambda_{ijp} + \phi_\iota \Lambda_{jip} < 0, \quad (i < j, \ \iota = 1, 2), \tag{8.15}
$$

with

$$\mathcal{P}_p \triangleq \sum_{q=1}^{n} \pi_{pq} P_q$$
$$\Lambda_{ijp} \triangleq -P_p + (1+\tau_1)(A_i + B_i K_{j,p}\Phi_p)^T \mathcal{P}_p (A_i + B_i K_{j,p}\Phi_p) + 2(1+\tau_1^{-1})\left[\varpi^2 \zeta_i I + m\varpi^2 \zeta_i \Phi_p^T \Phi_p\right],$$

then the closed-loop system (8.8) under the SS protocol is stochastically stable.

Proof Choose the Lyapunov function

$$V(x(k), \rho(k)) = x^T(k) P_{\rho(k)} x(k). \tag{8.16}$$

Let $\rho(k) = p$ and $\rho(k+1) = q$. Taking the difference of the Lyapunov function $V(x(k), \rho(k))$ in (8.16), one has

$$\begin{aligned}
&\mathbf{E}\{\Delta V(x(k), p) \mid x(k), p\} \\
&= \mathbf{E}\{V(x(k+1), \rho(k+1)) \mid x(k), p\} - V(x(k), p) \\
&= \mathbf{E}\{x(k+1)^T \mathcal{P}_p x(k+1) | x(k), p\} - x^T(k) P_p x(k).
\end{aligned} \tag{8.17}$$

Along with the closed-loop system (8.8), we can obtain

$$\begin{aligned}
&\mathbf{E}\{x(k+1)^T \mathcal{P}_p x(k+1) \mid x(k), p\} \\
&\leq \sum_{i=1}^{r}\sum_{j=1}^{r} \varsigma_j(k) m_i(x(k)) m_j(x(k)) \Big[(A_i + B_i K_{j,p}\Phi_p)x(k) \\
&+ B_i(d(x(k), k) - \varpi \|\bar{x}(k)\| \text{sign}(\bar{s}(k)))\Big]^T \\
&\times \mathcal{P}_p \Big[(A_i + B_i K_{j,p}\Phi_p)x(k) + B_i(d(x(k), k) - \varpi ||\bar{x}(k)|| \text{sign}(\bar{s}(k)))\Big] \\
&\leq \sum_{i=1}^{r}\sum_{j=1}^{r} \varsigma_j(k) m_i(x(k)) m_j(x(k)) \Big\{(1+\tau_1)x^T(k)\Big[(A_i + B_i K_{j,p}\Phi_p)^T \mathcal{P}_p (A_i \\
&+ B_i K_{j,p}\Phi_p)\Big]x(k) + 2(1+\tau_1^{-1})\varpi^2 x^T(k)\Big[\zeta_i I + m\zeta_l \Phi_p^T \Phi_p\Big]x(k)\Big\},
\end{aligned} \tag{8.18}$$

where $\tau_1 > 0$ is an adjustable scalar.

Substituting (8.18) into (8.17) yields

$$
\begin{aligned}
&\mathbf{E}\{\Delta V(x(k), p) \mid x(k), p\} \\
&\leq \sum_{i=1}^{r}\sum_{j=1}^{r} \varsigma_j(k) m_i(x(k)) m_j(x(k)) \Big\{(1+\tau_1)x^T(k)\big[(A_i + B_i K_{j,p}\Phi_p)^T \mathcal{P}_p \\
&\quad \times (A_i + B_i K_{j,p}\Phi_p)\big]x(k) \qquad (8.19)\\
&\quad + 2(1+\tau_1^{-1})\varpi^2 x^T(k)\big[\zeta_i I + m\zeta_l \Phi_p^T \Phi_p\big]x(k) - x^T(k) P_p x(k)\Big\} \\
&\leq \sum_{i=1}^{r} \varsigma_i(k) m_i^2(x(k)) x^T(k)\Lambda_{iip} x(k) \\
&\quad + \sum_{i=1}^{r-1}\sum_{j>i} \varsigma_j(k) m_i(x(k)) m_j(x(k)) x^T(k)\Big[\Lambda_{ijp} + \frac{\varsigma_i(k)}{\varsigma_j(k)}\Lambda_{jip}\Big]x(k). \qquad (8.20)
\end{aligned}
$$

According to the relation (8.12), it can be obtained that there exist two positive scalars ν_1 and ν_2 satisfying $\nu_1 + \nu_2 = 1$ and $\nu_1\phi_1 + \nu_2\phi_2 = \frac{\varsigma_l(k)}{\varsigma_r(k)}$.

Thus, we have

$$
\Lambda_{ijp} + \frac{\varsigma_i(k)}{\varsigma_j(k)}\Lambda_{jip} = \sum_{\iota=1}^{2} \nu_\iota[\Lambda_{ijp} + \phi_\iota \Lambda_{jip}]. \qquad (8.21)
$$

Substituting (8.21) into the expression (8.19) yields:

$$
\begin{aligned}
&\mathbf{E}\{\Delta V(x(k), p) \mid x(k), p\} \\
&\leq \sum_{i=1}^{r} \varsigma_i(k) m_i^2(x(k)) x^T(k)\Lambda_{iip} x(k) \\
&\quad + \sum_{i=1}^{r-1}\sum_{j>i} \varsigma_j(k) m_i(x(k)) m_j(x(k)) x^T(k)\left\{\sum_{\iota=1}^{2} \nu_\iota\left[\Lambda_{ijp} + \phi_\iota \Lambda_{jip}\right]\right\} x(k). \qquad (8.22)
\end{aligned}
$$

If the conditions (8.14)–(8.15) hold, we have from (8.22)

$$
\mathbf{E}\{\Delta V(x(k), p) | x(k), p\} \leq -\theta \|x(k)\|^2 < 0 \qquad (8.23)
$$

with a positive scalar $\theta = \min_{i,j,p,\iota}\{\lambda_{\min}(-\Lambda_{iip}), \lambda_{\min}(-\Lambda_{ijp} - \phi_\iota \Lambda_{jip})\} > 0$.

Taking the mathematical expectation for both sides of the expression (8.23) yields

$$
\mathbf{E}\{\Delta V(x(k), p)\} \leq -\theta \mathbf{E}\{\|x(k)\|^2\}. \qquad (8.24)
$$

Taking sum on the both sides of (8.24) with respect to k from 0 to T with $T > 0$, one has

$$\mathbf{E}\{V(x(T+1), \rho(T+1))\} - V(x(0), \rho(0)) \le -\theta \mathbf{E}\left\{\sum_{k=0}^{T} ||x(k)||^2\right\} \tag{8.25}$$

from which we can further obtain

$$\mathbf{E}\left\{\sum_{k=0}^{T} ||x(k)||^2\right\} \le \frac{V(x(0), \rho(0)) - \mathbf{E}\{V(x(T+1), \rho(T+1))\}}{\theta}. \tag{8.26}$$

As $T \to \infty$, one has from the above inequality

$$\mathbf{E}\left\{\sum_{k=0}^{\infty} ||x(k)||^2\right\} < \frac{\mathbf{E}\{V(x(0), \rho(0))\}}{\theta} < \infty, \tag{8.27}$$

which means that the closed-loop system (8.8) can be guaranteed to be stochastically stable according to Definition 8.1. □

8.4.2 Reachability of Sliding Surface

In this subsection, we will further derive the sufficient conditions for ensuring the reachability of the specified sliding surface $s(k) = 0$.

Theorem 8.2 *Consider the IT2 T-S fuzzy system (8.2) with given positive scalars τ_1, τ_2, ϕ_1, and ϕ_2, if there exist scalars $\zeta_i > 0$, $\xi_i > 0$, $\pi_{pq} \in [0, 1]$ with $\sum_{q=1}^{n} \pi_{pq} = 1$, matrices $P_p > 0$, $Q_p > 0$, G, and $K_{j,p}$, satisfying (8.13) and the following conditions (for $p, q \in \mathbb{M}$, $i, j = 1, \ldots, r$)*

$$B_i^T G^T \mathcal{Q}_p G B_i \le \xi_i I, \tag{8.28}$$

$$\hat{\Lambda}_{ijp} < 0, \quad (i = j), \tag{8.29}$$

$$\hat{\Lambda}_{ijp} + \phi_\iota \hat{\Lambda}_{jip} < 0, \quad (i < j, \iota = 1, 2), \tag{8.30}$$

with

$$\mathcal{Q}_p \triangleq \sum_{q=1}^{n} \pi_{pq} Q_q,$$

$$\hat{\Lambda}_{ijp} \triangleq \Lambda_{ijp} + (1 + \tau_2)(A_i + B_i K_{j,p} \Phi_p)^T G^T \mathcal{Q}_p G (A_i + B_i K_{j,p} \Phi_p),$$

then the scheduling signal-dependent fuzzy sliding mode controller (8.7) can drive state trajectories entering (in mean square) into the following domain

$$\mathcal{D} = \left\{ s(k) \mid ||s(k)|| \leq \max_{i,p} \sqrt{\frac{\mathcal{N}}{\lambda_{\min}(\mathcal{Q}_p)}} \right\} \tag{8.31}$$

with $\mathcal{N} \triangleq 2(1+\tau_2^{-1})\varpi^2\xi_i(||x(k)||^2 + m||\bar{x}(k)||^2)$.

Proof Combining the closed-loop system (8.8) and sliding function (8.6) obtains that

$$\begin{aligned} s(k+1) = & \sum_{i=1}^{r}\sum_{j=1}^{r} \varsigma_j(k) m_i(x(k)) m_j(x(k)) G\Big[(A_i + B_i K_{j,\rho(k)} \Phi_{\rho(k)}) x(k) \\ & + B_i\big(d(k, x(k)) - \varpi ||\bar{x}(k)|| \text{sign}(\bar{s}(k))\big)\Big]. \end{aligned} \tag{8.32}$$

Choose the Lyapunov function as

$$U(x(k), \rho(k)) = V(x(k), \rho(k)) + s(k)^T \mathcal{Q}_{\rho(k)} s(k). \tag{8.33}$$

For $\rho(k) = p$ and $\rho(k+1) = q$, one has from (8.33)

$$\begin{aligned} & \mathbf{E}\{\Delta U(x(k), p) \mid x(k), p\} \\ & = \mathbf{E}\{\Delta V(x(k), p) \mid x(k), p\} + \mathbf{E}\{s^T(k+1)\mathcal{Q}_p s(k+1) | s(k), p\} \\ & - s^T(k)\mathcal{Q}_p s(k). \end{aligned} \tag{8.34}$$

According to (8.32), we have

$$\begin{aligned} & \mathbf{E}\{s^T(k+1)\mathcal{Q}_p s(k+1) \mid x(k), p\} \\ & \leq \sum_{i=1}^{r}\sum_{j=1}^{r} \varsigma_j(k) m_i(x(k)) m_j(x(k)) \Big[(A_i + B_i K_{j,p}\Phi_p) x(k) \\ & + B_i\Big(d(x(k), k) - \varpi ||\bar{x}(k)|| \text{sign}(\bar{s}(k))\Big)\Big]^T G^T \mathcal{Q}_p G \Big[(A_i + B_i K_{j,p}\Phi_p) x(k) \\ & + B_i\Big(d(x(k), k) - \varpi ||\bar{x}(k)|| \text{sign}(\bar{s}(k))\Big)\Big] \\ & \leq \sum_{i=1}^{r}\sum_{j=1}^{r} \varsigma_j(k) m_i(x(k)) m_j(x(k)) \Big\{(1+\tau_2) x^T(k)[(A_i + B_i K_{j,p}\Phi_p)^T G^T \mathcal{Q}_p \\ & \times G(A_i + B_i K_{j,p}\Phi_p)] x(k) + 2(1+\tau_2^{-1})\varpi^2\xi_i\big(||x(k)||^2 + m||\bar{x}(k)||^2\big)\Big\}. \end{aligned} \tag{8.35}$$

Combining (8.34), (8.35) and (8.19), one has

$$\begin{aligned}
&\mathbf{E}\{\Delta U(x(k), p) \mid x(k), p\} \\
&\leq \sum_{i=1}^{r}\sum_{j=1}^{r} \varsigma_j(k) m_i(x(k)) m_j(x(k)) x^T(k) \hat{\Lambda}_{ijp} x(k) \\
&\quad - \left[s^T(k) Q_p s(k) - 2(1+\tau_2^{-1}) \varpi^2 \xi_i \left(||x(k)||^2 + m||\bar{x}(k)||^2 \right) \right].
\end{aligned} \tag{8.36}$$

Along the similar lines in (8.19)–(8.21), we can obtain

$$\begin{aligned}
&\mathbf{E}\{\Delta U(x(k), p) | x(k), p\} \\
&\leq \sum_{i=1}^{r} \varsigma_i(k) m_i^2(x(k)) x^T(k) \hat{\Lambda}_{iip} x(k) \\
&\quad + \sum_{i=1}^{r-1}\sum_{j>i} \varsigma_j(k) m_i(x(k)) m_j(x(k)) x^T(k) \left\{ \sum_{\iota=1}^{2} \nu_\iota \left[\hat{\Lambda}_{ijp} + \phi_\iota \hat{\Lambda}_{jip} \right] \right\} x(k) \\
&\quad - \left[s^T(k) Q_p s(k) - 2(1+\tau_2^{-1}) \varpi^2 \xi_i \left(||x(k)||^2 + m||\bar{x}(k)||^2 \right) \right].
\end{aligned} \tag{8.37}$$

Thus, if the state trajectories are outside the sliding domain, i.e.,

$$||s(k)|| > \sqrt{\frac{\mathcal{N}}{\lambda_{\min}(Q_i)}},$$

one has from the conditions (8.28)–(8.30)

$$\mathbf{E}\{\Delta U(x(k), p) | x(k), p\} < 0. \tag{8.38}$$

Taking the expectation for the both sides of the expression (8.38), we have $\mathbf{E}\{\Delta U(x(k), i)\} < 0$, i.e., $\mathbf{E}\{U(x(k+1), \rho(k+1))\} < \mathbf{E}\{U(x(k), \rho(k))\}$, which means that the Lyapunov function $V_2(k)$ is monotonically decreasing outside the domain $\mathcal{D}$ in mean-square sense. So, the state trajectories of closed-loop system (8.8) can be driven (in mean square) into the sliding domain $\mathcal{D}$ around the specified sliding surface $s(k) = 0$. □

8.5 Co-design Scheme and Performance Optimization

In this section, we will establish the co-design scheme of the SS protocol and fuzzy controller, and derive the sufficient conditions on the TPM Π and the scheduling signal-dependent control gains $K_{j,\rho}$. Meanwhile, an optimization solving algorithm is provided to search sliding gain matrix and token-dependent parameters for improving system performance.

8.5.1 Membership-Function-Dependent Sufficient Conditions

In order to guarantee the stochastic stability of the closed-loop system (8.8) and the reachability of the specified sliding surface $s(k) = 0$, the sufficient conditions in Theorems 8.1 and 8.2 should be guaranteed simultaneously, the following Theorem 8.3 is given.

Theorem 8.3 *Consider the IT2 T-S fuzzy system (8.2) with given positive scalars* τ_1, τ_2, ϕ_1, *and* ϕ_2, *if there exist positive scalars* μ, λ, ζ_i, ξ_i, γ_p, χ_p, ϵ_p, ε_p, $\pi_{pq} \in [0, 1]$ *with* $\sum_{q=1}^{n} \pi_{pq} = 1$, *matrices* $P_p > 0$, $Q_p > 0$, $X_p > 0$, $Y_p > 0$, G, *and* $K_{j,p}$ *satisfying the following conditions (for* $p, q \in \mathbb{M}, i, j = 1, \ldots, r$*)*

$$\begin{bmatrix} \Gamma_{11} & * \\ \Gamma_{21} & -\gamma_p \acute{X} \end{bmatrix} \leq 0, \tag{8.39}$$

$$\begin{bmatrix} \Omega_{11} & * \\ \Omega_{21} & -\chi_p \acute{Y} \end{bmatrix} \leq 0, \tag{8.40}$$

$$\tilde{\Lambda}_{ijp} < 0, \quad (i = j), \tag{8.41}$$

$$\tilde{\Lambda}_{ijp} + \phi_\iota \tilde{\Lambda}_{jip} < 0, \quad (i < j, \ \ \iota = 1, 2), \tag{8.42}$$

where

$$\tilde{\Lambda}_{ijp} = \begin{bmatrix} \tilde{\Lambda}_{ijp}^{11} & * & * & * & * & * & * \\ \tilde{\Lambda}_{ijp}^{21} & -\bar{X} & * & * & * & * & * \\ \tilde{\Lambda}_{ijp}^{31} & 0 & -\epsilon_p \bar{X} & * & * & * & * \\ \tilde{\Lambda}_{ijp}^{41} & 0 & 0 & -\bar{Y} & * & * & * \\ \tilde{\Lambda}_{ijp}^{51} & 0 & 0 & 0 & -\varepsilon_p \bar{Y} & * & * \\ \tilde{\Lambda}_{ijp}^{61} & 0 & 0 & 0 & 0 & -\bar{P} & * \\ \tilde{\Lambda}_{ijp}^{71} & 0 & 0 & 0 & 0 & 0 & -\bar{Q} \end{bmatrix},$$

$$\Gamma_{11} = -\zeta_i I - \gamma_p^{-1} \sum_{q=1}^{n} B_i^T P_q B_i,$$

$$\Gamma_{21} = \mathbf{1}_n \otimes B_i + \frac{\gamma_p}{2} [\pi_{p1}, \pi_{p2}, \ldots, \pi_{pn}]^T \otimes B_i,$$

$$\Omega_{11} = -\xi_i I - \chi_p^{-1} \sum_{q=1}^{n} (G B_i)^T Q_q G B_i,$$

$$
\begin{aligned}
\Omega_{21} &= \mathbf{1}_n \otimes GB_i + \frac{\chi_p}{2}[\pi_{p1}, \pi_{p2}, \ldots, \pi_{pn}]^T \otimes GB_i, \\
\bar{P} &= \frac{1}{1+\tau_1}\mathrm{diag}\{P_1, P_2, \ldots, P_n\}, \ \ \bar{Q} = \frac{1}{1+\tau_2}\mathrm{diag}\{Q_1, Q_2, \ldots, Q_n\}, \\
\bar{X} &= \frac{1}{1+\tau_1}\acute{X}, \ \acute{X} = \mathrm{diag}\{X_1, X_2, \ldots, X_n\}, \\
\bar{Y} &= \frac{1}{1+\tau_2}\acute{Y}, \ \acute{Y} = \mathrm{diag}\{Y_1, Y_2, \ldots, Y_n\}, \\
\tilde{\Lambda}_{ijp}^{11} &= -P_p - 2(1+\tau_1)\mu A_i^T A_i - 2(1+\tau_2)\lambda(GA_i)^T(GA_i) \\
&\quad - (1+\tau_1)\epsilon_p^{-1}\sum_{q=1}^{n} A_i^T P_q A_i \\
&\quad - (1+\tau_2)\varepsilon_p^{-1}\sum_{q=1}^{n}(GA_i)^T Q_q(QA_i) + 2(1+\tau_1^{-1})\varpi^2\zeta_i(I + m\Phi_p^T\Phi_p), \\
\tilde{\Lambda}_{ijp}^{21} &= \mathbf{1}_n \otimes B_i K_{j,p}\Phi_p + [\pi_{p1}, \pi_{p2}, \ldots, \pi_{pn}]^T \otimes A_i, \\
\tilde{\Lambda}_{ijp}^{31} &= \mathbf{1}_n \otimes A_i + \frac{\epsilon_p}{2}[\pi_{p1}, \pi_{p2}, \ldots, \pi_{pn}]^T \otimes A_i, \\
\tilde{\Lambda}_{ijp}^{41} &= \mathbf{1}_n \otimes GB_i K_{j,p}\Phi_p + [\pi_{p1}, \pi_{p2}, \ldots, \pi_{pn}]^T \otimes GA_i, \\
\tilde{\Lambda}_{ijp}^{51} &= \mathbf{1}_n \otimes GA_i + \frac{\varepsilon_i}{2}[\pi_{p1}, \pi_{p2}, \ldots, \pi_{pn}]^T \otimes GA_i, \\
\tilde{\Lambda}_{ijp}^{61} &= \mu\mathbf{1}_n \otimes A_i, \ \ \tilde{\Lambda}_{ijp}^{71} = \lambda\mathbf{1}_n \otimes GA_i,
\end{aligned}
$$

with $X_p = P_p^{-1}$ and $Y_p = Q_p^{-1}$, then the stochastic stability of the closed-loop system (8.8) and the reachability of the sliding domain can be guaranteed simultaneously.

Proof Due to $(1+\tau_2)(A_i + B_i K_{j,p}\Phi_p)^T G^T Q_p G(A_i + B_i K_{j,p}\Phi_i) > 0$, the conditions (8.14)–(8.15) in Theorem 8.1 can be ensured by the conditions (8.29)–(8.30) in Theorem 8.2.

Thus, we just need to prove that the conditions (8.39)–(8.42) can guarantee the inequalities (8.13), (8.28) and (8.29)–(8.30).

It follows from the condition (8.13) that

$$
\begin{aligned}
& B_i^T\left(\sum_{q=1}^{n}\pi_{pq}P_q\right)B_i \\
&\leq \sum_{q=1}^{n}\left(\frac{1}{2}\pi_{pq}B_i\right)^T P_q B_i + \sum_{q=1}^{n}B_i^T P_q\left(\frac{1}{2}\pi_{pq}B_i^T\right) + \gamma_p\sum_{q=1}^{n}\left(\frac{1}{2}\pi_{pq}B_i\right)^T P_q\left(\frac{1}{2}\pi_{pq}B_l\right) \\
&= \gamma_p^{-1}\Big[\sum_{q=1}^{n}\left(B_i + \frac{\gamma_p}{2}\pi_{pq}B_i\right)^T P_q\left(B_i + \frac{\gamma_p}{2}\pi_{pq}B_i\right) - \sum_{q=1}^{n}B_i^T P_q B_i\Big],
\end{aligned} \tag{8.43}
$$

where $\gamma_p > 0$ is an adjustable scalar. Thus, it can be easily obtained from the inequality (8.43) that the condition (8.39) can guarantee the condition (8.13) in Theorem

8.1. Along the similar lines as (8.43), the condition (8.28) can be ensured by the condition (8.40) with a positive scalar χ_p.

For the inequalities (8.29)–(8.30), we can obtain the following expressions by using "completing the square"

$$
\begin{aligned}
&(A_i + B_i K_{j,p}\Phi_p)^T \left(\sum_{q=1}^{n} \pi_{pq} P_q\right) (A_i + B_i K_{j,p}\Phi_p) \\
&= A_i^T \left(\sum_{q=1}^{n} \pi_{pq} P_q\right) A_i + A_i^T \left(\sum_{q=1}^{n} \pi_{pq} P_q\right) B_i K_{j,p}\Phi_p \\
&+ (B_i K_{j,p}\Phi_p)^T \left(\sum_{q=1}^{n} \pi_{pq} P_q\right) A_i + (B_i K_{j,p}\Phi_p)^T \left(\sum_{q=1}^{n} \pi_{pq} P_q\right) (B_i K_{j,p}\Phi_p) \\
&\le \sum_{q=1}^{n} \left(\frac{1}{2}\pi_{pq} A_i\right)^T P_q A_i + \sum_{q=1}^{n} A_i^T P_q \left(\frac{1}{2}\pi_{pq} A_i\right) \\
&+ \sum_{q=1}^{n} (\pi_{pq} A_i)^T P_q (B_i K_{j,p}\Phi_p) + \sum_{q=1}^{n} (B_i K_{j,p}\Phi_p)^T P_q (\pi_{pq} A_i) \\
&+ \sum_{q=1}^{n} (B_i K_{j,p}\Phi_p)^T P_q (B_i K_{j,p}\Phi_p) + \epsilon_p \sum_{q=1}^{n} \left(\frac{1}{2}\pi_{pq} A_i\right)^T P_q \left(\frac{1}{2}\pi_{pq} A_i\right)
\end{aligned}
\tag{8.44}
$$

and

$$
\begin{aligned}
&(GA_i + GB_i K_{j,p}\Phi_p)^T \left(\sum_{q=1}^{n} \pi_{pq} Q_q\right) (GA_i + GB_i K_{j,p}\Phi_p) \\
&= (GA_i)^T \left(\sum_{q=1}^{n} \pi_{pq} Q_q\right) GA_i + (GA_i)^T \left(\sum_{q=1}^{n} \pi_{pq} Q_q\right) GB_i K_{j,p}\Phi_i \\
&+ (GB_i K_{j,p}\Phi_p)^T \left(\sum_{q=1}^{n} \pi_{pq} Q_q\right) GA_i + (GB_i K_{j,p}\Phi_p)^T \left(\sum_{q=1}^{n} \pi_{pq} Q_q\right) (GB_i K_{j,p}\Phi_p) \\
&\le \sum_{q=1}^{n} \left(\frac{1}{2}\pi_{pq} GA_i\right)^T Q_q GA_i + \sum_{q=1}^{n} (GA_i)^T Q_q \left(\frac{1}{2}\pi_{pq} GA_i\right) \\
&+ \sum_{q=1}^{n} (\pi_{pq} GA_i)^T Q_q (GB_i K_{j,p}\Phi_p) + \sum_{q=1}^{n} (GB_i K_{j,p}\Phi_p)^T Q_q (\pi_{pq} GA_i) \\
&+ \sum_{q=1}^{n} (GB_i K_{j,p}\Phi_p)^T Q_q (GB_i K_{j,p}\Phi_p) + \varepsilon_p \sum_{q=1}^{n} \left(\frac{1}{2}\pi_{pq} GA_i\right)^T Q_q \left(\frac{1}{2}\pi_{pq} GA_i\right)
\end{aligned}
\tag{8.45}
$$

with $\epsilon_p > 0$ and $\varepsilon_p > 0$.

In the light of (8.44)–(8.45), the conditions (8.29)–(8.30) can be guaranteed by the following inequalities

$$\bar{\Lambda}_{iip} < 0, (i = j), \tag{8.46}$$

$$\bar{\Lambda}_{ijp} + \phi_\iota \bar{\Lambda}_{jip} < 0, \quad (i < j, \ \iota = 1, 2), \tag{8.47}$$

with

$$\bar{\Lambda}_{ijp} = \begin{bmatrix} \bar{\Lambda}_{ijp}^{11} & * & * & * & * \\ \tilde{\Lambda}_{ijp}^{21} & -\bar{X} & * & * & * \\ \tilde{\Lambda}_{ijp}^{31} & 0 & -\epsilon_p \bar{X} & * & * \\ \tilde{\Lambda}_{ijp}^{41} & 0 & 0 & -\bar{Y} & * \\ \tilde{\Lambda}_{ijp}^{51} & 0 & 0 & 0 & -\varepsilon_p \bar{Y} \end{bmatrix},$$

and

$$\begin{aligned} \bar{\Lambda}_{ijp}^{11} = & -P_p - (1+\tau_1)\sum_{q=1}^{n}(\pi_{pq}A_i)^T P_q(\pi_{pq}A_i) - (1+\tau_1)\epsilon_p^{-1}\sum_{q=1}^{n} A_i^T P_q A_i \\ & - (1+\tau_2)\sum_{q=1}^{n}(\pi_{pq}GA_i)^T Q_q(\pi_{pq}GA_i) - (1+\tau_2)\varepsilon_p^{-1}\sum_{q=1}^{n}(GA_i)^T Q_q(QA_i) \\ & + 2(1+\tau_1^{-1})\varpi^2\zeta_i(I + m\Phi_p^T\Phi_p). \end{aligned}$$

For the terms $-\sum_{q=1}^{n}(\pi_{pq}A_i)^T P_q(\pi_{pq}A_i)$ and $-\sum_{q=1}^{n}(\pi_{pq}GA_i)^T Q_q(\pi_{pq}GA_i)$, we can further obtain

$$-\sum_{q=1}^{n}(\pi_{pq}A_i)^T P_q \pi_{pq} A_i \leq \mu^2 \sum_{q=1}^{n} A_i^T P_q^{-1} A_i - 2\mu A_i^T A_i, \tag{8.48}$$

$$-\sum_{q=1}^{n}(\pi_{pq}GA_i)^T Q_q(\pi_{pq}GA_i) \leq \lambda^2 \sum_{q=1}^{n}(GA_i)^T Q_q^{-1} GA_i - 2\lambda(GA_i)^T GA_i, \tag{8.49}$$

where μ and λ are two positive scalars.

In view of (8.48)–(8.49), it can be obtained that the conditions (8.46)–(8.47) can be ensured by the conditions (8.41)–(8.42). Thus, the conditions (8.29)–(8.30) hold if the inequalities (8.41)–(8.42) are feasible. □

Remark 8.2 It is worth mentioning that the conditions in Theorem 8.3 can ensure that the system state trajectories are only driven into the domain $\mathcal{D}$ around the specified sliding surface $s(k) = 0$, rather than staying on the sliding surface, which is usually termed as the quasi-sliding mode (QSM). This is just the characteristic of the discrete-time SMC. Moreover, a smaller convergence domain implies a better control performance. To achieve this, we will propose an optimization algorithm within the co-design framework to minimize the sliding domain, thereby enhancing control performance.

8.5.2 PSO-Based Solving Algorithm

It should be pointed out that some coupling terms (e.g., $\gamma_p^{-1} \sum_{q=1}^{n} B_i^T P_q B_i$), are dependent on token-dependent parameters $\{\gamma_p, \chi_p, \epsilon_p, \varepsilon_p\}$ and matrix G such that it is difficult to solve the conditions (8.39)–(8.42) in Theorem 8.3. Fortunately, if these token-dependent parameters and sliding matrix could be chosen *a priori*, these coupling terms would be reduced into the linear ones. Notice that the choice of these parameters will affect the solution of the conditions (8.39)–(8.42), and will further have an impact on the sliding domain $\mathcal{D}$. Therefore, we shall propose an iterative optimization algorithm to search the desirable token-dependent parameters $\{\gamma_p, \chi_p, \epsilon_p, \varepsilon_p\}$ and matrix G via reducing the sliding domain. To this end, the following optimization problem is formulated as:

$$\begin{aligned} \min_{\gamma_p,\chi_p,\epsilon_p,\varepsilon_p,G} \quad & F_{obj} \\ \text{s.t. conditions}(8.38) -& -(8.41) \end{aligned} \tag{8.50}$$

with the objective function $F_{obj} \triangleq \sqrt{2(1+\tau_2^{-1})\varpi^2 \max_i\{\xi_i\}/\min_p \lambda_{\min}(Q_p)}$.

Due to the distinctive ability of handling the optimization problems with the non-linear constraints, some remarkable intelligent optimization algorithms, e.g., genetic algorithm, and PSO algorithm, have been widely used in many practical applications. In this chapter, the **PSO-based solving algorithm** 8.1 is proposed to solve the above optimization problem (8.50):

8.5.3 Co-design Scheme Under Structure Constraints of Protocol

In practical application, since the TPM of the SS protocol determines the transmission of the communication network, it may inevitably be subject to some constraints. For instance, the Gossip Markov chain shown in Fig. 8.2 may be integrated on the network-on-chip for allocating the data packet flow rates over the network communication [1], whose TPM obeys the following structure:

$$\Pi = \begin{bmatrix} p & q & 1-(p+q) & 0 \\ 1-(p+q) & p & 0 & q \\ q & 0 & p & 1-(p+q) \\ 0 & 1-(p+q) & q & p \end{bmatrix} \tag{8.52}$$

with $p \in [0, 1]$ and $q \in [0, 1]$. It can be found that the structure of the TPM (8.52) is highly constrained, in which only two parameters p and q can be adjusted flexibly and some elements are fixed a priori.

Algorithm 8.1 PSO-based solving algorithm

Particle Encoding and Initialization

1: The set of token-dependent parameters $\{\gamma_p, \chi_p, \epsilon_p, \varepsilon_p\}$, and matrix $G \in \mathbb{R}^{m\times n}$ include $4n$ and $m \times n$ independent variables, respectively. Those variables can be encoded as a particle as follows:

$$w \triangleq [\gamma_1, \cdots, \gamma_n, \chi_1, \cdots, \chi_n, \epsilon_1, \cdots, \epsilon_n, \varepsilon_1, \cdots, \varepsilon_n,\\ g_{11}, \cdots, g_{1n}, g_{21}, \cdots, g_{mn}] \in \mathbb{R}^{1\times\bar{n}},$$

with $\bar{n} = (m+4)n$.

2: In the initial step $l = 0$, a swarm with N particles is generated as $W^l = \{w_1^l, \cdots, w_N^l\}$. For the t-th particle w_t^l, the initial position $w_t^l = [w_{1t}^l, \cdots, w_{\bar{n}t}^l]$ and its velocity $v_t^l = [v_{1t}^l, \cdots, v_{\bar{n}t}^l]$ are developed over the appointed search space $S = \{w_{ht}^l, v_{ht}^l | w_{ht}^l \in [\underline{w}_h, \bar{w}_h], v_{ht}^l \in [\underline{v}_h, \bar{v}_h], h = 1, \cdots, \bar{n}\}$.

Evaluate Fitness Function

3: Solve the conditions (8.39)–(8.42), and calculate the fitness function F_{obj} of each particle.

4: For each particle, compare the current fitness value with the best value. Obtain the individual best value $pBest_t^l$ of the t-th particle and the global best value $gBest^l$ of the entire swarm. Enter the next step and update the positions and velocities for all particles.

Update Position and Velocity

5: The position w_t and velocity v_t of t-th particle will be updated according to the following rules:

$$v_t^{l+1} = c_0 v_t^l + c_1 rand_1(pbest_{t'}^l - w_t^l) + c_2 rand_2(gbest^l - w_t^l), \quad (8.50)$$

$$w_t^{l+1} = w_t^l + v_t^l, \quad (8.51)$$

where c_0 denotes inertia weight factor, c_1 and c_2 represent individual and social learning factors, respectively. $rand_1$ and $rand_2 \in [0, 1]$ are two random numbers. For the previous l iterations, the best position among the entire particle swarm is denoted as $gbest^l$. $pbest_{t'}^l$ is the individual best fitness value of the t'-th particle, which is determined by the rule $M_t = a_1 + a_2\sqrt{R_t^{-1}}$, where R_t denotes the order of the t-th particle according to the fitness value from small to large, a_1 and $a_2 \in [0, 1]$ are two learning parameters. If a random number $r > M_t$, the t-th particle will learn from itself, i.e., $pbest_{t'} = pbest_t$. Otherwise, the t-th particle learns from its neighbor's best individual information around N_p particles, i.e., $t' = Mod(t + rands * N_p, N)$, where random number $rands \in [-1, 1]$.

Terminal Condition:

6: If the terminal condition is met, i.e., the number of iteration l reaches the maximum iterations $T_{\max}$, stop and return optimal solutions. Otherwise, update position and velocity again.

7: **return** K_{jp}, G, Ω, τ_j, and ρ.

In addition, since the transition probability can reflect the allocation for the data flow in the communication network, it could be constrained within a specified scope according to the practical requirements. For example, the transition probability π_{ij} is subject to the following constraint

$$\underline{\pi}_{ij} \leq \pi_{ij} \leq \overline{\pi}_{ij}, \quad (8.53)$$

where the conditional probabilities $\underline{\pi}_{ij}, \overline{\pi}_{ij} \in [0, 1]$ are given scalars.

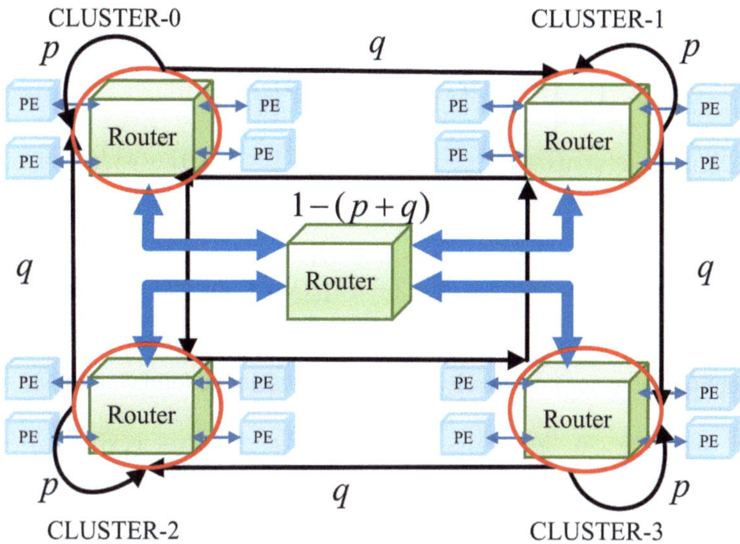

Fig. 8.2 The stochastic communication under Gossip Markov chain [1]

Apparently, the constraint condition (8.52) or (8.53) can be directly integrated into the conditions in Theorem 8.3, that is, the proposed PSO-based optimization solving algorithm can be utilized to solve the conditions (8.39)–(8.42) with constraint (8.52) or (8.53). In the following section, the Example 8.2 is provided to verify the co-design scheme with the constraints (8.52) and (8.53).

8.6 Simulation Example

In this section, two examples are provided to illustrate the proposed co-design scheme of fuzzy sliding mode controller and the SS protocol.

Example 8.1 Consider the tunnel diode circuit system [6] as follows:

$$\begin{cases} C\dot{V}_C(t) = -\frac{1}{R_L}V_C(t) - \frac{1}{R_D(t)}V_C(t) + i_L(t), \\ L\dot{i}_L(t) = -V_C(t) - R_E i_L(t) + V_s(t) + u(t), \end{cases} \tag{8.54}$$

where C is the capacitance, L is the inductance, $R_D(t) = \frac{V_D(t)}{i_D(t)} = \frac{1}{0.02+\delta V_C^2(t)}$ with the uncertain parameter $\delta \in [\delta_1, \delta_2]$ and $\delta_1 = 0.001,\ \delta_2 = 0.003$, and $V_s(t) = 1.5V_C(t)$.

Define $x_1(t) \triangleq V_C(t)$, $x_2(t) \triangleq i_L(t)$, and $x(t) \triangleq [x_1(t), x_2(t)]^T$. The nonlinear system (8.56) can be further represented as the following system model:

$$\dot{x}(t) = \check{A}x(t) + \check{B}u(t) \tag{8.55}$$

with

$$\check{A} = \begin{bmatrix} g(x_1(t)) & \frac{1}{C} \\ \frac{0.5}{L} & -\frac{R_E}{L} \end{bmatrix}, \quad \check{B} = \begin{bmatrix} 0 \\ 1 \end{bmatrix}$$

$$g(x_1(t)) = -\frac{1}{CR_L} - \frac{1}{CR_D(t)} = -\frac{1}{CR_L} - \frac{0.02 + \delta x_1^2(t)}{C}.$$

Let the corresponding parameters as $L = 1H, C = 0.5F, R_E = 0.1\Omega$, and $R_L = 20\Omega$. Suppose the system state $x_1(t) \in [-2, 2]$. Denote the maximum and minimum values of the nonlinear function $g(x_1(t))$ as $g_{\max}$ and $g_{\min}$, respectively. Choose $\hat{g}_{\min} = g_{\min} - 0.02$ and $\hat{g}_{\max} = g_{\max} + 0.02$. Thus, one has $g(x_1(t)) \in [\hat{g}_{\min}, \hat{g}_{\max}]$. By adopting the sector nonlinearity method, we obtain the following MFs:

$$m_1(x_1(t)) = \frac{g(x_1(t)) - \hat{g}_{\min}}{\hat{g}_{\max} - \hat{g}_{\min}}, \quad m_2(x_1(t)) = \frac{\hat{g}_{\max} - g(x_1(t))}{\hat{g}_{\max} - \hat{g}_{\min}}.$$

However, the uncertain parameter δ leads to the above MFs $m_1(x_1(t))$ and $m_2(x_1(t))$ to be uncertain. Based on the range of the uncertain parameter $\delta \in [\delta_1, \delta_2]$, we can further derive the upper and lower MFs as:

$$\overline{m}_1(x_1(t)) = \frac{g(x_1(t))|_{\delta=\delta_1} - \hat{g}_{\min}}{\hat{g}_{\max} - \hat{g}_{\min}}, \quad \underline{m}_1(x_1(t)) = \frac{g(x_1(t))|_{\delta=\delta_2} - \hat{g}_{\min}}{\hat{g}_{\max} - \hat{g}_{\min}},$$

and $\overline{m}_2(x_1(t)) = 1 - \underline{m}_1(x_1(t)), \underline{m}_2(x_1(t)) = 1 - \overline{m}_1(x_1(t))$.

As stated in Remark 8.1, the MFs of the fuzzy model should be greater than zero. To this end, we enlarge the modeling range of nonlinearity $g(x_1(t))$ from $[g_{\min}, g_{\max}]$ to $[\hat{g}_{\min}, \hat{g}_{\max}]$ so that the lower MFs $\overline{m}_1(x_1(t))$ and $\underline{m}_2(x_1(t))$ are greater than zero over the entire practical operation domain $x_1(t) \in [-2, 2]$.

Then, the nonlinear system (8.55) can be modeled as the following 2-rule IT2 T-S fuzzy system

$$\dot{x}(t) = \sum_{i=1}^{2} m_i(x_i(t))[\check{A}_i x(t) + \check{B}_i u(t)] \tag{8.56}$$

with

$$\check{A}_1 = \begin{bmatrix} \hat{g}_{\max} & \frac{1}{C} \\ \frac{0.5}{L} & -\frac{R_E}{L} \end{bmatrix}, \check{A}_2 = \begin{bmatrix} \hat{g}_{\min} & \frac{1}{C} \\ \frac{0.5}{L} & -\frac{R_E}{L} \end{bmatrix}, \check{B}_1 = \check{B}_2 = \check{B}.$$

By discretizing the above continuous-time IT2 T-S fuzzy system (8.56) with sampling period $T = 0.2$ and taking the external disturbance $d(x(k), k) = \varpi\sqrt{2|x_1(k)x_2(k)|}$ ($\varpi = 0.1$) into consideration, we obtain the following discrete-time IT2 T-S fuzzy system as

$$x(k+1) = \sum_{i=1}^{2} m_i(x_i(k))[A_i x(k) + B_i(u(k) + d(x(k), k))]$$

with the following parameters

$$A_1 = \begin{bmatrix} 0.9833 & 0.3914 \\ 0.0979 & 0.9998 \end{bmatrix}, \quad A_2 = \begin{bmatrix} 0.9959 & 0.3939 \\ 0.0985 & 0.9998 \end{bmatrix},$$
$$B_1 = \begin{bmatrix} 0.0394 \\ 0.1993 \end{bmatrix}, \quad B_2 = \begin{bmatrix} 0.0395 \\ 0.1993 \end{bmatrix}.$$

Notice that only the upper and lower MFs $\overline{m}_i(x_i(k))$ and $\underline{m}_i(x_i(k))$ are known, and the actual MFs $m_i(x_i(k))$ are unavailable for the controller. In order to remove the obstacle, by using the received state signal $\bar{x}(k)$, the MFs in IT2 T-S fuzzy controller (8.7) are constructed as $h_1(\bar{x}(k)) = 0.6\underline{m}_1(\bar{x}(k)) + 0.4\overline{m}_1(\bar{x}(k))$ and $h_2(\bar{x}(k)) = 1 - h_1(\bar{x}(k))$. Meanwhile, parameters $\eta_1 = \eta_2 = 0.08$ are chosen to satisfying the expression (8.9), we can obtain the parameters $\phi_1 = 0.5924$ and $\phi_2 = 1.6881$ for solving the MFD conditions.

In the PSO-based solving algorithm, the particle population and maximum iteration number are set as $N = 60$ and $T_{\max} = 50$, respectively. The searching range of token-dependent parameters $\{\gamma_p, \chi_p, \epsilon_p, \varepsilon_p\}$ is $[0.001, 2.000]$, and $g_{11}, \ g_{12} \in [-1, 1]$. Other parameters are chosen as: $c_0 = 0.8$, $c_1 = c_2 = 1.5$, $a_1 = 0.1$, $a_2 = 0.5$ and $N_p = 2$. By performing the proposed PSO-based solving algorithm, we can obtain the token-dependent parameters $\gamma_1 = 0.1473, \ \gamma_2 = 0.2210, \chi_1 = 1.1372, \ \chi_2 = 0.8279, \ \epsilon_1 = 0.5144, \ \epsilon_2 = 1.4203, \ \varepsilon_1 = 1.54416, \ \varepsilon_2 = 1.3542$, and the TPM Π, the sliding matrix G, and the control gains $K_{j,p}$ as follows:

$$\Pi = \begin{bmatrix} 0.6613 & 0.3387 \\ 0.6505 & 0.3495 \end{bmatrix}, \quad G = \begin{bmatrix} -0.3464 & -0.6765 \end{bmatrix},$$
$$K_{11} = \begin{bmatrix} -0.8661 & 0 \end{bmatrix}, \quad K_{21} = \begin{bmatrix} -0.8697 & 0 \end{bmatrix},$$
$$K_{12} = \begin{bmatrix} 0 & -3.1015 \end{bmatrix}, \quad K_{22} = \begin{bmatrix} 0 & -3.3798 \end{bmatrix}.$$

Moreover, the fitness value during the optimization solving process is shown in Fig. 8.3.

In the simulation, suppose the weighting coefficients $\overline{\alpha}_1(x(k)) = \sin(x_1(k))$ and $\underline{\alpha}_1(x(k)) = 1 - \overline{\alpha}_1(x(k))$ of upper and lower MFs in the IT2 T-S fuzzy system (8.2). Then, one can obtain the system's MFs as $m_1(x(k))$ and $m_2(x(k)) = 1 - m_1(x(k))$. Under the initial conditions $x(0) = [-1.2, \ 0.5]^T$, the simulation results are shown in Fig. 8.4a–d. The selected sensor node under the SS protocol is depicted in Fig. 8.4d, in which for $\rho(k) = 1$, the state $x_1(k)$ is transmitted at the current instant, and the state $x_2(k)$ is transmitted for $\rho(k) = 2$. It can be seen from Fig. 8.4d that $x_1(k)$ has more transmission times than $x_2(k)$. Although only one sensor is permitted to transmit its signal at each instant, it is shown in Fig. 8.4a that the stability of the tunnel diode circuit system can still be guaranteed. In addition, the control signal $u(k)$ and

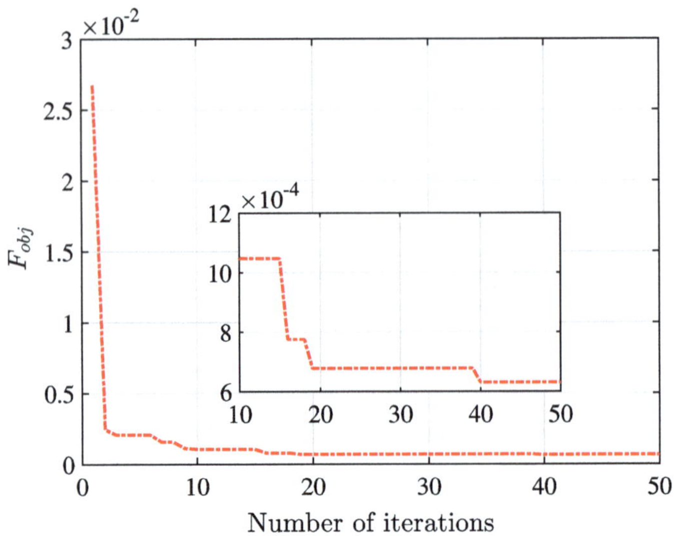

Fig. 8.3 Iteration of fitness function

sliding variable $s(k)$ are shown in Fig. 8.4b–c, from which it can be observed that the reachability of the sliding surface $s(k) = 0$ can be ensured under the proposed SMC scheme. The above simulation results verifies the effectiveness of the proposed co-design strategy of the fuzzy sliding mode controller and the SS protocol.

Example 8.2 In this example, suppose that the TPM of the SS protocol has the structure (8.52) with $0.2 \leq p \leq 0.8$. Consider a 4-order system with the following parameters:

$$A_1 = \begin{bmatrix} 1.0 & 0.4 & 0 & 0 \\ 0 & 0.3 & 0 & 0 \\ 0.1 & 1.0 & 0.6 & 1.0 \\ 0.1 & 1.0 & 0 & 1.0 \end{bmatrix}, \quad B_1 = \begin{bmatrix} 1.0 \\ 0.6 \\ 1.0 \\ 0.4 \end{bmatrix},$$

$$A_2 = \begin{bmatrix} 1.2 & 0.5 & 0 & 0 \\ 0 & 0.2 & 0 & 0 \\ 0 & 0.5 & 0.2 & 1.0 \\ 0.1 & 0.8 & 0 & 1.0 \end{bmatrix}, B_2 = \begin{bmatrix} 0.8 \\ 0.7 \\ 1.0 \\ 0.5 \end{bmatrix},$$

$$d(x(k), k) = \varpi\sqrt{|2x_1(k)x_4(k)| + x_2^2(k)}, \quad (\varpi = 0.1).$$

The upper and lower MFs of the fuzzy model are shown in Table 8.1.

The scheduling signal-dependent fuzzy SMC law (8.7) is designed by choosing the MFs $h_1(\bar{x}(k)) = 0.7\underline{m}_1(\bar{x}(k)) + 0.3\overline{m}_1(\bar{x}(k))$ and $h_2(\bar{x}(k)) = 1 - h_1(\bar{x}(k))$. Besides, in the expression (8.9), the parameters $\eta_1 = 0.5$ and $\eta_2 = 0.2$ are chosen to obtain $\phi_1 = 0.5048$ and $\phi_2 = 1.9809$ for solving the MFD conditions.

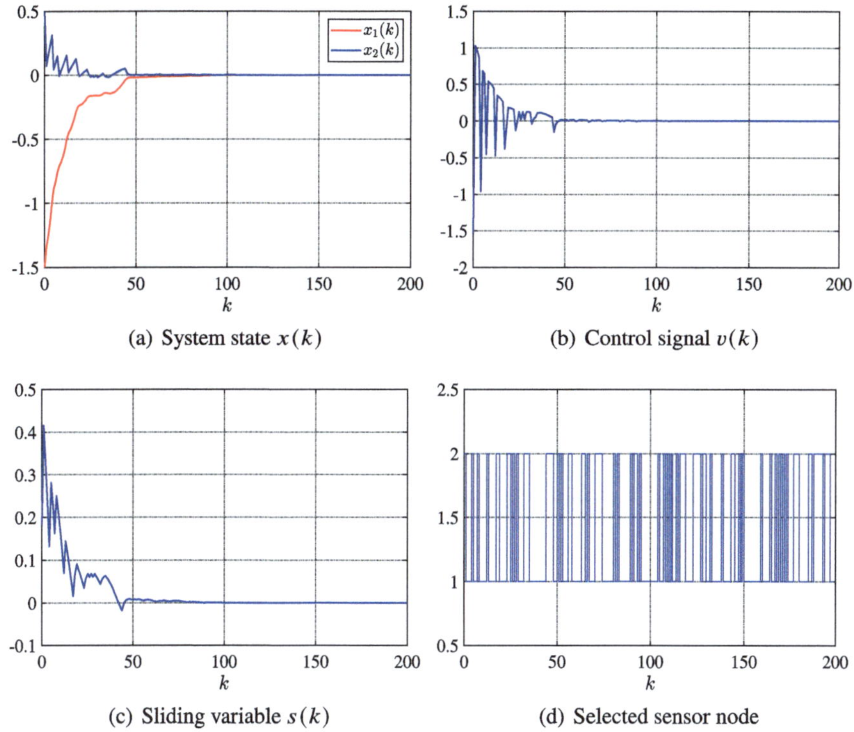

Fig. 8.4 Simulation results under the co-design scheme of controller and SS protocol

Table 8.1 The upper and lower MFs of fuzzy model

Upper MFs	Lower MFs
$\overline{m}_1(x(k)) = \frac{4-0.2x_1^2(k)}{5}$	$\underline{m}_1(x(k)) = \frac{4-0.6x_1^2(k)}{5}$
$\overline{m}_2(x(k)) = 1 - \underline{m}_1(x(k))$	$\underline{m}_2(x(k)) = 1 - \overline{m}_1(x(k))$

By performing the proposed POS-based solving algorithm, we can obtain the sliding matrix and control gains as

$$
\begin{aligned}
G &= \begin{bmatrix} 0.9379 \; 0.5189 \; 0.8687 \; 0.9856 \end{bmatrix}, \\
K_{11} &= \begin{bmatrix} -0.0301 \; 0 \; 0 \; 0 \end{bmatrix}, \quad K_{21} = \begin{bmatrix} -0.0320 \; 0 \; 0 \; 0 \end{bmatrix}, \\
K_{12} &= \begin{bmatrix} 0 \; -0.0055 \; 0 \; 0 \end{bmatrix}, \quad K_{22} = \begin{bmatrix} 0 \; -0.0096 \; 0 \; 0 \end{bmatrix}, \\
K_{13} &= \begin{bmatrix} 0 \; 0 \; -0.0355 \; 0 \end{bmatrix}, \quad K_{23} = \begin{bmatrix} 0 \; 0 \; -0.0109 \; 0 \end{bmatrix}, \\
K_{14} &= \begin{bmatrix} 0 \; 0 \; 0 \; -0.0172 \end{bmatrix}, \quad K_{24} = \begin{bmatrix} 0 \; 0 \; 0 \; -0.0290 \end{bmatrix}.
\end{aligned}
$$

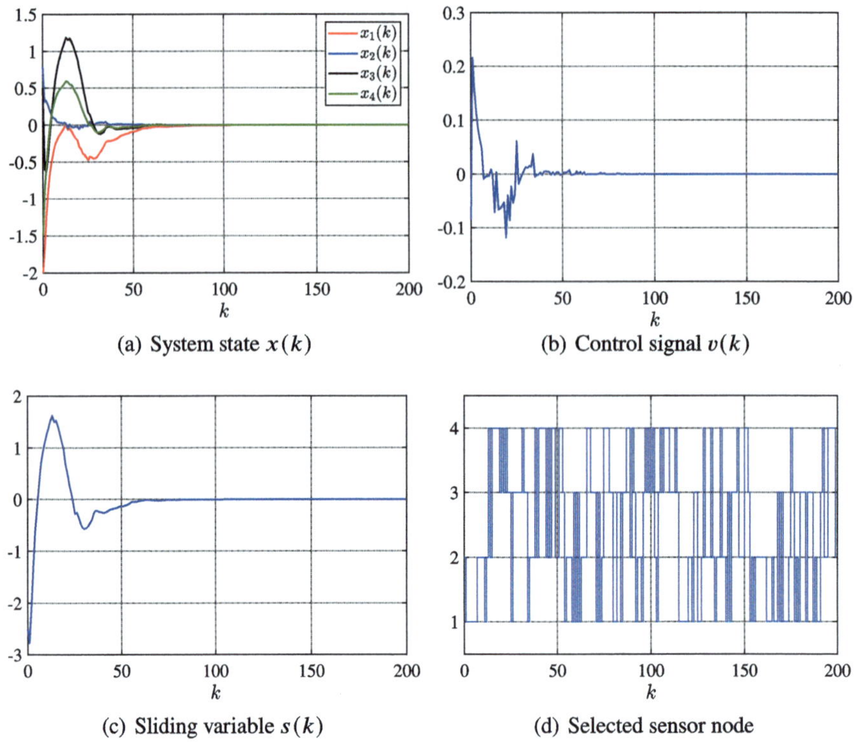

(a) System state $x(k)$ (b) Control signal $v(k)$

(c) Sliding variable $s(k)$ (d) Selected sensor node

Fig. 8.5 Simulation results under SS protocol with structure constraint

In the simulation, we assume that $\overline{\alpha}_1(x(k)) = \sin(x_1(k))$ and $\underline{\alpha}_1(x(k)) = 1 - \overline{\alpha}_1(x(k))$, thereby obtaining the system's MFs as $m_1(x(k))$ and $m_2(x(k)) = 1 - m_1(x(k))$. The simulation results under the initial conditions $x(0) = [-2, 0.8, 0.5, -1.5]^T$ are presented in Fig. 8.5a–d. As shown in Fig. 8.5d, for the selected sensor node governed by the Markov chain with constraints (8.52) and (8.53), the first sensor node (i.e., $x_1(k)$) and the fourth sensor node (i.e., $x_4(k)$) cannot be selected at two consecutive instants. This is because that the transition probability between them is zero. Likewise, the state $x_2(k)$ and $x_3(k)$ cannot also be transmitted at two consecutive instants. Besides, it can be seen from Fig. 8.5a that the stability of the closed-loop system can be guaranteed despite the structure limitation of the SS protocol.

8.7 Conclusion

This chapter has addressed the co-design issue of the SS protocol and the fuzzy sliding mode controller for IT2 T-S fuzzy system under the network communication. The transition probability matrix of the SS protocol is co-designed with the fuzzy sliding mode controller, and the SS protocol has been implemented to schedule sensor nodes. A scheduling signal-dependent fuzzy sliding mode controller has been designed by using the received state information, and the relation between the MFs in the fuzzy system and the controller has been established. Moreover, the PSO-based solving algorithm has been proposed for obtaining the optimal TPM of Markov chain and the fuzzy sliding mode controller.

References

1. Sehgal, V.K.: Markovian models based stochastic communication in networks-in-package. IEEE Trans. Parallel Distrib. Syst. **26**(10), 2806–2821 (2015)
2. Song, J., Niu, Y., Lam, J., Shu, Z.: A hybrid design approach for output feedback exponential stabilization of Markovian jump systems. IEEE Trans. Autom. Control **63**(5), 1404–1417 (2018)
3. Feng, J., Lam, J., Shu, Z.: Stabilization of Markovian systems via probability rate synthesis and output feedback. IEEE Trans. Autom. Control **55**(3), 773–777 (2010)
4. Lam, H.K., Leung, F.: Stability analysis of fuzzy control systems subject to uncertain grades of membership. IEEE Transactions On Systems, Man, and Cybernetics-Part B: Cybernetics **35**(6), 1322–1325 (2005)
5. Ghaoui, L.E., Oustry, F., AitRami, M.: A cone complementarity linearization algorithm for static output-feedback and related problems. IEEE Trans. Autom. Control **42**(8), 1171–1176 (1997)
6. Wang, Y., Shi, P., Yan, H.: Reliable control of fuzzy singularly perturbed systems and its application to electronic circuits. IEEE Trans. Circuits Syst. I Regul. Pap. **65**(10), 3519–3528 (2018)

Chapter 9
Co-design of Output-Feedback Controller and Multi-node Stochastic Scheduling Protocol

9.1 Introduction

Chapters 7 and 8 systematically investigated the fuzzy SMC problem for IT2 T-S fuzzy systems under the MN-SS protocols and proposed a co-design scheme of the IT2 fuzzy sliding mode controller and the SS protocol. However, these investigations are based on an implicit assumption that the full state of the controlled systems is measurable and can be effectively regulated via the SS protocols. Nevertheless, practical control systems often encounter scenarios where system states are not fully measurable, and only output signals can be measured. This motivates the development of an output feedback control strategy to ensure desired control performance.

This chapter focuses on the design of output-feedback SMC for IT2 T-S fuzzy systems with unmeasured states under resource-constrained communication networks. To address the bandwidth limitations in the S/C channels, a MN-SS protocol introduced in Chap. 7 is implemented to govern the transmission of measured output signals. Based on the received measurement output under the scheduling protocol, an output-feedback fuzzy sliding mode controller is designed with distinct fuzzy rules and MFs from those in the fuzzy model. Besides, the co-design scheme established in Chap. 8 is further extended to address the optimization design of the output-feedback fuzzy sliding mode controller and the Markov-chain-based MN-SS protocol. Moreover, by leveraging global boundary information of mismatched MFs between the fuzzy system and the controller, sufficient conditions are derived to ensure the stochastic stability of the closed-loop system and the reachability of the specified sliding surface. Finally, an optimization-based solving algorithm is developed to enhance control performance by minimizing the convergence domain of the system state.

Y. Yang et al., *Analysis and Synthesis of Interval Type-2 Fuzzy Systems Under Communication Protocols*, Studies in Systems, Decision and Control 636,
https://doi.org/10.1007/978-981-95-2355-9_9

9.2 Problem Description

9.2.1 System Model

Consider the following IT2 T-S fuzzy systems with the i-th rule:

Rule i: IF $f_1(x(k))$ is $\mathscr{F}_1^i$ and ... and $f_\hbar(x(k))$ is $\mathscr{F}_\hbar^i$, THEN

$$\begin{cases} x(k+1) = A_i x(k) + B_i(u(k) + d(x(k), k)), \\ y(k) = Cx(k) \end{cases} \tag{9.1}$$

where $x(k) \in \mathbb{R}^n$, $y(k) \in \mathbb{R}^p$, and $u(k) \in \mathbb{R}^m$ represent the system state, the measurement output, an the control input, respectively. $f_\iota(x(k))$ $(\iota = 1, \dots, \hbar)$ is the premise variable and $\mathscr{F}_\iota^i$ $(i = 1, \dots, r)$ is the IT2 fuzzy sets of the i-th fuzzy rule. A_i and B_i are known matrices. $d(x(k), k)$ is the external disturbance satisfying $\|d(x(k), k)\| \le \varpi$ with a known upper bound $\varpi \ge 0$.

The firing strength of the i-th rule is represented as

$$\overrightarrow{m}_i(x(k)) = \left[\underline{m}_i(x(k)), \overline{m}_i(x(k))\right],$$

with upper and lower MFs $\underline{m}_i(x(k)) = \prod_{\iota=1}^{\hbar} \underline{\mu}_{\mathscr{F}_\iota^i}(f_\iota(x(k))) > 0$, $\overline{m}_i(x(k)) = \prod_{\iota=1}^{\hbar} \overline{\mu}_{\mathscr{F}_\iota^i}(f_\iota(x(k))) > 0$ and $0 \le \underline{m}_i(x(k)) \le \overline{m}_i(x(k))$; $\overline{\mu}_{\mathscr{F}_\iota^i}(f_\iota(x(k)))$ and $\underline{\mu}_{\mathscr{F}_\iota^i}(f_\iota(x(k)))$ denote the lower and upper grades of the membership function, respectively.

Then, the IT2 T-S fuzzy system (9.1) can be inferred as

$$\begin{cases} x(k+1) = \sum_{i=1}^{r} m_i(x(k))[A_i x(k) + B_i(u(k) + d(x(k), k))], \\ y(k) = Cx(k) \end{cases} \tag{9.2}$$

where $m_i(x(k))$ is the membership function satisfying $\sum_{i=1}^{r} m_i(x(k)) = 1$, and $m_i(x(k)) = \underline{\alpha}_i(x(k))\underline{m}_i(x(k)) + \overline{\alpha}_i(x(k))\overline{m}_i(x(k))$. Besides, $\overline{\alpha}_i(x(k))$ and $\underline{\alpha}_i(x(k))$ are the weighting coefficients satisfying $\underline{\alpha}_i(x(k)), \overline{\alpha}_i(x(k)) \in [0, 1]$ and $\underline{\alpha}_i(x(k)) + \overline{\alpha}_i(x(k)) = 1$.

9.2.2 Multi-node Stochastic Scheduling Protocol

Consider the S/C channel, which permits only c $(1 \le c \le p)$ sensor nodes to transmit their measurement signals. That is, the MN-SS protocol can select c sensor nodes from the p-dimensional measurement output signal $y(k) = [y_1(k), \dots, y_p(k)]^T$ to access the communication channels. Define the binary-valued function $\psi_i(k) : \mathbb{Z}^+ \to \{0, 1\}$ representing the access status of the i-th sensor node at instant k. When the sensor node i is permitted to access the channel (i.e., $\psi_i(k) = 1$), the controller will receive the state information $y_i(k)$. Otherwise, when $\psi_i(k) = 0$, the controller treats $y_i(k)$ as zero. Thus, the available measurement signal of the i-th sensor node for the controller at instant k is denoted as

$$\bar{y}_i(k) = \begin{cases} y_i(k), & \text{if } \psi_i(k) = 1\,; \\ 0, & \text{otherwise.} \end{cases} \tag{9.3}$$

Let the token $\rho(k) \in \mathbb{M} \triangleq \{1, 2, \ldots, \bar{p}\}$ with $\bar{p} = C_p^c$ represent the selected sensor node at instant k, which obeys a discrete-time Markov chain with the following conditional probability

$$\pi_{pq} = \mathbf{Pr}\{\rho(k+1) = \wp | \rho(k) = \ell\}, \quad \ell, \wp \in \mathbb{M}, \tag{9.4}$$

with $\pi_{\ell\wp} \in [0, 1]$ and $\sum_{\wp=1}^{\bar{p}} \pi_{\ell\wp} = 1$. Define $\Pi = [\pi_{\ell\wp}]_{\bar{p}\times\bar{p}}$ as the TPM.

Define $\bar{y}(k) \triangleq [\bar{y}_1(k), \ldots, \bar{y}_p(k)]^T$. The available information for the controller can be denoted as

$$\bar{y}(k) = \Phi_{\rho(k)} y(k), \tag{9.5}$$

where $\Phi_{\rho(k)} = \text{diag}\{\delta(\rho(k) - 1), \ldots, \delta(\rho(k) - \bar{p})\}$ with $\delta(\cdot)$ being the Kronecker delta function, and $\Phi_{\rho(k)} \in \{\Phi_1, \ldots, \Phi_{\bar{p}}\}$.

9.3 Output-Feedback Fuzzy Sliding Mode Control Design

Firstly, choose the following sliding function

$$s(k) = Gy(k), \tag{9.6}$$

where G is the sliding matrix to be designed later.

Under the MN-SS protocol, only a part of measurement output signals can be transmitted, and only the scheduled measurement signal $\bar{y}(k)$ in (9.5) is available for the controller. Then, the following scheduling signal-dependent fuzzy sliding mode controller is designed as

$$u(k) = \sum_{j=1}^{c} h_j(\bar{y}(k)) K_{j,\rho(k)} \bar{y}(k) - \varpi \,\text{sign}(\bar{s}(k)), \tag{9.7}$$

where $K_{r,\rho(k)}$ is the token-dependent control gain to be designed together with the TPM Π of SS protocol later, and $\bar{s}(k)$ is the sliding variable by replacing $y(k)$ in (9.6) with $\bar{y}(k)$. Besides, new upper and lower MFs dependent on the available signal $\bar{y}(k)$ are constructed as $\overline{h}_j(\bar{y}(k))$ and $\underline{h}_j(\bar{y}(k))$ such that controller's MFs can be obtained as $h_j(\bar{y}(k)) = \underline{\beta}_j(\bar{y}(k))\underline{h}_j(\bar{y}(k)) + \overline{\beta}_j(\bar{y}(k))\overline{h}_j(\bar{y}(k))$ satisfying $h_j(\bar{y}(k)) \in [0, 1]$ and $\sum_{j=1}^{r} h_j(\bar{y}(k)) = 1$ with $\underline{\beta}_j(\bar{y}(k))$ with $\overline{\beta}_j(\bar{y}(k))$ as weighting coefficients.

Substituting the sliding mode controller (9.7) into the IT2 T-S fuzzy system (9.2), the closed-loop system can be obtained as

$$x(k+1) = \sum_{i=1}^{r}\sum_{j=1}^{c} m_i(x(k))h_j(\bar{x}(k))\Big[(A_i + B_i K_{j,\rho(k)}\Phi_{\rho(k)}C)x(k) + B_i\big(d(x(k), k) - \varpi\,\mathrm{sign}(\bar{s}(k))\big)\Big]. \tag{9.8}$$

The following definition is provided for analyzing the control performance of closed-loop system (9.8).

Definition 9.1 The closed-loop system (9.8) is said to be exponentially ultimately bounded (EUB) in the mean-square sense, if there exist scalars $\gamma \in [0, 1)$, $c_1 > 0$, and $c_2 \geq 0$ satisfying

$$\mathbf{E}\left\{\|x(k)\|^2\right\} \leq c_1\gamma^k\|x(0)\|^2 + c_2,\ \forall k > 0 \tag{9.9}$$

where $x(0)$ is the initial condition, and c_2 is the convergent upper bound (CUB) of $\mathbf{E}\left\{\|x(k)\|^2\right\}$.

The following objective is to develop the co-design scheme of output-feedback controller and Markov chain-based multi-node scheduling protocol such that the EUB of closed-loop system (9.8) in the mean-square sense and the reachability of the specified sliding surface $s(k) = 0$ can be ensured simultaneously.

9.4 Stability Analysis Under Multi-node Stochastic Scheduling Protocol

In this section, the stochastic stability of the closed-loop system (9.8) and the reachability of the sliding surface $s(k) = 0$ will be analyzed, and then, the co-design scheme will be provided to determine the control gain $K_{j,\rho(k)}$ and TPM Π.

9.4.1 Stochastic Stability of Closed-Loop System

Theorem 9.1 *For the IT2 T-S fuzzy system (9.2) with given positive scalars τ_1, τ_2, ϕ_1, and ϕ_2, if there exist scalars $\zeta_i > 0$, $\pi_{\ell\wp} \in [0, 1]$ with $\sum_{\wp=1}^{\bar{p}} \pi_{\ell\wp} = 1$, matrices $P_\ell > 0$, G, $K_{j,\ell}$ satisfying the following condition ($\ell, \wp = 1, \ldots, \bar{p}$ and $i, j = 1, \ldots, r$)*

$$\sum_{i=1}^{r}\sum_{j=1}^{c} m_i(x(k))h_j(\bar{x}(k))\Lambda_{ij\ell} < 0, \tag{9.10}$$

with

$$\mathcal{P}_\ell \triangleq \sum_{\wp=1}^{\bar{p}} \pi_{\ell\wp} P_\wp,$$

$$\Lambda_{ij\ell} \triangleq -\gamma P_\ell + (1+\tau_1)(A_i + B_i K_{j,\ell}\Phi_\ell C)^T \mathcal{P}_\ell (A_i + B_i K_{j,\ell}\Phi_\ell C),$$

then the closed-loop system (9.8) under the MN-SS protocol is stochastically bounded, and system state can converge into the following domain:

$$\mathcal{B} = \left\{ x(k) \mid ||x(k)|| \le \sqrt{\frac{\psi}{\underline{\lambda}_{\min}(P_\ell)(1-\gamma)}} \right\} \tag{9.11}$$

with $\psi \triangleq 2(1+\tau_1^{-1})(1+m)\varpi^2\bar{\lambda}_{\max}(B_i^T P_\ell B_i)$, $\bar{\lambda}_{\max}(B_i^T P_\ell B_i) = \max_{i,\ell}\{\lambda_{\max}(B_i^T P_\ell B_i)\}$, *and* $\underline{\lambda}_{\min}(P_\ell) = \min_{\ell}\{\lambda_{\min}(P_\ell)\}$.

Proof Choose the Lyapunov function

$$V(x(k), \rho(k)) = x^T(k) P_{\rho(k)} x(k). \tag{9.12}$$

Let $\rho(k) = \ell$ and $\rho(k+1) = \wp$. Taking the difference of the Lyapunov function $V(x(k), \rho(k))$ in (9.12), one has

$$\begin{aligned}
&\mathbf{E}\{V(x(k+1), \rho(k+1)) \mid x(k), \ell\} - \gamma V(x(k), \ell) \\
&\quad = \mathbf{E}\{x(k+1)^T \mathcal{P}_\ell x(k+1) | x(k), \ell\} - \gamma x^T(k) P_\ell x(k) \\
&\quad \le \sum_{i=1}^{r}\sum_{j=1}^{c} m_i(x(k))h_j(\bar{x}(k)) \Big[(A_i + B_i K_{j,\ell}\Phi_\ell C)x(k) + B_i(d(x(k), k) \\
&\qquad - \varpi \operatorname{sign}(\bar{s}(k)))\Big]^T \mathcal{P}_\ell \Big[(A_i + B_i K_{j,\ell}\Phi_\ell C)x(k) + B_i(d(x(k), k) \\
&\quad - \varpi \operatorname{sign}(\bar{s}(k)))\Big] - \gamma x^T(k) P_\ell x(k) \\
&\le \sum_{i=1}^{r}\sum_{j=1}^{c} m_i(x(k))h_j(\bar{x}(k)) \Big\{(1+\tau_1) x^T(k)\big[(A_i + B_i K_{j,\ell}\Phi_\ell C)^T \mathcal{P}_\ell \\
&\times (A_i + B_i K_{j,\ell}\Phi_\ell C)\big]x(k) - \gamma x^T(k) P_\ell x(k)\Big\} \\
&+ 2(1+\tau_1^{-1})(1+m)\varpi^2 \lambda_{\max}(B_i^T \mathcal{P}_\ell B_i) \\
&\le \sum_{i=1}^{r}\sum_{j=1}^{c} m_i(x(k))h_j(\bar{x}(k)) x^T(k)\Lambda_{ij\ell} x(k) + \psi
\end{aligned} \tag{9.13}$$

where $0 < \gamma < 1$, $0 < \tau_1$ are adjustable scalars and $\psi \triangleq 2(1+\tau_1^{-1})(1+m)\varpi^2\lambda_{\max}(B_i^T P_\ell B_i)$.

If conditions (9.10) holds, we have from (9.13)

$$\mathbf{E}\{V(x(k+1), \rho(k+1)) \mid x(k), \ell\} \leq \lambda V(x(k), \ell) + \psi. \tag{9.14}$$

Taking mathematical expectation for both sides of expression (9.14) yields

$$\mathbf{E}\{V(x(k+1), \rho(k+1))\} \leq \gamma \mathbf{E}\{V(x(k), \ell)\} + \psi, \tag{9.15}$$

from which it can be further obtained

$$\begin{aligned} \mathbf{E}\{V(x(k), \rho(k))\} \leq& \gamma \mathbf{E}\{V(x(k-1), \ell)\} + \psi \\ \leq& \gamma^k V(x(0), \rho(0)) + \psi \frac{1-\gamma^k}{1-\gamma}. \end{aligned} \tag{9.16}$$

According to the definition of $V(x(k), \rho(k))$, one has

$$\mathbf{E}\left\{||x(k)||^2\right\} \leq \frac{\gamma^k \lambda_{\max}(P_\ell)}{\lambda_{\min}(P_\ell)} \|x(0)\|^2 + \psi \frac{1-\gamma^k}{\lambda_{\min}(P_\ell)(1-\gamma)}. \tag{9.17}$$

Thus, the closed-loop system (9.8) is EUB in mean square. □

9.4.2 Reachability of Sliding Surface

In the sequel, we will further give the sufficient conditions to ensure the reachability of the specified sliding surface $s(k) = 0$.

Theorem 9.2 *Consider the IT2 T-S fuzzy system (9.2) with given positive scalars τ_1, τ_2, ϕ_1, and ϕ_2, if there exist scalars $\zeta_i > 0$, $\xi_i > 0$, $\pi_{\ell\wp} \in [0, 1]$ with $\sum_{\wp=1}^{\bar{p}} \pi_{\ell\wp} = 1$, matrices $P_\ell > 0$, $Q_\ell > 0$, G, and $K_{j,\ell}$, satisfying the following conditions (for $\ell, \wp = 1, \ldots, \bar{p}$ and $i, j = 1, \ldots, r$)*

$$\sum_{i=1}^{r}\sum_{j=1}^{c} m_i(x(k)) h_j(\bar{x}(k)) \hat{\Lambda}_{ij\ell} < 0, \tag{9.18}$$

with

$$\mathcal{Q}_\ell \triangleq \sum_{\wp=1}^{\bar{p}} \pi_{\ell\wp} Q_\wp,$$

$$\hat{\Lambda}_{ij\ell} \triangleq \Lambda_{ij\ell} + (1+\tau_2)(A_i + B_i K_{j,\ell}\Phi_\ell C)^T C^T G^T \mathcal{Q}_\ell G C (A_i + B_i K_{j,\ell}\Phi_\ell C),$$

then the designed token-dependent output-feedback fuzzy sliding mode controller (9.7) can drive state trajectories entering (in mean square) into the following domain

$$\mathcal{D} = \left\{ s(k) \mid ||s(k)|| \le \sqrt{\frac{\bar{\psi}}{\gamma \underline{\lambda}_{\min}(Q)}} \right\}, \tag{9.19}$$

with $\bar{\psi} \triangleq \psi + 2(1+\tau_2^{-1})\varpi^2(1+m)\bar{\lambda}_{\max}$, $\bar{\lambda}_{\max} = \max\limits_{i,\ell}\{\lambda_{\max}(B_i^T C^T G^T Q_\ell G C B_i)\}$, *and* $\underline{\lambda}_{\min}(Q) = \min\limits_{\ell}\{\lambda_{\min}(Q_\ell)\}$

Proof Combining the closed-loop system (9.8) and the sliding function (9.6) obtains that

$$\begin{aligned} s(k+1) = & \sum_{i=1}^{r}\sum_{j=1}^{c} m_i(x(k))h_j(\bar{x}(k))GC\Big[(A_i + B_i K_{j,\rho(k)}\Phi_{\rho(k)}C)x(k) \\ & + B_i\big(d(x(k),k) - \varpi \operatorname{sign}(\bar{s}(k))\big)\Big]. \end{aligned} \tag{9.20}$$

Choose the Lyapunov function as

$$U(x(k),\rho(k)) = V(x(k),\rho(k)) + s(k)^T Q_{\rho(k)} s(k). \tag{9.21}$$

For $\rho(k) = \ell$ and $\rho(k+1) = \wp$, one has from (9.21)

$$\begin{aligned} & \mathbf{E}\{U(x(k+1),\rho(k+1)) \mid x(k),\ell\} - \gamma U(x(k),\rho(k)) \\ = & \mathbf{E}\{V(x(k+1),\ell) \mid x(k),\ell\} + \mathbf{E}\{s^T(k+1)\mathcal{Q}_\ell s(k+1) | s(k),\ell\} \\ & - \gamma x^T(k) P_\ell x(k) - \gamma s^T(k) Q_\ell s(k). \end{aligned} \tag{9.22}$$

According to (9.20), we have

$$\begin{aligned} & \mathbf{E}\{s^T(k+1)\mathcal{Q}_\ell s(k+1) \mid x(k),\ell\} \\ & \le \sum_{i=1}^{r}\sum_{j=1}^{c} m_i(x(k))h_j(\bar{x}(k))\Big[(A_i + B_i K_{j,\ell}\Phi_\ell C)x(k) + B_i\Big(d(x(k),k) \\ & - \varpi \operatorname{sign}(\bar{s}(k))\Big)\Big]^T C^T G^T \mathcal{Q}_\ell G C\Big[(A_i + B_i K_{j,\ell}\Phi_\ell C)x(k) \\ & + B_i\Big(d(x(k),k) - \varpi \operatorname{sign}(\bar{s}(k))\Big)\Big] \\ & \le \sum_{i=1}^{r}\sum_{j=1}^{c} m_i(x(k))h_j(\bar{x}(k))\Big\{(1+\tau_2)x^T(k)[(A_i + B_i K_{j,\ell}\Phi_\ell C)^T C^T G^T \\ & \times \mathcal{Q}_\ell G C(A_i + B_i K_{j,\ell}\Phi_\ell C)]x(k) + 2(1+\tau_2^{-1})\varpi^2(1+m)\bar{\lambda}_{\max}\Big\}. \end{aligned} \tag{9.23}$$

Combining (9.22), (9.23) and (9.13), one has

$$\begin{aligned}
&\mathbf{E}\{U(x(k+1),\rho(k+1)) \mid x(k),\ell\} - \gamma U(x(k),\rho(k)) \\
&\leq \sum_{i=1}^{r}\sum_{j=1}^{c} m_i(x(k))h_j(\bar{x}(k))x^T(k)\hat{\Lambda}_{ij\ell}x(k) \\
&\quad - \left[\gamma s^T(k)Q_\ell s(k) - \psi - 2(1+\tau_2^{-1})\varpi^2(1+m)\bar{\lambda}_{\max}\right] \\
&\leq \sum_{i=1}^{r}\sum_{j=1}^{c} m_i(x(k))h_j(\bar{x}(k))x^T(k)\hat{\Lambda}_{ij\ell}x(k) - \gamma\lambda_{\min}(\mathcal{Q}_\ell)\|s(k)\|^2 + \bar{\psi}. \qquad (9.24)
\end{aligned}$$

Thus, if the state trajectories are outside the sliding domain, i.e.,

$$||s(k)|| > \sqrt{\frac{\bar{\psi}}{\gamma\lambda_{\min}(\mathcal{Q}_\ell)}}, \qquad (9.25)$$

one has from the conditions (9.18) that

$$\mathbf{E}\{U(x(k+1),\rho(k+1))|x(k),\ell\} - \gamma U(x(k),\ell) < 0. \qquad (9.26)$$

Taking the expectation for the both sides of the expression (9.26), we have $\mathbf{E}\{U(x(k+1),\rho(k+1))\} < \gamma\mathbf{E}\{U(x(k),\rho(k))\}$, which means that the Lyapunov function $U(x(k),\rho(k))$ is monotonically decreasing outside the domain $\mathcal{D}$ in mean-square sense. So, the system trajectories of closed-loop system (9.8) can be driven (in mean square) into the sliding domain $\mathcal{D}$ around the specified sliding surface $s(k) = 0$. □

9.5 Co-design Scheme and Performance Optimization

9.5.1 Global-Boundary-Information-Based Sufficient Conditions

In order to attain the stochastic stability of the closed-loop system (9.8) and the reachability of the sliding surface $s(k) = 0$, the conditions in Theorems 9.1 and 9.2 should be guaranteed simultaneously. To this end, the following Theorem 9.3 is further given.

Theorem 9.3 *Consider the IT2 T-S fuzzy system (9.2) with given positive scalars* τ_1, τ_2, ϕ_1, *and* ϕ_2, *if there exist positive scalars* μ, λ, ζ_i, ξ_i, γ_ℓ, χ_ℓ, ϵ_ℓ, ε_ℓ, $\pi_{\ell\wp} \in [0, 1]$ *with* $\sum_{\wp=1}^{\bar{p}} \pi_{\ell\wp} = 1$, *matrices* $P_\ell > 0$, $Q_\ell > 0$, $X_\ell > 0$, $Y_\ell > 0$, $M_{ij} > 0$, $N_{ij} > 0$, G, *and* $K_{j,\ell}$ *satisfying the following condition (for* $\ell, \wp = 1, \ldots, \bar{p}$ *and* $i, j = 1, \ldots, r$*)*

$$\begin{bmatrix} \tilde{\Lambda}_{ij\ell}^{11} & * & * & * & * & * & * \\ \tilde{\Lambda}_{ij\ell}^{21} & -\bar{X} & * & * & * & * & * \\ \tilde{\Lambda}_{ij\ell}^{31} & 0 & -\epsilon_\ell \bar{X} & * & * & * & * \\ \tilde{\Lambda}_{ij\ell}^{41} & 0 & 0 & -\bar{Y} & * & * & * \\ \tilde{\Lambda}_{ij\ell}^{51} & 0 & 0 & 0 & -\varepsilon_\ell \bar{Y} & * & * \\ \tilde{\Lambda}_{ij\ell}^{61} & 0 & 0 & 0 & 0 & -\bar{P} & * \\ \tilde{\Lambda}_{ij\ell}^{71} & 0 & 0 & 0 & 0 & 0 & -\bar{Q} \end{bmatrix} < 0, \tag{9.27}$$

where

$$\begin{aligned}
\bar{P} =& \frac{1}{1+\tau_1}\text{diag}\{P_1, P_2, \ldots, P_{\bar{p}}\}, \quad \bar{Q} = \frac{1}{1+\tau_2}\text{diag}\{Q_1, Q_2, \ldots, Q_{\bar{p}}\}, \\
\bar{X} =& \frac{1}{1+\tau_1}\text{diag}\{X_1, X_2, \ldots, X_{\bar{p}}\}, \bar{Y} = \frac{1}{1+\tau_2}\text{diag}\{Y_1, Y_2, \ldots, Y_{\bar{p}}\}, \\
\tilde{\Lambda}_{ij\ell}^{11} =& -\gamma P_\ell - 2(1+\tau_1)\mu A_i^T A_i - 2(1+\tau_2)\lambda (GCA_i)^T (GCA_i) \\
& - (1+\tau_1)\epsilon_\ell^{-1} \sum_{\wp=1}^{\bar{p}} A_i^T P_\wp A_i - (1+\tau_2)\varepsilon_\ell^{-1} \sum_{\wp=1}^{\bar{p}} (GCA_i)^T Q_\wp (GCA_i) \\
& + \sum_{\imath=1}^{r} \sum_{\jmath=1}^{c} \left(\overline{h}_{\imath\jmath} M_{\imath\jmath} - \underline{h}_{\imath\jmath} N_{\imath\jmath} \right) - M_{ij} + N_{ij}, \\
\tilde{\Lambda}_{ij\ell}^{21} =& \mathbf{1}_n \otimes B_i K_{j,\ell} \Phi_\ell C + [\pi_{\ell 1}, \pi_{\ell 2}, \ldots, \pi_{\ell \bar{p}}]^T \otimes A_i, \\
\tilde{\Lambda}_{ij\ell}^{31} =& \mathbf{1}_n \otimes A_i + \frac{\epsilon_\ell}{2} [\pi_{\ell 1}, \pi_{\ell 2}, \ldots, \pi_{\ell \bar{p}}]^T \otimes A_i, \\
\tilde{\Lambda}_{ij\ell}^{41} =& \mathbf{1}_n \otimes GCB_i K_{j,\ell} \Phi_\ell C + [\pi_{\ell 1}, \pi_{\ell 2}, \ldots, \pi_{\ell \bar{p}}]^T \otimes GCA_i, \\
\tilde{\Lambda}_{ij\ell}^{51} =& \mathbf{1}_n \otimes GCA_i + \frac{\varepsilon_\ell}{2} [\pi_{\ell 1}, \pi_{\ell 2}, \ldots, \pi_{\ell \bar{p}}]^T \otimes GCA_i, \\
\tilde{\Lambda}_{ij\ell}^{61} =& \mu \mathbf{1}_n \otimes A_i, \quad \tilde{\Lambda}_{ij\ell}^{71} = \lambda \mathbf{1}_n \otimes GCA_i,
\end{aligned}$$

with $X_\ell = P_\ell^{-1}$ and $Y_\ell = Q_\ell^{-1}$, then the stochastic stability of the closed-loop system (9.8) and the reachability of the sliding domain can be guaranteed simultaneously.

Proof Due to $(1+\tau_2)(A_i + B_i K_{j,\ell} \Phi_\ell C)^T C^T G^T Q_\ell GC(A_i + B_i K_{j,\ell} \Phi_i C) > 0$, the condition (9.10) in Theorem 9.1 can be ensured by the condition (9.18) in Theorem 9.2. Thus, we just need to prove that the condition (9.27) can guarantee the inequality (9.18).

It should be pointed out that condition (9.18) contains MFs, which cannot be directly solved. Here, the global-boundary-information of MFs is utilized to deal with the time-varying MFs. Similar to the method in Chapter 5, we can calculate the upper and lower bounds of membership function $m_i(x(k))h_j(\bar{x}(k))$ as $\overline{h}_{ij}$ and $\underline{h}_{ij}$ satisfying $0 \le \underline{h}_{ij} \le m_i(x(k))h_j(\bar{x}(k)) \le \overline{h}_{ij} \le 1$.

By introducing slack matrices $M_{ij}^T = M_{ij} > 0$ and $N_{ij}^T = N_{ij} > 0$, one can obtain the following inequalities

$$\sum_{i=1}^{r}\sum_{j=1}^{c}(\overline{h}_{ij} - m_i(x(k))h_j(\bar{x}(k)))M_{ij} > 0 \tag{9.28}$$

$$\sum_{i=1}^{r}\sum_{j=1}^{c}(m_i(x(k))h_j(\bar{x}(k)) - \underline{h}_{ij})N_{ij} > 0. \tag{9.29}$$

By means of expressions (9.28) and (9.29), it can be obtained from condition (9.18) that

$$\begin{aligned}
&\sum_{i=1}^{r}\sum_{j=1}^{c} m_i(x(k))h_j(\bar{x}(k))\hat{\Lambda}_{ij\ell} \\
&\leq \sum_{i=1}^{r}\sum_{j=1}^{c} m_i(x(k))h_j(\bar{x}(k))\Big\{\hat{\Lambda}_{ij\ell} + (\overline{h}_{ij} - m_i(x(k))h_j(\bar{x}(k)))M_{ij} \\
&+ (m_i(x(k))h_j(\bar{x}(k)) - \underline{h}_{ij})N_{ij}\Big\} \\
&= \sum_{i=1}^{r}\sum_{j=1}^{c} m_i(x(k))h_j(\bar{x}(k))\Big\{\sum_{\imath=1}^{r}\sum_{\jmath=1}^{c}\Big(\overline{h}_{\imath\jmath}M_{\imath\jmath} - \underline{h}_{\imath\jmath}N_{\imath\jmath}\Big) \\
&+ \hat{\Lambda}_{ij\ell} - M_{ij} + N_{ij}\Big\}.
\end{aligned} \tag{9.30}$$

It can be obtained that condition (9.18) is ensured if the following inequality holds:

$$\sum_{\imath=1}^{r}\sum_{\jmath=1}^{c}\Big(\overline{h}_{\imath\jmath}M_{\imath\jmath} - \underline{h}_{\imath\jmath}N_{\imath\jmath}\Big) + \hat{\Lambda}_{ij\ell} - M_{ij} + N_{ij} < 0. \tag{9.31}$$

For the matrix $\hat{\Lambda}_{ijp}$ in the inequality (9.31), we can obtain the following expressions by using "completing the square"

$$\begin{aligned}
&(A_i + B_i K_{j,\ell}\Phi_\ell C)^T\Big(\sum_{\wp=1}^{\bar{p}}\pi_{\ell\wp}P_\wp\Big)(A_i + B_i K_{j,\ell}\Phi_\ell C) \\
&= A_i^T\Big(\sum_{\wp=1}^{\bar{p}}\pi_{\ell\wp}P_\wp\Big)A_i + A_i^T\Big(\sum_{\wp=1}^{\bar{p}}\pi_{\ell\wp}P_\wp\Big)B_i K_{j,\ell}\Phi_\ell C \\
&+ (B_i K_{j,\ell}\Phi_\ell C)^T\Big(\sum_{\wp=1}^{\bar{p}}\pi_{\ell\wp}P_\wp\Big)A_i + (B_i K_{j,\ell}\Phi_\ell C)^T\Big(\sum_{\wp=1}^{\bar{p}}\pi_{\ell\wp}P_\wp\Big)(B_i K_{j,\ell}\Phi_\ell C) \\
&\leq \sum_{\wp=1}^{\bar{p}}\Big[(\pi_{\ell\wp}A_i + B_i K_{j,\ell}\Phi_\ell C)^T P_\wp(\pi_{\ell\wp}A_i + B_i K_{j,\ell}\Phi_\ell C) - (\pi_{\ell\wp}A_i)^T P_\wp(\pi_{\ell\wp}A_i)\Big] \\
&+ \sum_{\wp=1}^{\bar{p}}\Big[\epsilon_\ell^{-1}\Big(\frac{\epsilon_\ell}{2}\pi_{\ell\wp}A_i + A_i\Big)^T P_\wp\Big(\frac{\epsilon_\ell}{2}\pi_{\ell\wp}A_i + A_i\Big) - \epsilon_\ell^{-1}A_i^T P_\wp A_i\Big]
\end{aligned} \tag{9.32}$$

and

$$
\begin{aligned}
&(GCA_i + GCB_iK_{j,\ell}\Phi_\ell C)^T\Big(\sum_{\wp=1}^{\bar{p}}\pi_{\ell\wp}Q_\wp\Big)(GCA_i + GCB_iK_{j,\ell}\Phi_\ell C)\\
&= (GCA_i)^T\Big(\sum_{\wp=1}^{\bar{p}}\pi_{pq}Q_\wp\Big)GCA_i + (GCA_i)^T\Big(\sum_{\wp=1}^{\bar{p}}\pi_{\ell\wp}Q_\wp\Big)GCB_iK_{j,\ell}\Phi_i C\\
&\quad + (GCB_iK_{j,\ell}\Phi_\ell C)^T\Big(\sum_{\wp=1}^{\bar{p}}\pi_{\ell\wp}Q_\wp\Big)GCA_i\\
&\quad + (GCB_iK_{j,\ell}\Phi_\ell C)^T\Big(\sum_{\wp=1}^{\bar{p}}\pi_{\ell\wp}Q_\wp\Big)(GCB_iK_{j,\ell}\Phi_\ell C)\\
&\le \sum_{\wp=1}^{\bar{p}}[(\pi_{\ell\wp}GCA_i + GCB_iK_{j,\ell}\Phi_\ell C)^TQ_\wp(\pi_{\ell\wp}GCA_i + GCB_iK_{j,\ell}\Phi_\ell C)\\
&\quad - (\pi_{\ell\wp}GCA_i)^TQ_\wp(\pi_{\ell\wp}GCA_i)]\\
&\quad + \sum_{\wp=1}^{\bar{p}}\Big[\varepsilon_\ell^{-1}\Big(\frac{\varepsilon_\ell}{2}\pi_{\ell\wp}GCA_i + GCA_i\Big)^TQ_\wp\Big(\frac{\varepsilon_\ell}{2}\pi_{\ell\wp}GCA_i + GCA_i\Big)\\
&\quad - \varepsilon_\ell^{-1}(GCA_i)^TQ_\wp(GCA_i)\Big]
\end{aligned} \tag{9.33}
$$

with $\epsilon_\ell > 0$ and $\varepsilon_\ell > 0$.

In the light of (9.32)–(9.33), the conditions (9.31) can be guaranteed by the following inequalities

$$\bar{\Lambda}_{ii\ell} < 0, \tag{9.34}$$

with

$$
\bar{\Lambda}_{ij\ell} = \begin{bmatrix}
\bar{\Lambda}_{ij\ell}^{11} & * & * & * & *\\
\tilde{\Lambda}_{ij\ell}^{21} & -\bar{X} & * & * & *\\
\tilde{\Lambda}_{ij\ell}^{31} & 0 & -\epsilon_\ell\bar{X} & * & *\\
\tilde{\Lambda}_{ij\ell}^{41} & 0 & 0 & -\bar{Y} & *\\
\tilde{\Lambda}_{ij\ell}^{51} & 0 & 0 & 0 & -\varepsilon_\ell\bar{Y}
\end{bmatrix},
$$

and

$$\begin{aligned}\bar{\Lambda}_{ij\ell}^{11} =& -\gamma P_\ell - (1+\tau_1)\sum_{\wp=1}^{\bar{p}}(\pi_{\ell\wp}A_i)^T P_\wp(\pi_{\ell\wp}A_i) - (1+\tau_1)\epsilon_\ell^{-1}\sum_{\wp=1}^{\bar{p}} A_i^T P_\wp A_i \\ & - (1+\tau_2)\sum_{\wp=1}^{\bar{p}}(\pi_{\ell\wp}GCA_i)^T Q_\wp(\pi_{\ell\wp}GCA_i) \\ & - (1+\tau_2)\varepsilon_\ell^{-1}\sum_{\wp=1}^{\bar{p}}(GCA_i)^T Q_\wp(GCA_i) \\ & + \sum_{\imath=1}^{r}\sum_{\jmath=1}^{c}\left(\bar{h}_{\imath\jmath}M_{\imath\jmath} - \underline{h}_{\imath\jmath}N_{\imath\jmath}\right) - M_{ij} + N_{ij}.\end{aligned}$$

For the terms $-\sum_{\wp=1}^{\bar{p}}(\pi_{\ell\wp}A_i)^T P_\wp(\pi_{\ell\wp}A_i)$ and $-\sum_{\wp=1}^{\bar{p}}(\pi_{\ell\wp}CGA_i)^T Q_\wp(\pi_{\ell\wp}CGA_i)$, we can further obtain

$$-\sum_{\wp=1}^{\bar{p}}(\pi_{\ell\wp}A_i)^T P_\wp\pi_{\ell\wp}A_i \leq \mu^2\sum_{\wp=1}^{\bar{p}} A_i^T P_\wp^{-1}A_i - 2\mu A_i^T A_i, \tag{9.35}$$

$$\begin{aligned}-\sum_{\wp=1}^{\bar{p}}(\pi_{\ell\wp}GCA_i)^T Q_\wp(\pi_{\ell\wp}GCA_i) \leq& \lambda^2\sum_{\wp=1}^{\bar{p}}(GCA_i)^T Q_\wp^{-1}GCA_i \\ & - 2\lambda(GCA_i)^T GCA_i,\end{aligned} \tag{9.36}$$

where μ and λ are two positive scalars.

In view of (9.35)–(9.36), it is easily shown that the condition (9.34) can be ensured by the condition (9.27). Thus, the condition (9.18) hold if the inequality (9.27) is feasible. The results of Theorems 9.1 and 9.2 can be ensured simultaneously. □

9.5.2 Optimization Solving on Control Gains and Scheduling Protocol

It is critical to note that some coupling terms in condition (9.27) in Theorem 9.3 depend on token-dependent parameters $\{\epsilon_\ell, \varepsilon_\ell\}$, sliding matrix G, positive definite matrices P_ℓ, Q_ℓ and their inverse matrices X_ℓ, Y_ℓ, which poses challenges to solve the control design parameters. However, if these token-dependent parameters $\{\epsilon_\ell, \varepsilon_\ell\}$ and matrices G, P_ℓ, and Q_ℓ are predetermined *a priori*, these nonlinear coupling terms can be reduced into linear relations. Notably, the selection of parameters $\{\epsilon_\ell, \varepsilon_\ell\}$, matrices G, P_ℓ, and Q_ℓ directly influences the feasibility of (9.27), the sliding domain $\mathcal{D}$ in (9.19) and the CUB in (9.11). To address this difficulty, an iterative optimization algorithm will be established to search the desirable token-dependent parameters $\{\epsilon_\ell, \varepsilon_\ell\}$ and matrices G, P_ℓ, Q_ℓ by solving the following optimization problem:

$$\min_{\mathcal{S}} \; F_{obj} \quad (9.37)$$
$$\text{s.t. condition} \quad (9.27)$$

where a set of appropriate variables $\mathcal{S} \triangleq \{\epsilon_\ell, \varepsilon_\ell, G, P_\ell, Q_\ell\}$ are searched to optimize the objective function F_{obj} representing the desired performance of the controlled systems.

In Theorems 9.1 and 9.2, it has been proved that the system state converges to a bounded domain with a CUB of $\sqrt{\psi/[\underline{\lambda}_{\min}(P_\ell)(1-\gamma)]}$, while state trajectories converge to a sliding domain bounded by $\sqrt{\bar{\psi}/[\gamma\underline{\lambda}_{\min}(Q)]}$. To enhance the dynamic performance of the closed-loop systems, the objective function F_{obj} can be defined as

$$F_{obj} = w_1\sqrt{\frac{\psi}{\underline{\lambda}_{\min}(P_\ell)(1-\gamma)}} + w_2\sqrt{\frac{\bar{\psi}}{\gamma\underline{\lambda}_{\min}(Q)}}, \quad (9.38)$$

where $w_1, w_2 \geq 0$ satisfying $w_1 + w_2 = 1$ are tunable weighting coefficients, offering a balanced trade-off between stability and transient performance.

Furthermore, the TPM Π governing the multi-node stochastic scheduling protocol may be inherently constrained by the network's limited communication capacity, such as bandwidth restrictions or hardware limitations. To ensure practical implementability, structural constraints on the TPM Π should be integrated into the protocol design. These constraints may include sparsity patterns, symmetry requirements, or bounds on transition probabilities, reflecting realistic communication scenarios. By explicitly incorporating such constraints, the proposed framework makes the scheduling protocol adheres to physical network capabilities while maintaining stochastic stability guarantees.

For the optimization problem (9.37), various optimization algorithms can be utilized to search the parameter set $\mathcal{S}$ for achieving the desired system performance. In this chapter, the PSO algorithm is employed to optimize these parameters and matrices, thereby enhancing system control performance. The detailed procedure can be referred to Algorithm 8.1 in Chapter 8.

9.6 Simulation Example

Consider an IT2 T-S fuzzy system (9.2) with the following parameters:

$$A_1 = \begin{bmatrix} 0.9 & 0.1 & -0.2 & 0.1 \\ 0.1 & 0.8 & 0.1 & 0.1 \\ 0.2 & 0.1 & 0.6 & 0.1 \\ 0.1 & 0.1 & 0.1 & 0.7 \end{bmatrix}, A_2 = \begin{bmatrix} 1.0 & 0.1 & 0.2 & 0.4 \\ 0.2 & -0.1 & 0.6 & 0.1 \\ 0.3 & 0.1 & 0.2 & 0.3 \\ 0.1 & 0.3 & 0.1 & 0.1 \end{bmatrix}, A_3 = \begin{bmatrix} 1.1 & 0.2 & 0.3 & 0.1 \\ 0.1 & 0.8 & 0.7 & 0.4 \\ 0.2 & 0.5 & -0.2 & 0.5 \\ 0.1 & 0.4 & 0.3 & 0.1 \end{bmatrix},$$

$$B_1 = \begin{bmatrix} 1.0 \\ 0.6 \\ 1.0 \\ 0.4 \end{bmatrix}, B_2 = \begin{bmatrix} 0.8 \\ 0.7 \\ 1.0 \\ 0.5 \end{bmatrix}, B_3 = \begin{bmatrix} 1.0 \\ 0.5 \\ 1.0 \\ 0.8 \end{bmatrix}, C = \begin{bmatrix} 1 & 0 & 0 & 0 \\ 0 & 1 & 0 & 0 \\ 0 & 0 & 1 & 0 \end{bmatrix},$$

$$d(x(k), k) = 0.05 \sin(x_1(k)x_4(k))$$

and the upper and lower MFs:

$$\overline{m}_1(x(k)) = 1 - \frac{1}{1.5 + e^{-1.5(x_1(k)+0.5)}}, \ \underline{m}_1(x(k)) = 1 - \frac{1}{1.5 + e^{-1.5(x_1(k)+1.5)}},$$

$$\overline{m}_3(x(k)) = \frac{1}{1.5 + e^{-1.5(x_1(k)-0.5)}}, \ \underline{m}_3(x(k)) = \frac{1}{1.5 + e^{-1.5(x_1(k)-1.5)}},$$

and $\overline{m}_2(x(k)) = 1 - \underline{m}_1(x(k)) - \underline{m}_3(x(k)), \underline{m}_2(x(k)) = 1 - \overline{m}_1(x(k)) - \overline{m}_3(x(k))$.

Due to the limited network resources, it is assumed that only two nodes among three sensor nodes may be chosen to access the communication network channels and transmit their measurement outputs at each instant. Meanwhile, the structure of the TPM is constrained as follows:

$$\Pi = \begin{bmatrix} \ell & \wp & 1-\ell-\wp \\ 1-\ell-\wp & \ell & \wp \\ \wp & \ell & 1-\ell-\wp \end{bmatrix}. \tag{9.39}$$

Based on the received system state, the following upper and lower MFs can be chosen to design a fuzzy sliding mode controller (9.7):

$$\overline{h}_1(\bar{x}(k)) = \frac{1}{1 + e^{-(\bar{x}_1(k)+0.5)}}, \ \underline{h}_1(\bar{x}(k)) = \frac{1}{1 + e^{-(\bar{x}_1(k)+1.5)}}$$

with the weighting coefficients $\overline{\beta}_1(\bar{x}(k)) = 0.6$. Then, the MFs can be obtained as $h_1(\bar{x}(k)) = 0.6\overline{h}_1(\bar{x}(k)) + 0.4\underline{h}_1(\bar{x}(k))$ and $h_2(\bar{x}(k)) = 1 - h_1(\bar{x}(k))$.

By performing the PSO-based optimization solving algorithm, we obtain the following design parameters on the TPM $\ell = 0.2994$, $\wp = 0.3566$, the sliding gain G, and the control gains $K_{j\ell}$ as

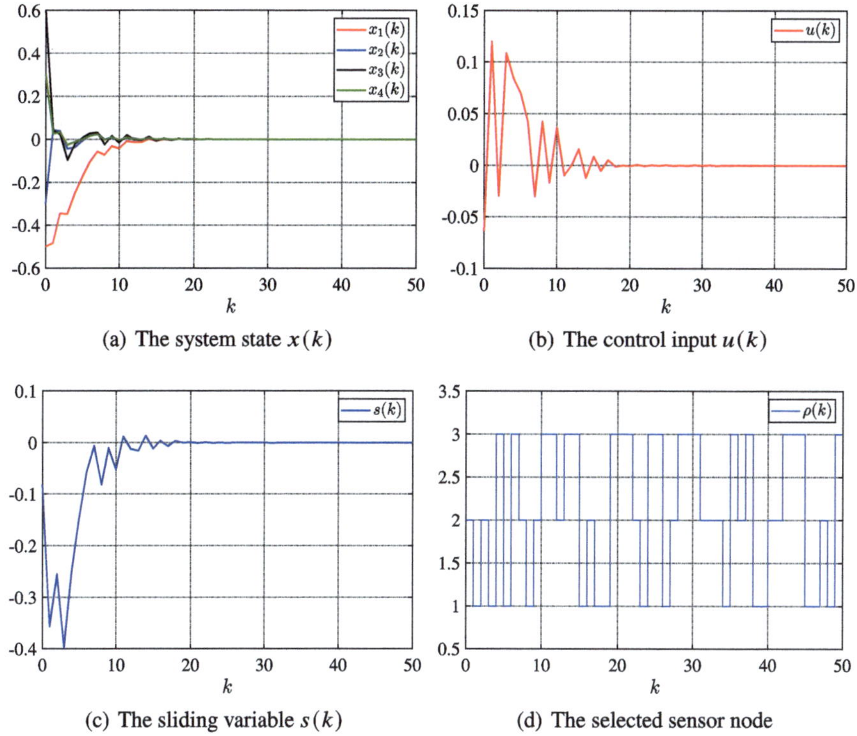

Fig. 9.1 The simulation results under the multi-node SS protocol

$$
\begin{aligned}
G &= \begin{bmatrix} 0.8523 \ 0.5142 \ 0.8307 \end{bmatrix}, \\
K_{11} &= \begin{bmatrix} -0.1677 \ -0.1460 \ 0 \end{bmatrix}, \quad K_{21} = \begin{bmatrix} -0.1677 \ -0.1460 \ 0 \end{bmatrix}, \\
K_{12} &= \begin{bmatrix} 0 \ -0.0709 \ -0.0677 \end{bmatrix}, \quad K_{22} = \begin{bmatrix} 0 \ -0.0712 \ -0.0674 \end{bmatrix}, \\
K_{13} &= \begin{bmatrix} -0.1496 \ 0 \ -0.1708 \end{bmatrix}, \quad K_{23} = \begin{bmatrix} -0.1496 \ 0 \ -0.1708 \end{bmatrix}.
\end{aligned}
$$

In simulation, the initial condition is set as $x_0 = [-0.4\ -0.3\ 0.6\ 0.3]^T$, the weighting functions are chosen as $\overline{\alpha}_1(x(k)) = (\sin(2x_1(k)) + 1)/2$ and $\underline{\alpha}_1(x(k)) = 1 - \overline{\alpha}_1(x(k))$ for obtaining $m_1(x(k))$, and $\overline{\alpha}_3(x(k)) = (\cos(2x_1(k)) + 1)/2$ and $\underline{\alpha}_3(x(k)) = 1 - \overline{\alpha}_1(x(k))$ for obtaining $m_3(x(k))$, and $m_2(x(k)) = 1 - m_1(x(k)) - m_3(x(k))$. The simulation results are shown in Fig. 9.1a–d. It is seen from Fig. 9.1a, c that the stability of closed-loop system and the reachability of sliding surface can be ensured. The control input is plotted in Fig. 9.1b, and the status of scheduling signal is shown in Fig. 9.1d. It can be concluded that, for IT2 T-S fuzzy system with unmeasured system state, the proposed co-design of output-feedback fuzzy SMC scheme and multi-node stochastic protocol can ensure a desired control performance.

9.7 Conclusion

This chapter addresses output-feedback SMC design for IT2 T-S fuzzy systems with unmeasurable states under bandwidth-constrained networks. To mitigate communication limitations in S/C channels, a stochastic scheduling protocol has been implemented to regulate output signal transmission, employing the multi-node scheduling protocol to enable concurrent data transmission from multiple sensors. A fuzzy sliding mode controller has been designed with distinct fuzzy rules and MFs from the fuzzy model, which can improve the design flexibility. The co-design scheme of the output-feedback controller and Markov-chain-governed scheduling protocol has been proposed. By exploiting global boundary information of mismatched MFs between the fuzzy system and controller, sufficient conditions have been derived to guarantee the stochastic stability of closed-loop systems and the reachability of specified sliding surfaces. Additionally, an optimization-based solving algorithm has been developed to enhance control performance.

Chapter 10
Sliding Mode Control Under Event-Triggered WTOD Protocol

10.1 Introduction

In Chaps. 2–9, the ETS and the SS protocol, as two important methods, have been investigated to regulate communication processes and improve the utilization efficiency of constrained network bandwidth. Besides, some relevant advancements have been considered for RR scheduling protocols [1, 2] and WTOD scheduling protocols [3, 4]. It should be emphasised that although both ETSs and channel-scheduling protocols (e.g., SS, RR, WTOD protocols) serve as effective approaches for alleviating communication burdens through transmission reduction, their main ideas and principles exhibit fundamental distinctions. Conventional ETSs (such as in [5, 6], Chaps. 2, 4, and 5) always utilize an event detector, also termed the ETM, to monitor whether state deviations overstep the specified threshold, where once the specified triggering condition is satisfied, all components of the current state will be transmitted simultaneously. Different from the transmission principle of the ETS, the WTOD protocol as in [4] adopts a competitive scheduling paradigm that exclusively authorizes transmission from the sensor node with the maximum state deviation at each instant, thereby ensuring persistent single-node transmission. Notably, the event-triggered protocol evaluates the deviation of system states with a predefined threshold, whereas the WTOD strategy performs relative comparisons exclusively among component-wise deviations of system states. From an implementation perspective, the synergistic integration of threshold-based evaluation and component-wise relative assessment could potentially optimize dynamic performance while conserving transmission resources. This necessitates the development of a comprehensive scheduling strategy that selectively authorizes transmission for sensor nodes exceeding predefined thresholds, while enforcing transmission dormancy when the deviations of all sensor nodes remain sub-threshold.

This chapter investigates the SMC issue for IT2 T-S fuzzy systems under limited communication resources. An event-triggered WTOD (ET-WTOD) protocol is formulated by introducing two thresholds to determine the state transmission of the

Y. Yang et al., *Analysis and Synthesis of Interval Type-2 Fuzzy Systems Under Communication Protocols*, Studies in Systems, Decision and Control 636,
https://doi.org/10.1007/978-981-95-2355-9_10

sensor nodes. The ET-WTOD protocol can dynamically adjust the selected nodes and permits only a part of components with larger errors to be sent at each triggering instant, which constitutes the key feature different from the existing ETSs and WTOD protocols. Under the imperfect premise matching framework, the controller's MFs are reconstructed via the received state and the known upper and lower MFs in fuzzy systems. Meanwhile, by establishing a new scheduling signal set, the scheduling-signal-dependent fuzzy sliding mode controller is designed. With the aid of the MFD approach, the mismatched premise variables of the fuzzy model and the controller are effectively handled by introducing some slack matrices, while the relaxed stability conditions are derived to ensure the stability of the closed-loop system and the reachability of the specified sliding surface. Moreover, by using the genetic algorithm (GA), an optimization-based SMC solving algorithm is proposed to attain a smaller sliding domain and improve the control performance.

10.2 Problem Description

10.2.1 System Model

Consider the following discrete-time IT2 T-S fuzzy system:

Rule i: IF $f_1(x(k))$ is $\mathscr{F}_1^i$ AND $\cdots$ AND $f_\hbar(x(k))$ is $\mathscr{F}_\hbar^i$, THEN

$$x(k+1) = A_i x(k) + B_i(u(k) + d(x(k), k)) \tag{10.1}$$

where $x(k) \in \mathbb{R}^n$ is the system state, $u(k) \in \mathbb{R}^m$ is the control input. $f_\iota(x(k))$ ($\iota = 1, \cdots, \hbar$) is the premise variable, and $\mathscr{F}_\iota^i$ ($i = 1, \cdots, r$) is the IT2 fuzzy set. A_i and B_i are known matrices with compatible dimensions. $d(x(k), k)$ is the external disturbance satisfying $\|d(x(k), k)\| \leq \varpi \|x(k)\|$ with a known constant $\varpi \geq 0$. The firing strength for the i-th rule is given as:

$$\overrightarrow{m}_i(x(k)) = \left[\underline{m}_i(x(k)), \overline{m}_i(x(k))\right],$$

where $\underline{\mu}_{\mathscr{F}_\iota^i}(f_\iota(x(k)))$ and $\overline{\mu}_{\mathscr{F}_\iota^i}(f_\iota(x(k)))$ are the membership grades satisfying $0 < \underline{\mu}_{\mathscr{F}_\iota^i}(f_\iota(x(k))) \leq \overline{\mu}_{\mathscr{F}_j^i}(f_\iota(x(k))) < 1$, and the upper and lower MFs $\overline{m}_i(x(k))$ and $\underline{m}_i(x(k))$ can be obtained as $\underline{m}_i(x(k)) = \prod_{\iota=1}^{\hbar} \underline{\mu}_{\mathscr{F}_\iota^i}(f_\iota(x(k)))$ and $\overline{m}_i(x(k)) = \prod_{\iota=1}^{\hbar} \overline{\mu}_{\mathscr{F}_\iota^i}(f_\iota(x(k)))$ satisfying $0 \leq \overline{m}_i(x(k)) \leq \underline{m}_i(x(k)) \leq 1$.

According to the fuzzy rules in (10.1), the IT2 T-S fuzzy system can be obtained as

$$x(k+1) = \sum_{i=1}^{r} m_i(x(k))[A_i x(k) + B_i(u(k) + f(x(k), k))], \tag{10.2}$$

where the membership function $m_i(x(k))$ can be represented as

$$m_i(x(k)) = \underline{m}_i(x(k))\underline{\alpha}_i(x(k)) + \overline{m}_i(x(k))\overline{\alpha}_i(x(k)), \tag{10.3}$$

with $\sum_{i=1}^{r} m_i(x(k)) = 1$ and the weighting functions $\underline{\alpha}_i(x(k))$ and $\overline{\alpha}_i(x(k))$ satisfying $\underline{\alpha}_i(x(k)), \overline{\alpha}_i(x(k)) \in [0, 1]$, and $\underline{\alpha}_i(x(k)) + \overline{\alpha}_i(x(k)) = 1$, which can extract the parameter uncertainties of nonlinear systems.

For the IT2 T-S fuzzy systems (10.2) under the constrained communication resources, the following objective is to develop the ET-WTOD protocol and propose an effective SMC scheme for ensuring a desired control performance.

10.2.2 Event-Triggered WTOD Protocol

As depicted in Fig. 10.1, the scheduling protocol is implemented in the S/C channel network to determine the transmission of the system states $x(k) \triangleq [x_1(k), x_2(k), \cdots, x_n(k)]^T \in \mathbb{R}^n$. Denote the available signals for the controller as $\breve{x}(k) \triangleq [\breve{x}_1(k), \breve{x}_2(k), \cdots, \breve{x}_n(k)]^T$, where $\breve{x}_i(k)$ is the i-th state component transmitted through the S/C communication channel.

Define the index $\delta_i(k)$ for the i-th state component as

$$\delta_i(k) \triangleq (\breve{x}_i(k-1) - x_i(k))\phi_i(\breve{x}_i(k-1) - x_i(k)), \tag{10.4}$$

where $\phi_i > 0$ is the specified weighting coefficient. The index $\delta_i(k)$ reflects the deviation of the current state from the previous transmitted one. Under the conventional WTOD protocol as in [3, 4], only the sensor node with the largest difference $\delta_i(k)$, i.e., the state component $x_i(k)$, will be permitted to access the network channels at instant k.

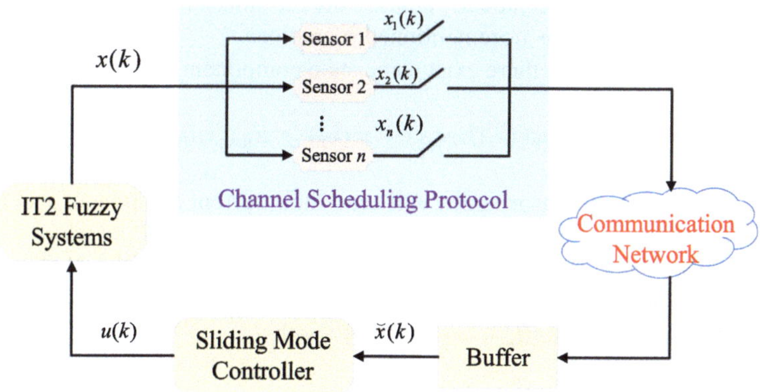

Fig. 10.1 Control system under the channel scheduling protocol

Denote $\sigma(k)$ as the scheduling signal representing the sensor node that is selected to transmit its signal at the instant k. Under the conventional WTOD protocol, one has

$$\sigma(k) = \arg\max_{1\leq i\leq n}(\delta_i(k)). \tag{10.5}$$

Different from the above protocol (10.5), this chapter will integrate event-triggering technique into WTOD protocol and propose the following ET-WTOD protocol:

Given two positive threshold values $E_{\min}$ and $E_{\max}$ satisfying $E_{\min} < E_{\max}$ and a positive scalar λ_i, the transmission of state components will be determined according to the following strategy:

(i) If there exist some $\delta_i(k)$ satisfying $\delta_i(k) \geq E_{\max}x_i(k)\lambda_i x_i(k)$, all those corresponding components $x_i(k)$ will be transmitted at the instant k.
(ii) If not any $\delta_i(k)$ satisfies the case (i) but there exist some $\delta_i(k)$ satisfying $\delta_i(k) \geq E_{\min}x_i(k)\lambda_i x_i(k)$, only the component $x_i(k)$ with the largest index $\delta_i(k)$ will be transmitted. Under the case, the selected sensor node is determined according to the scheduling signal $\sigma(k)$ in the WTOD protocol (10.5).
(iii) If all $\delta_i(k)$ $(i = 1, \cdots, n)$ satisfy $\delta_i(k) < E_{\min}x_i(k)\lambda_i x_i(k)$, no components $x_i(k)$ will be transmitted at the instant k.

According to the above three transmission cases (i)–(iii) under the ET-WTOD protocol, there exist three sets on transmission instant:

(a) Set $\mathcal{I}_1$ denotes those instants at which all those components with $\delta_i(k) \geq E_{\max}x_i(k)\lambda_i x_i(k)$ will be transmitted, that is, more than one sensor nodes may obtain the access to transmit their signals.
(b) Set $\mathcal{I}_2$ represents those instants at which only the sensor with the largest difference $\delta_i(k)$ can transmit its signal in accordance with the WTOD protocol.
(c) Set $\mathcal{I}_3$ denotes those instants at which no signals are transmitted.

The following will systematically analyze the transmission characteristics of state component under the three aforementioned scenarios:

- At the instant $k \in \mathcal{I}_1$, there exist some state components satisfying

$$(\breve{x}_i(k-1) - x_i(k))\phi_i(\breve{x}_i(k-1) - x_i(k)) \geq E_{\max}x_i(k)\lambda_i x_i(k), \tag{10.6}$$

and the corresponding sensor nodes will send their current component $x_i(k)$ to the controller.

By adopting the ZOH strategy, i.e., for those untransmitted nodes, their previous transmitted value will be held, the available signals for the controller are denoted as $\breve{x}(k) = x(l_m) \triangleq [x_1(l_m^1), x_2(l_m^2), \cdots, x_n(l_m^n)]^T$ with

$$x_i(l_m^i) = \begin{cases} x_i(k), & \text{if } x_i(k) \text{ satisfying (10.6)}; \\ x_i(l_{m-1}^i), & \text{otherwise} \end{cases} \tag{10.7}$$

where l^i_{m-1} denotes the previous transmission instant.

According to the expression (10.7), all state components $x_i(k)$ satisfy $(x_i(l^i_m) - x_i(k))^T \phi_i (x_i(l^i_m) - x_i(k)) \leq E_{\max} x_i^T(k)\lambda_i x_i(x)$.

• At the instant $k \in \mathcal{I}_2$, only one sensor node can obtain the access to transmit its corresponding state component. Under the case, denote the available signals for the controller at previous instant as $\bar{x}(k-1) \triangleq [\bar{x}_1(k-1), \bar{x}_2(k-1), \cdots, \bar{x}_n(k-1)]^T$. In light of the conventional WTOD protocol and the scheduling signal $\sigma(k)$ in (10.5), one has

$$\bar{x}_i(k) = \begin{cases} x_i(k), & \text{if } i = \sigma(k)\,; \\ \bar{x}_i(k-1), & \text{otherwise}\,. \end{cases} \tag{10.8}$$

At the current instant k, the available signal for the controller can be represented as $\breve{x}(k) = \bar{x}(k) = \Psi_{\sigma(k)}x(k) + (I - \Psi_{\sigma(k)})\bar{x}(k-1)$, where $\Psi_{\sigma(k)} \in \{\Psi_1, \cdots, \Psi_n\}$ is the scheduling matrix with $\Psi_i = \text{diag}\{\delta_{i1}, \delta_{i2}, \cdots, \delta_{in}\}$ and Kronecker delta function $\delta_{ij} = \delta(i-j)$.

• At the instant $k \in \mathcal{I}_3$, all indexes $\delta_i(k)$ are less than the lower threshold, i.e., $(\breve{x}_i(k-1) - x_i(k))^T \phi_i (\breve{x}_i(k-1) - x_i(k)) < E_{\min} x_i^T(k)\lambda_i x_i(x)$. Under this case, all sensor nodes will not transmit their signals and the controller will utilize the previous received values via the buffer, i.e., $\breve{x}(k) = \breve{x}(k-1)$.

Remark 10.1 For the sake of analysis, two intermediate variables $x(l_m)$ and $\bar{x}(k)$ are introduced in expressions (10.7)–(10.8). Actually, they are only the different expression of the available signals $\breve{x}(k)$ for the controller in different transmission cases. In the design later, only the signal $\breve{x}(k)$ will be utilized.

Remark 10.2 In this chapter, the ETS and the WTOD protocol are integrated to formulate the new ET-WTOD protocol by introducing two thresholds $E_{\min}$ and $E_{\max}$. The key feature of the ET-WTOD protocol is that it can flexibly adjust the number of the transmitted state components according to the relation between the threshold and each state component. To be more specific, the number of the transmitted nodes will be more than one when the deviation of multiple nodes exceeds the pre-specified threshold. Besides, there exist some instants at which no signals are transmitted. This selection strategy will be more conducive to promoting the control performance under the limited transmission resources, which has also been shown in the simulation results later.

10.3 Sliding Mode Control Design Under the ET-WTOD Protocol

In this section, a fuzzy sliding mode controller will be designed under the proposed ET-WTOD protocol.

First, choose the following sliding function

$$s(k) = Gx(k), \tag{10.9}$$

where $G \in \mathbb{R}^{m \times n}$ is the sliding matrix to be determined later.

It is worth mentioning that under the proposed ET-WTOD protocol, the controller can only receive the transmitted signals $\check{x}(k)$ that will be determined according to three different transmission cases. Thus, a scheduling-signal-dependent IT2 fuzzy SMC law is designed as:

$$u(k) = \sum_{j=1}^{r} h_j(\check{x}(k)) K_{j\sigma(k)} \check{x}(k) - \varpi ||\check{x}(k)|| \text{sign}(\check{s}(k)), \tag{10.10}$$

where $K_{j\sigma(k)}$ is the control gain, and $\hat{s}(k)$ is the sliding variable $s(k)$ in (10.9) with $x(k)$ replaced by $\hat{x}(k)$. The MFs $h_j(\check{x}(k))$ in (10.10) are constructed as $h_j(\check{x}(k)) = \underline{\beta}_j(\check{x}(k))\underline{m}_j(\check{x}(k)) + \overline{\beta}_j(\check{x}(k))\overline{m}_j(\check{x}(k))$ by introducing two weighting coefficients $\underline{\beta}_j(\check{x}(k))$ and $\overline{\beta}_j(\check{x}(k))$. Here, the reconstructed MFs $h_j(\check{x}(k))$ can ensure the practical implementation of the sliding mode controller (10.10).

It should be pointed out that the control gain $K_{j\sigma(k)}$ reflects the impact of the scheduling signal $\sigma(k)$. In the case (ii), according to the conventional WTOD protocol, the scheduling signal $\sigma(k)$ can take value in the set $\{1, \ldots, n\}$. Besides, by defining the scheduling signal $\sigma(k) = 0$ for the case (iii), and $\sigma(k) = n + 1$ for the case (i), the scheduling-signal-dependent controller can be switched according to the selection of transmission channels. Thus, under the proposed ET-WTOD protocol, the scheduling signal $\sigma(k)$ will take value in the set $\mathcal{T} \triangleq \{0, 1, \ldots, n + 1\}$ for different transmission scenarios.

Substituting the SMC law (10.10) into (10.2) yields the closed-loop system as:

$$\begin{aligned} x(k+1) = &\sum_{i=1}^{r}\sum_{j=1}^{r} m_i(x(k))h_j(\check{x}(k))[A_i x(k) + B_i K_{j\sigma(k)}\check{x}(k) \\ &+ B_i(d(x(k), k) - \varpi ||\check{x}(k)|| \text{sign}(\check{s}(k)))]. \end{aligned} \tag{10.11}$$

10.4 Performance Analysis Under Event-Triggered WTOD Protocol

In this section, the stability of closed-loop system (10.11) and the reachability of sliding surface $s(k) = 0$ will analyzed under the ET-WTOD protocol, respectively.

10.4.1 Stability of Closed-Loop System

The following theorem will give sufficient conditions for ensuring the asymptotic stability of the closed-loop system (10.11) under the ET-WTOD protocol.

Theorem 10.1 *For given positive parameters $E_{\max}$, $E_{\min}$ with $E_{\max} > E_{\min}$, ϕ_i, λ_i, ϵ_1 and θ_{vq} with $\sum_{q=1}^{n}\theta_{vq} = 1$, if there exist matrices $P > 0$, $R > 0$, K_{j0}, K_{jv}, $K_{j(n+1)}$, and scalars $\xi_i > 0$, $\varphi > 0$ satisfying the following conditions (for $i, j = 1, \ldots, r$ and $v, q = 1, \ldots, n$)*

$$B_i^T P B_i \le \xi_i I, \tag{10.12}$$

$$\sum_{i=1}^{r}\sum_{j=1}^{r} m_i(x(k))h_j(\breve{x}(k))\Omega_{ij} < 0, \tag{10.13}$$

$$\sum_{i=1}^{r}\sum_{j=1}^{r} m_i(x(k))h_j(\breve{x}(k))\Theta_{ij} < 0, \tag{10.14}$$

$$\sum_{i=1}^{r}\sum_{j=1}^{r} m_i(x(k))h_j(\breve{x}(k))\Pi_{ij} < 0, \tag{10.15}$$

with

$$\Omega_{ij} = \begin{bmatrix} \Omega_{ij}^{11} & * \\ \Omega_{ij}^{21} & \Omega_{ij}^{22} \end{bmatrix}, \quad \Theta_{ij} = \begin{bmatrix} \Theta_{ij}^{11} & * \\ \Theta_{ij}^{21} & \Theta_{ij}^{22} \end{bmatrix}, \quad \Pi_{ij} = \begin{bmatrix} \Pi_{ij}^{11} & * \\ \Pi_{ij}^{21} & \Pi_{ij}^{22} \end{bmatrix},$$

and

$$\begin{aligned}
\Omega_{ij}^{11} &= -P + 2(1+\epsilon_1^{-1})(m+1)\mu^2\xi_i I + \varphi E_{\max}\Lambda \\
&\quad + (1+\epsilon_1)(A_i + B_i K_{j(n+1)})^T P(A_i + B_i K_{j(n+1)}) + R, \\
\Omega_{ij}^{21} &= (1+\epsilon_1)(B_i K_{j(n+1)})^T P(A_i + B_i K_{j(n+1)}) + 2(1+\epsilon_1^{-1})m\mu^2\xi_i I + R, \\
\Omega_{ij}^{22} &= -\varphi\Phi + (1+\epsilon_1)(B_i K_{j(n+1)})^T P B_i K_{j(n+1)} + 2(1+\epsilon_1^{-1})m\mu^2\xi_i I + R, \\
\Theta_{ij}^{11} &= -P + 2(1+\epsilon_1^{-1})\mu^2\xi_i I + (1+\epsilon_1)(A_i + B_i K_{jv}\Psi_v)^T P(A_i + B_i K_{jv}\Psi_v) \\
&\quad + \Psi_v^T (R + 2(1+\epsilon_1^{-1})m\mu^2\xi_i I)\Psi_v + \Delta_v, \\
\Theta_{ij}^{21} &= (1+\epsilon_1)(B_i K_{jv}(I - \Psi_v))^T P(A_i + B_i K_{jv}\Psi_v) \\
&\quad + (I - \Psi_v)^T (R + 2(1+\epsilon_1^{-1})m\mu^2\xi_i I)\Psi_v - \Delta_v, \\
\Theta_{ij}^{22} &= -R + (1+\epsilon_1)(B_i K_{jv})^T P B_i K_{jv} \\
&\quad + (I - \Psi_v)^T (R + 2(1+\epsilon_1^{-1})m\mu^2\xi_i I)(I - \Psi_v) + \Delta_v, \\
\Pi_{ij}^{11} &= -P + 2(1+\epsilon_1^{-1})(m+1)\mu^2\xi_i I + \varphi E_{\min}\Lambda \\
&\quad + (1+\epsilon_1)(A_i + B_i K_{j0})^T P(A_i + B_i K_{j0}), \\
\Pi_{ij}^{21} &= (1+\epsilon_1)(B_i K_{j0})^T P(A_i + B_i K_{j0}) + 2(1+\epsilon_1^{-1})m\mu^2\xi_i I, \\
\Pi_{ij}^{22} &= -\varphi\Phi + (1+\epsilon_1)(B_i K_{j0})^T P B_i K_{j0} + 2(1+\epsilon_1^{-1})m\mu^2\xi_i I,
\end{aligned}$$

$$\Delta_v = \Phi\sum_{q=1}^{n}\theta_{vq}(\Psi_v - \Psi_q), \;\; \Phi = \mathrm{diag}\{\phi_1, \phi_2, \cdots, \phi_n\}, \;\; \Lambda = \mathrm{diag}\{\lambda_1, \lambda_2, \cdots, \lambda_n\}$$

then, the closed-loop system (10.11) under the ET-WTOD protocol is asymptotically stable.

Proof Choose the Lyapunov function as

$$V(x(k), \breve{x}(k-1)) \triangleq x^T(k)Px(k) + \breve{x}^T(k-1)R\breve{x}(k-1). \tag{10.16}$$

Taking the difference of $V(x(k), \breve{x}(k-1))$ along the closed-loop system (10.11), one has

$$\begin{aligned}
&\Delta V(x(k), \breve{x}(k-1)) \\
&\quad = V(x(k+1), \breve{x}(k)) - V(x(k), \breve{x}(k-1)) \\
&\quad = x^T(k+1)Px(k+1) + \breve{x}^T(k)R\breve{x}(k) - x^T(k)Px(k) - \breve{x}^T(k-1)R\breve{x}(k-1) \\
&\quad \le \sum_{i=1}^{r}\sum_{j=1}^{r} m_i(x(k))w_j(\breve{x}(k))\Big\{[A_i x(k) + B_i K_{j\sigma(k)}\breve{x}(k) \\
&\qquad + B_i(d(x(k), k) - \varpi||\breve{x}(k)||\text{sign}(\breve{s}(k)))]^T P[A_i x(k) + B_i K_{j\sigma(k)}\breve{x}(k) \\
&\qquad + B_i(d(x(k), k) - \varpi||\breve{x}(k)||\text{sign}(\breve{s}(k)))]\Big\} + \breve{x}^T(k)R\breve{x}(k) \\
&\quad - x^T(k)Px(k) - \breve{x}^T(k-1)R\breve{x}(k-1).
\end{aligned} \tag{10.17}$$

In the following, the difference $\Delta V_1(x(k), \breve{x}(k-1))$ will be analyzed for different transmission cases:

• For $\forall k \in \mathcal{I}_1$, i.e., there exist the state component $x_i(k)$ satisfying $\delta_i(k) \ge E_{\max}x_i(k)\lambda_i x_i(k)$, and the scheduling signal $\sigma(k) = n+1$. At the current transmission instant, the available signals for the controller can be represented as $\breve{x}(k) = x(l_m) = [x_1(l_m^1), \cdots, x_n(l_m^n)]^T$, from which and the expression (10.17) it can be obtained that

$$\begin{aligned}
&\Delta V(x(k), \breve{x}(k-1)) \\
&\le \sum_{i=1}^{r}\sum_{j=1}^{r} m_i(x(k))h_j(\breve{x}(k))\Big\{[A_i x(k) + B_i K_{j(n+1)}\breve{x}(k) + B_i(d(x(k), k) \\
&\quad - \varpi||\breve{x}(k)||\text{sign}(\breve{s}(k)))]^T P[A_i x(k) + B_i K_{j(n+1)}\breve{x}(k) + B_i(d(x(k), k) \\
&\quad - \varpi||\breve{x}(k)||\text{sign}(\breve{s}(k)))]\Big\} + x^T(l_m)Rx(l_m) - x^T(k)Px(k) \\
&\le \sum_{i=1}^{r}\sum_{j=1}^{r} m_i(x(k))h_j(\breve{x}(k))\Big\{(1+\epsilon_1)[A_i x(k) + B_i K_{j(n+1)}x(l_m)]^T \\
&\quad \times P[A_i x(k) + B_i K_{j(n+1)}x(l_m)] + 2(1+\epsilon_1^{-1})\varpi^2\xi_i x^T(k)x(k) \\
&\quad + 2(1+\epsilon_1^{-1})m\varpi^2\xi_i x^T(l_m)x(l_m)\Big\} + x(l_m)^T Rx(l_m) - x^T(k)Px(k).
\end{aligned} \tag{10.18}$$

In accordance with the case (i) in the proposed ET-WTOD protocol, the following expression holds:

$$\sum_{\ell=1}^{n}[E_{\max}x_\ell(k)\lambda_\ell x_\ell(k)-e_\ell(k)\phi_\ell e_\ell(k)]\geq 0 \tag{10.19}$$

with $e_\ell(k)\triangleq x_\ell(l_m^\ell)-x_\ell(k)$.

Define $e(k)\triangleq[e_1(k),e_2(k),\cdots,e_n(k)]^T$. It can be obtained from the inequalities (10.18) and (10.19) that

$$\begin{aligned}
&\Delta V(x(k),\breve{x}(k-1))\\
\leq&\sum_{i=1}^{r}\sum_{j=1}^{r}m_i(x(k))h_j(\breve{x}(k))\Big\{(1+\epsilon_1)[(A_i+B_iK_{j(n+1)})x(k)+B_iK_{j(n+1)}e(k)]^T\\
&\times P[(A_i+B_iK_{j(n+1)})x(k)+B_iK_{j(n+1)}e(k)]+2(1+\epsilon_1^{-1})\varpi^2\xi_ix^T(k)x(k)\\
&+2(1+\epsilon_1^{-1})m\varpi^2\xi_i[x(k)+e(k)]^T[x(k)+e(k)]\Big\}\\
&+[x(k)+e(k)]^TR[x(k)+e(k)]\\
&-x^T(k)Px(k)+\varphi\left[E_{\max}x^T(k)\Lambda x(k)-e^T(k)\Phi e(k)\right]\\
\leq&\eta_1^T(k)\sum_{i=1}^{r}\sum_{j=1}^{r}m_i(x(k))h_j(\breve{x}(k))\Omega_{ij}\eta_1(k),
\end{aligned}\tag{10.20}$$

with scalar $\varphi>0$ and $\eta_1(k)\triangleq[x^T(k),e^T(k)]^T$.

• For $\forall k\in\mathcal{I}_2$, the scheduling signal $\sigma(k)$ is granted to the sensor node with the largest difference $\delta_i(k)$, and $\sigma(k)\in\{1,\cdots,n\}$. In this case, the available signals for the controller can be represented as $\breve{x}(k)=\bar{x}(k)=\Psi_{\sigma(k)}x(k)+(I-\Psi_{\sigma(k)})\bar{x}(k-1)$, from which and the expression (10.17) we can obtain

$$\begin{aligned}
&\Delta V(x(k),\breve{x}(k-1))\\
\leq&\sum_{i=1}^{r}\sum_{j=1}^{r}m_i(x(k))h_j(\breve{x}(k))\Big\{(1+\epsilon_1)[A_ix(k)+B_iK_{j\sigma(k)}\Psi_{\sigma(k)}x(k)\\
&+B_iK_{j\sigma(k)}(I-\Psi_{\sigma(k)})\bar{x}(k-1)]^TP[A_ix(k)+B_iK_{j\sigma(k)}\Psi_{\sigma(k)}x(k)\\
&+B_iK_{j\sigma(k)}(I-\Psi_{\sigma(k)})\bar{x}(k-1)]+2(1+\epsilon_1^{-1})\varpi^2\xi_ix^T(k)x(k)\\
&+2(1+\epsilon_1^{-1})m\varpi^2\xi_i\bar{x}^T(k)\bar{x}(k)+[\Psi_{\sigma(k)}x(k)+(I-\Psi_{\sigma(k)})\bar{x}(k-1)]^T\\
&\times R[\Psi_{\sigma(k)}x(k)+(I-\Psi_{\sigma(k)})\bar{x}(k-1)]\Big\}\\
&-x^T(k)Px(k)-\bar{x}^T(k-1)R\bar{x}(k-1).
\end{aligned}\tag{10.21}$$

In this case, according to the selection principle of the WTOD protocol, one has from (10.4) that:

$$(\bar{x}(k-1)-x(k))^T\Phi(\Psi_{\sigma(k)}-\Psi_q)(\bar{x}(k-1)-x(k))\geq 0 \tag{10.22}$$

for all $q=1,\ldots,n$.

Let $\sigma(k)=\nu\in\{1,\ldots,n\}$. It follows from (10.21) and (10.22) that

$$\begin{aligned}
&\Delta V(x(k),\check{x}(k-1))\\
\leq&V(x(k+1),\check{x}(k))-V(x(k),\check{x}(k-1))\\
&+(\bar{x}(k-1)-x(k))^T\sum_{j=1}^{n}\Phi\theta_{\iota j}(\Psi_\iota-\Psi_j)(\bar{x}(k-1)-x(k))\\
\leq&\sum_{i=1}^{r}\sum_{j=1}^{r}m_i(x(k))h_j(\check{x}(k))\{(1+\epsilon_1)[(A_i+B_iK_{j\iota}\Psi_\iota)x(k)+B_iK_{j\iota}(I-\Psi_\iota)\bar{x}(k-1)]^T\\
&\times P[(A_i+B_iK_{j\iota}\Psi_\iota)x(k)+B_iK_{j\iota}(I-\Psi_\iota)\bar{x}(k-1)]+2(1+\epsilon_1^{-1})\varpi^2\xi_ix^T(k)x(k)\\
&+2(1+\epsilon_1^{-1})m\varpi^2\xi_i\bar{x}^T(k)\bar{x}(k)+[\Psi_\iota x(k)+(I-\Psi_\iota)\bar{x}(k-1)]^TR[\Psi_\iota x(k)\\
&+(I-\Psi_\iota)\bar{x}(k-1)]\}-x^T(k)Px(k)-\bar{x}^T(k-1)R\bar{x}(k-1)\\
&+(\bar{x}(k-1)-x(k))^T\Delta_\iota(\bar{x}(k-1)-x(k))\\
\leq&\eta_2^T(k)\sum_{i=1}^{r}\sum_{j=1}^{r}m_i(x(k))h_j(\check{x}(k))\Theta_{ij}\eta_2(k),
\end{aligned}\tag{10.23}$$

with $\eta_2(k)\triangleq[x^T(k),\ \bar{x}^T(k-1)]^T$.

• For $\forall k\in\mathcal{I}_3$, i.e., no sensor node transmits signal at the instant k, and the scheduling signal $\sigma(k)$ is taken as $\sigma(k)=0$. In this case, the controller will utilize the previous received state, i.e., $\check{x}(k)=\check{x}(k-1)$, by which and (10.17) one has

$$\begin{aligned}
&\Delta V(x(k),\check{x}(k-1))\\
&=x^T(k+1)Px(k+1)-x^T(k)Px(k)\\
&\leq\sum_{i=1}^{r}\sum_{j=1}^{r}m_i(x(k))h_j(\check{x}(k))\{[A_ix(k)+B_iK_{j0}\check{x}(k)\\
&\quad+B_i(d(k,x(k))-\varpi||\check{x}(k)||\text{sign}(\check{s}(k)))]^TP[A_ix(k)+B_iK_{j0}\check{x}(k)\\
&\quad+B_i(d(k,x(k))-\varpi||\check{x}(k)||\text{sign}(\check{s}(k)))]\}-x^T(k)Px(k)\\
&\leq\sum_{i=1}^{r}\sum_{j=1}^{r}m_i(x(k))h_j(\check{x}(k))\{(1+\epsilon_1)[A_ix(k)+B_iK_{j0}x(l_m)]^T\\
&\quad\times P[A_ix(k)+B_iK_{j0}x(l_m)]+2(1+\epsilon_1^{-1})\varpi^2\xi_ix^T(k)x(k)\\
&\quad+2(1+\epsilon_1^{-1})m\varpi^2\xi_ix(l_m)^Tx(l_m)\}-x^T(k)Px(k),
\end{aligned}\tag{10.24}$$

where $\check{x}(k)=x(l_m)$ is the previous received value.

According to the ET-WTOD protocol, if $\delta_i(k) < E_{\min} x_i(k)\lambda_i x_i(i)$ for all $i = 1, \ldots, n$, we have

$$\sum_{\ell=1}^{n} [E_{\min} x_\ell(k)\lambda_\ell x_\ell(k) - e_\ell(k)\phi_\ell e_\ell(k)] > 0. \tag{10.25}$$

From expressions (10.24) and (10.25), one can obtain

$$\begin{aligned}
&\Delta V(x(k), \breve{x}(k-1)) \\
\le & \sum_{i=1}^{r}\sum_{j=1}^{r} m_i(x(k))h_j(\breve{x}(k))\{(1+\epsilon_1)[(A_i + B_i K_{j0})x(k) + B_i K_{j0} e(k)]^T \\
& \times P[(A_i + B_i K_{j0})x(k) + B_i K_{j0} e(k)] + 2(1+\epsilon_1^{-1})\varpi^2 \xi_1 x^T(k)x(k) \\
& + 2(1+\epsilon_1^{-1}) m\varpi^2 \xi_i [x(k) + e(k)]^T [x(k) + e(k)]\} \\
& - x^T(k)Px(k) + \varphi[E_{\min} x^T(k)\Lambda x(k) - e^T(k)\Phi e(k)] \\
\le & \eta_1^T(k) \sum_{i=1}^{r}\sum_{j=1}^{r} m_i(x(k))h_j(\breve{x}(k))\Pi_{ij}\eta_1(k).
\end{aligned} \tag{10.26}$$

According to the above three transmission cases, it can be obtained from expressions (10.20), (10.23) and (10.26) that the conditions (10.12)–(10.15) can always guarantee $\Delta V(x(k), \breve{x}(k-1)) < 0$ for $\eta_1(k), \eta_2(k) \neq 0$. Thus, the closed-loop system (10.11) under the ET-WTOD protocol is asymptotically stable. □

10.4.2 Reachability of Sliding Surface

In the following, the reachability of the specified sliding surface $s(k) = 0$ will be further analyzed according to three transmission cases.

Theorem 10.2 *For given positive parameters $E_{\max}$, $E_{\min}$ with $E_{\max} > E_{\min}$, ϕ_i, λ_i, ϵ_1, ϵ_2, and θ_{vq} with $\sum_{q=1}^{n} \theta_{vq} = 1$, if there exist matrices $P > 0$, $R > 0$, $Q > 0$, K_{j0}, K_{jv}, $K_{j(n+1)}$, G, and scalars $\xi_i > 0$, $\chi > 0$, $\varphi > 0$ satisfying the condition (10.12) and the following conditions (for $i, j = 1, \ldots, r$, and $v, q = 1, \ldots, n$)*

$$B_i^T G^T Q G B_i \le \chi I, \tag{10.27}$$

$$\sum_{i=1}^{r}\sum_{j=1}^{r} m_i(x(k))w_j(\breve{x}(k))\bar{\Omega}_{ij} < 0, \tag{10.28}$$

$$\sum_{i=1}^{r}\sum_{j=1}^{r} m_i(x(k))w_j(\breve{x}(k))\bar{\Theta}_{ij} < 0, \tag{10.29}$$

$$\sum_{i=1}^{r}\sum_{j=1}^{r} m_i(x(k))w_j(\breve{x}(k))\bar{\Pi}_{ij} < 0, \tag{10.30}$$

with

$$\bar{\Omega}_{ij} = \begin{bmatrix} \Omega_{ij}^{11}+\Gamma & * \\ \Omega_{ij}^{21}+\Gamma & \Omega_{ij}^{22}+\Gamma \end{bmatrix}, \quad \bar{\Theta}_{ij} = \begin{bmatrix} \bar{\Theta}_{ij}^{11} & * \\ \bar{\Theta}_{ij}^{21} & \bar{\Theta}_{ij}^{22} \end{bmatrix}, \quad \bar{\Pi}_{ij} = \begin{bmatrix} \Pi_{ij}^{11}+\Xi & * \\ \Pi_{ij}^{21}+\Xi & \Pi_{ij}^{22}+\Xi \end{bmatrix},$$

$$\begin{aligned}
\Gamma &= (1+\epsilon_2)(GA_i+GB_iK_{j(n+1)})^TQ(GA_i+GB_iK_{j(n+1)})+2(1+\epsilon_2^{-1})m\varpi^2\chi I,\\
\bar{\Theta}_{ij}^{11} &= \Theta_{ij}^{11}+(1+\epsilon_2)(GA_i+GB_iK_{j\nu}\Psi_\nu)^TQ(GA_i+GB_iK_{j\nu}\Psi_\nu)\\
&\quad +2(1+\epsilon_2^{-1})m\varpi^2\chi\Psi_\nu^T\Psi_\nu,\\
\bar{\Theta}_{ij}^{21} &= \Theta_{ij}^{21}+(1+\epsilon_2)(GB_iK_{j\nu}(I-\Psi_\nu))^TQ(GB_iK_{j\nu}\Psi_\nu)\\
&\quad +2(1+\epsilon_2^{-1})m\varpi^2\chi(I-\Psi_\nu)^T\Psi_\nu,\\
\bar{\Theta}_{ij}^{22} &= \Theta_{ij}^{22}+(1+\epsilon_2)(GB_iK_{j\nu}(I-\Psi_\nu))^TQ(GB_iK_{j\nu}(I-\Psi_\nu))\\
&\quad +2(1+\epsilon_2^{-1})m\varpi^2\chi(I-\Psi_\nu)^T(I-\Psi_\nu),\\
\Xi &= (1+\epsilon_2)(GA_i+GB_iK_{j0})^TQ(GA_i+GB_iK_{j0})+2(1+\epsilon_2^{-1})m\varpi^2\chi I,
\end{aligned}$$

then, the state trajectories of the IT2 T-S fuzzy system (10.11) will be driven into the following domain $\mathcal{D}$

$$\mathcal{D} \triangleq \left\{ s(k) \mid \|s(k)\| \le \sqrt{\frac{2(1+\epsilon_2^{-1})\mu^2\chi}{\lambda_{\min}(Q)}}\|x(k)\| \right\}. \tag{10.31}$$

Proof According to (10.9) and (10.11), one has

$$\begin{aligned}
s(k+1) = \sum_{i=1}^{r}\sum_{j=1}^{r} m_i(x(k))h_j(\breve{x}(k))\big[&GA_ix(k)+GB_iK_{j\sigma(k)}\breve{x}(k)\\
&+GB_i(d(x(k),k)-\varpi\|\breve{x}(k)\|\text{sign}(\breve{s}(k)))\big].
\end{aligned} \tag{10.32}$$

Choose the Lyapunov function as

$$U(x(k),\breve{x}(k-1)) = V(x(k),\breve{x}(k-1)) + V_s(k) \tag{10.33}$$

with $V(x(k),\breve{x}(k-1))$ defined in (10.16) and $V_s(k) \triangleq s^T(k)Qs(k)$.

One has

$$
\begin{aligned}
&\Delta V_s(k)\\
&= s^T(k+1)Qs(k+1) - s^T(k)Qs(k)\\
&\leq \sum_{i=1}^{r}\sum_{j=1}^{r} m_i(x(k))h_j(\check{x}(k))(1+\epsilon_2)[GA_ix(k) + GB_iK_{j\sigma(k)}\check{x}(k))]^T\\
&\quad \times Q[GA_ix(k) + GB_iK_{j\sigma(k)}\check{x}(k))] + 2(1+\epsilon_2^{-1})m\varpi^2\chi||\check{x}(k)||^2\\
&\quad + 2(1+\epsilon_2^{-1})\varpi^2\chi||x(k)||^2 - s^T(k)Qs(k).
\end{aligned}
\tag{10.34}
$$

In the sequel, the difference of $U(x(k), \check{x}(k-1))$ will be analyzed for different transmission cases.

- For $\forall k \in \mathcal{I}_1$, one has from the expressions (10.20) and (10.34) that

$$
\begin{aligned}
&\Delta U(x(k), \check{x}(k-1))\\
\leq&\Delta V(x(k), \check{x}(k-1)) + \Delta V_s(k) + \varphi\Big[E_{\max}x^T(k)\Lambda x(k) - e^T(k)\Phi e(k)\Big]\\
\leq&\eta_1^T(k)\sum_{i=1}^{r}\sum_{j=1}^{r} m_i(x(k))h_j(\check{x}(k))\bar{\Omega}_{ij}\eta_1(k)\\
&- [\lambda_{\min}(Q)||s(k)||^2 - 2(1+\epsilon_2^{-1})\varpi^2\chi||x(k)||^2].
\end{aligned}
\tag{10.35}
$$

- For $\forall k \in \mathcal{I}_2$, by combining the expressions (10.23) and (10.34), one has

$$
\begin{aligned}
&\Delta U(x(k), \check{x}(k-1))\\
\leq&\Delta V(x(k), \check{x}(k-1)) + \Delta V_s(k)\\
&+ (\bar{x}(k-1) - x(k))^T\sum_{j=1}^{n}\Phi\theta_{ij}(\Psi_i - \Psi_j)(\bar{x}(k-1) - x(k))\\
\leq&\eta_2^T(k)\sum_{i=1}^{r}\sum_{j=1}^{r} m_i(x(k))h_j(\check{x}(k))\bar{\Theta}_{ij}\eta_2(k)\\
&- [\lambda_{\min}(Q)||s(k)||^2 - 2(1+\epsilon_2^{-1})\varpi^2\chi||x(k)||^2].
\end{aligned}
\tag{10.36}
$$

- For $\forall k \in \mathcal{I}_3$, one has from (10.26) and (10.34) that

$$
\begin{aligned}
&\Delta U(x(k), \check{x}(k-1))\\
\leq&\Delta V(x(k), \check{x}(k-1)) + \Delta V_s(k) + \varphi\Big[E_{\min}x^T(k)\Lambda x(k) - e^T(t)\Phi e(t)\Big]\\
\leq&\eta_1^T(k)\sum_{i=1}^{r}\sum_{j=1}^{r} m_i(x(k))h_j(\check{x}(k))\bar{\Pi}_{ij}\eta_1(k)\\
&- \Big[\lambda_{\min}(Q)||s(k)||^2 - 2(1+\epsilon_2^{-1})\varpi^2\chi||x(k)||^2\Big].
\end{aligned}
\tag{10.37}
$$

Thus, when the state trajectories lie outside the domain $\mathcal{D}$, i.e.,

$$||s(k)|| > \sqrt{\frac{2(1+\epsilon_2^{-1})\mu^2\chi}{\lambda_{\min}(Q)}}||x(k)||, \tag{10.38}$$

one has from (10.35)–(10.37) and (10.27)–(10.30) that $\Delta U(x(k), \breve{x}(k-1)) < 0$ for all of three transmission cases, that is, the Lyapunov functional $U(x(k), \breve{x}(k-1))$ monotonically deceases outside the domain $\mathcal{D}$. Thus, the state trajectories of the closed-loop system (10.11) will be driven into the domain $\mathcal{D}$. □

10.5 Sufficient Conditions and Optimization Solving Algorithm

It is noted that the time-varying MFs $m_i(x(k))$ and $h_i(\breve{x}(k))$ are involved in the conditions in Theorems 10.1 and 10.2, which makes the number of the matrix inequalities be infinite such that these conditions are hard to solve. To remove this difficulty, the following theorem is further derived by means of the relation between $m_i(x(k))$ and $h_i(\breve{x}(k))$ via the MFD analysis techniques.

10.5.1 Membership-Function-Dependent Sufficient Conditions

Theorem 10.3 *For given positive parameters $E_{\max}$, $E_{\min}$ with $E_{\max} > E_{\min}$, λ_i, ϕ_i, ϵ_1, ϵ_2, θ_{vq} with $\sum_{q=1}^{n}\theta_{vq} = 1$ and τ_i satisfying $h_i(\breve{x}(k)) - \tau_i m_i(x(k)) \geq 0$, if there exist matrices $M_i = M_i^T$, $N_i = N_i^T$, $S_i = S_i^T$, $P > 0$, $X > 0$, $R > 0$, $\bar{Q} > 0$, G, K_{j0}, K_{jv}, $K_{j(n+1)}$, and scalars $\xi_i > 0$, $\chi > 0$, $\varphi > 0$ satisfying the following conditions (for $i, j = 1, \ldots, r$, and $v, q = 1, \ldots, n$)*

$$\begin{bmatrix} -\xi_i I & * \\ B_i & -X \end{bmatrix} \leq 0, \quad \begin{bmatrix} -\chi I & * \\ GB_i & -\bar{Q} \end{bmatrix} \leq 0, \tag{10.39}$$

$$PX = I, \tag{10.40}$$

$$\hat{\Omega}_{ij} - M_i < 0, \tag{10.41}$$

$$\tau_i\hat{\Omega}_{ii} + M_i - \tau_i M_i < 0, \tag{10.42}$$

$$\tau_j\hat{\Omega}_{ij} + M_i - \tau_j M_i + \tau_i\hat{\Omega}_{ji} + M_j - \tau_i M_j < 0, \ i < j, \tag{10.43}$$

$$\hat{\Theta}_{ij} - N_i < 0, \tag{10.44}$$

$$\tau_i \hat{\Theta}_{ii} + N_i - \tau_i N_i < 0, \tag{10.45}$$

$$\tau_j \hat{\Theta}_{ij} + N_i - \tau_j N_i + \tau_i \hat{\Theta}_{ji} + N_j - \tau_i N_j < 0,\ i < j, \tag{10.46}$$

$$\hat{\Pi}_{ij} - S_i < 0, \tag{10.47}$$

$$\tau_i \hat{\Pi}_{ii} + S_i - \tau_i S_i < 0, \tag{10.48}$$

$$\tau_j \hat{\Pi}_{ij} + S_i - \tau_j S_i + \tau_i \hat{\Pi}_{ji} + S_j - \tau_i S_j < 0,\ i < j, \tag{10.49}$$

with

$$\hat{\Omega}_{ij} = \begin{bmatrix} \hat{\Omega}_{ij}^{11} & * & * \\ \hat{\Omega}_{ij}^{21} & \hat{\Omega}_{ij}^{22} & * \\ \hat{\Omega}_{ij}^{31} & 0 & \hat{\Omega}_{ij}^{33} \end{bmatrix},\ \hat{\Theta}_{ij} = \begin{bmatrix} \hat{\Theta}_{ij}^{11} & * & * \\ \hat{\Theta}_{ij}^{21} & \hat{\Theta}_{ij}^{22} & * \\ \hat{\Theta}_{ij}^{31} & 0 & \hat{\Theta}_{ij}^{33} \end{bmatrix},\ \hat{\Pi}_{ij} = \begin{bmatrix} \hat{\Pi}_{ij}^{11} & * & * \\ \hat{\Pi}_{ij}^{21} & \hat{\Pi}_{ij}^{22} & * \\ \hat{\Pi}_{ij}^{31} & 0 & \hat{\Pi}_{ij}^{33} \end{bmatrix},$$

$$\hat{\Omega}_{ij}^{11} = \begin{bmatrix} -P + \bar{\xi}_i I + \varphi E_{\max}\Lambda + R & * \\ R & -\varphi\Phi + R \end{bmatrix},$$

$$\hat{\Omega}_{ij}^{21} = \begin{bmatrix} A_i + B_i K_{j(n+1)} & B_i K_{j(n+1)} \\ GA_i + GB_i K_{j(n+1)} & GB_i K_{j(n+1)} \end{bmatrix},$$

$$\hat{\Omega}_{ij}^{31} = \begin{bmatrix} \sqrt{2(1+\epsilon_i^{-1})m\mu\xi_i I} & \sqrt{2(1+\epsilon_i^{-1})m\mu\xi_i I} \\ \sqrt{2(1+\epsilon_i^{-1})m\mu\chi I} & \sqrt{2(1+\epsilon_i^{-1})m\mu\chi I} \end{bmatrix},$$

$$\hat{\Omega}_{ij}^{22} = \mathrm{diag}\left\{-\frac{1}{1+\epsilon_1}X, -\frac{1}{1+\epsilon_2}\bar{Q}\right\},\ \hat{\Omega}_{ij}^{33} = \mathrm{diag}\{-\xi_i I, -\chi I\}$$

$$\hat{\Theta}_{ij}^{11} = \begin{bmatrix} -P + 2(1+\epsilon_1^{-1})\mu^2\xi_i I + \Delta_\nu & * \\ \Delta_\nu & -R + \Delta_\nu \end{bmatrix},$$

$$\hat{\Theta}_{ij}^{21} = \begin{bmatrix} A_i + B_i K_{j\nu}\Psi_\nu & B_i K_{j\nu}\Psi_\nu \\ R\Psi_\nu & R\Psi_\nu \\ GA_i + GB_i K_{j\nu}\Psi_\nu & GB_i K_{j\nu}\Psi_\nu \end{bmatrix},\ \hat{\Pi}_{ij}^{21} = \begin{bmatrix} A_i + B_i K_{j0} & B_i K_{j0} \\ GA_i + GB_i K_{j0} & GB_i K_{j0} \end{bmatrix},$$

$$\hat{\Pi}_{ij}^{11} = \mathrm{diag}\{-P + 2(1+\epsilon_1^{-1})\mu^2\xi_i I + \varphi E_{\min}\Lambda,\ -\varphi\Phi\},$$

$$\hat{\Theta}_{ij}^{31} = \hat{\Pi}_{ij}^{31} = \hat{\Omega}_{ij}^{31},\ \hat{\Theta}_{ij}^{22} = \hat{\Pi}_{ij}^{22} = \hat{\Omega}_{ij}^{22},\ \hat{\Theta}_{ij}^{33} = \hat{\Pi}_{ij}^{33} = \hat{\Omega}_{ij}^{33},\ \bar{\xi}_i = 2(1+\epsilon_1^{-1})\mu^2\xi_i,$$

then, both the stability of the closed-loop system (10.11) and the reachability of the sliding domain $\mathcal{D}$ can be guaranteed simultaneously.

Proof Let $\bar{Q} = Q^{-1}$, $X = P^{-1}$. By virtue of Schur complement, the condition (10.39) can guarantee conditions (10.12) and (10.27). Meanwhile, the conditions (10.13) and (10.28) can be ensured by the following inequality:

$$\sum_{i=1}^{r}\sum_{j=1}^{r} m_i(x(k))h_j(\check{x}(k))\hat{\Omega}_{ij} < 0, \tag{10.50}$$

Similarly, the conditions (10.14), (10.29) and (10.15), (10.30) can be guaranteed, if the following inequalities hold

$$\sum_{i=1}^{r}\sum_{j=1}^{r} m_i(x(k))h_j(\breve{x}(k))\hat{\Theta}_{ij} < 0, \tag{10.51}$$

and

$$\sum_{i=1}^{r}\sum_{j=1}^{r} m_i(x(k))h_j(\breve{x}(k))\hat{\Pi}_{ij} < 0. \tag{10.52}$$

To cope with the mismatched MFs in the condition (10.50), we introduce the slack matrix M_i via the following equality

$$\begin{aligned}&\sum_{i=1}^{r}\sum_{j=1}^{r} m_i(x(k))(m_j(x(k)) - h_j(\breve{x}(k)))M_i \\ &= \sum_{i=1}^{r} m_i(x(k))\left[\sum_{j=1}^{r} m_j(x(k)) - \sum_{j=1}^{r} h_j(\breve{x}(k))\right] M_i = 0.\end{aligned} \tag{10.53}$$

By means of the condition (10.53), we can obtain from (10.50)

$$\begin{aligned}&\sum_{i=1}^{r}\sum_{j=1}^{r} m_i(x(k))h_j(\breve{x}(k))\hat{\Omega}_{ij} \\ &= \sum_{i=1}^{r}\sum_{j=1}^{r} m_i(x(k))\Big(h_j(\breve{x}(k)) - \tau_j m_j(x(k)) + \tau_j m_j(x(k))\Big)\hat{\Omega}_{ij} \\ &\quad + \sum_{i=1}^{r}\sum_{j=1}^{r} m_i(x(k))\Big(m_j(x(k)) - h_j(\breve{x}(k)) + \tau_j m_j(x(k)) - \tau_j m_j(x(k))\Big)M_i \\ &= \sum_{i=1}^{r}\sum_{j=1}^{r} m_i(x(k))\left(h_j(\breve{x}(k)) - \tau_j m_j(x(k)))\right)\left[\hat{\Omega}_{ij} - M_i\right] \\ &\quad + \sum_{i=1}^{r} m_i^2(x(k))\left[\tau_i\hat{\Omega}_{ii} + M_i - \tau_i M_i\right] \\ &\quad + \sum_{i=1}^{r}\sum_{j>i} m_i(x(k))m_j(x(k))\Big[\tau_j\hat{\Omega}_{ij} + M_i - \tau_j M_i + \tau_i\hat{\Omega}_{ji} + M_j - \tau_i M_j\Big].\end{aligned} \tag{10.54}$$

Note that $h_i(\breve{x}(k)) - \tau_i m_i(x(k)) \geq 0$ for given τ_i. Thus, the conditions (10.39)–(10.41) can ensure the condition (10.54).

Along the similar lines, the slack matrices N_i and S_i can be introduced for the inequalities (10.51) and (10.52) as follows:

$$\sum_{i=1}^{r}\sum_{j=1}^{r} m_i(x(k))(m_j(x(k)) - h_j(\breve{x}(k)))N_i = 0, \tag{10.55}$$

and

$$\sum_{i=1}^{r}\sum_{j=1}^{r} m_i(x(k))(m_j(x(k)) - h_j(\breve{x}(k)))S_i = 0. \tag{10.56}$$

Similar to the derivation in (10.54), it can be obtained that the conditions (10.44) and (10.49) can guarantee the conditions (10.51) and (10.52). Thus, the proof is completed. □

In the above analysis, in order to deal with the mismatched problem between $m_i(x(k))$ and $h_i(\breve{x}(k))$, slack matrices M_i, N_i, and S_i are introduced to obtain the equations (10.53), (10.55) and (10.56). By combining the expression (10.53) and the condition $h_i(\breve{x}(k)) - \tau_i m_i(x(k)) \geq 0$, a series of relaxed conditions (10.41)–(10.43) can be obtained from (10.54). Compared with the MFI method, i.e., directly solving $\hat{\Omega}_{ij} < 0$, $\hat{\Theta}_{ij} < 0$ and $\hat{\Pi}_{ij} < 0$, for all $i, j = 1, \ldots, r$, this MFD technique can integrate the information of MFs into the stability conditions and reduce the conservatism.

10.5.2 GA-Based Solving Algorithm

It is noted that the conditions (10.39)–(10.49) in Theorem 10.3 involving nonlinear coupling terms GB_iK_{j0}, $GB_iK_{j\upsilon}\Psi_\upsilon$, and $GB_iK_{j(n+1)}$ will reduce to the liner matrix inequalities if the matrices X and G can be determined aforehand. Hence, we shall propose a solving approach to search the feasible matrices X and G via minimizing the sliding domain $\mathcal{D}$ according to the following optimization problem:

$$\begin{aligned} &\min_{X,G} \quad F_{obj} \\ &\text{s.t.} \quad \text{conditons } (10.39) - (10.49), \end{aligned} \tag{10.57}$$

with $F_{obj} = 1/\sqrt{\frac{2(1+\epsilon_2^{-1})\varpi^2\chi}{\lambda_{\min}(Q)}}$.

By means of the binary-encoded GA, the GA-based control design solving algorithm 10.1 is proposed for dealing with the above optimization problem (10.57).

Algorithm 10.1 GA-based control design solving algorithm

Initialization

1: For the symmetric matrix $X_{n\times n}$ and the matrix $G_{m\times n}$, we encode $n(n+1)/2$ independent elements in $X_{n\times n}$ and $m \times n$ independent elements in the matrix $G_{m\times n}$ into the following binary strings with the length L_1 and L_2, respectively:

$$[X,\ G] \rightarrow [\mathfrak{p}_{11}, \cdots, \mathfrak{p}_{1n}, \mathfrak{p}_{22}, \cdots, \mathfrak{p}_{2n}, \mathfrak{p}_{33}, \cdots, \mathfrak{p}_{nn},$$

$$g_{11} \cdots, g_{1n}, g_{21}, \cdots, g_{mn}].$$

2: Randomly generate an initial population with N chromosomes.

Evaluation

3: Solve the conditions (10.39)–(10.49) by the decoded real matrices X and G, and calculate the fitness function F_{obj}. If the matrix $X \leq 0$ or the conditions (10.39) and (10.41)–(10.49) are infeasible, the fitness function is assigned artificially as a small value (e.g., 10^{-6}) for reducing its possibility to survive in the next generation.

Reproduction

4: Perform *Selection*, *Crossover*, and *Mutation* operations and reproduce the offsprings (the details can be seen in [7]).

5: If the terminal condition is met, i.e., the number of generations reaches the maximum number $T_{\max}$ of generations, stop and return the optimal solutions. Otherwise, go to the **Evaluation** step.

6: **return** K_{j0}, K_{j1}, ..., $K_{j(n+1)}$, and G.

10.6 Simulation Example

Consider a mass-spring-damper system as follows:

$$m\ddot{q}(t) + F_s(t) + F_f(t) = u(t), \tag{10.58}$$

where $q(t)$ and $u(t)$ denote the position and the control force, respectively. $F_f(t)$ and $F_s(t)$ are the spring restoring force and the force of friction, with $F_f(t) = c\dot{q}(t)$ and $F_s(t) = \kappa q(t) + \kappa a^2 q^3(t)$. The physical meaning and their values are shown in Table 10.1, and the uncertain parameter κ satisfies $\kappa \in [\kappa_1,\ \kappa_2]$ with $\kappa_1 = 5N/m$, $\kappa_2 = 8N/m$.

Table 10.1 The symbols in the model

Symbol	Physical meaning	Value
m	the mass of the rigid body	1 kg
κ	the elasticity factor	[5 N/m, 8N/m]
a	auxiliary variable	$0.3\,\mathrm{m}^{-1}$
c	the friction factor	2 N/(m/s)

Define $x(t) \triangleq [x_1(t), x_2(t)]^T \triangleq [q(t), \dot{q}(t)]^T$. It can be obtained from (10.58) that the following nonlinear system model

$$\dot{x}(t) = \begin{bmatrix} 0 & 1 \\ g(t) & -\frac{c}{m} \end{bmatrix} x(t) + \begin{bmatrix} 0 \\ 1 \end{bmatrix} u(t) \tag{10.59}$$

with $g(t, \kappa) = (-\kappa - \kappa a^2 x_1^2(t))/m$.

Assuming $x_1(t) \in [-2,\ 2]$, we can obtain $g_{\max} = -\kappa_1$ and $g_{\min} = -\kappa_2 - 4a^2\kappa_2$, and $g(t) \in [\bar{g}_{\min},\ \bar{g}_{\max}]$ with $\bar{g}_{\min} = g_{\min} - 0.02$ and $\bar{g}_{\max} = g_{\max} + 0.02$. By adapting the sector nonlinearity method, one has

$$\begin{cases} m_1(x(t)) + m_2(x(t)) = 1 \\ \bar{g}_{\min} m_1(x(t)) + \bar{g}_{\max} m_2(x(t)) = g(t, \kappa) \end{cases} \tag{10.60}$$

with the MFs as:

$$m_1(x(t)) = \frac{\bar{g}_{\max} - g(t, \kappa)}{\bar{g}_{\max} - \bar{g}_{\min}}, \quad m_2(x(t)) = \frac{g(t, \kappa) - \bar{g}_{\min}}{\bar{g}_{\max} - \bar{g}_{\min}}.$$

The upper and lower MFs can be calculated as

$$\underline{m}_1(x(k)) = \frac{g(t, \kappa_1) + \bar{g}_{\max}}{\bar{g}_{\max} - \bar{g}_{\min}}, \overline{m}_1(x(k)) = \frac{g(t, \kappa_2) + \bar{g}_{\max}}{\bar{g}_{\max} - \bar{g}_{\min}},$$
$$\underline{m}_2(x(k)) = 1 - \overline{m}_1(x(k)),\ \overline{m}_2(x(k)) = 1 - \underline{m}_1(x(k)).$$

And then, the original system (10.59) can be represented as the following IT2 fuzzy model:

$$\dot{x}(t) = \sum_{i=1}^{2} m_i(x(t)) \left[\bar{A}_i x(t) + \bar{B}_i u(t)\right] \tag{10.61}$$

with

$$\bar{A}_1 = \begin{bmatrix} 0 & 1 \\ \bar{g}_{\min} & -\frac{c}{m} \end{bmatrix},\ \bar{A}_2 = \begin{bmatrix} 0 & 1 \\ \bar{g}_{\max} & -\frac{c}{m} \end{bmatrix},\ \bar{B}_1 = \bar{B}_2 = \begin{bmatrix} 0 \\ 1 \end{bmatrix}.$$

After discretizing the system (10.61) with the sampling period 0.25, we obtain

$$x(k+1) = \sum_{i=1}^{2} m_i(x(k))[A_i x(k) + B_i(u(k) + d(x(k), k))] \tag{10.62}$$

with

$$A_1 = \begin{bmatrix} 0.7248 & 0.1752 \\ -1.9131 & 0.3744 \end{bmatrix}, A_2 = \begin{bmatrix} 0.8714 & 0.1868 \\ -0.9246 & 0.4978 \end{bmatrix},$$
$$B_1 = \begin{bmatrix} 0.0252 \\ 0.1752 \end{bmatrix},\ B_2 = \begin{bmatrix} 0.0260 \\ 0.1868 \end{bmatrix},$$

and the disturbance $d(x(k), k) = 0.03\sqrt{|2x_1(k)x_2(k)|}$.

The upper and lower MFs in controller (10.10) are chosen as $\overline{m}_i(\check{x}(k))$ and $\underline{m}_i(\check{x}(k))$ for $i = 1, 2$. In this example, the MFs $m_i(x(k))$ and $w_i(\check{x}(k))$ $(i = 1, 2)$ can be obtained by choosing $\underline{\alpha}_1(x(k)) = \sin^2(x_1(k))$ (acting as uncertainty for demonstration purposes), $\overline{\alpha}_1(x(k)) = 1 - \underline{\alpha}_1(x(k))$, $\underline{\beta}_1(\check{x}(k)) = 0.4$ and $\overline{\beta}_1(\check{x}(k)) = 1 - \underline{\beta}_1(\check{x}(k))$. Referring to Theorem 10.3, choose $\tau_1 = 0.2343$ and $\tau_2 = 0.4183$ satisfying $h_i(\check{x}(k)) - \tau_i m_i(x(k)) \geq 0,\ (i = 1, 2)$.

By executing the GA-based solving algorithm, the optimized solutions can be obtained as:

$$X = \begin{bmatrix} 0.0173 & -0.0169 \\ -0.0169 & 0.1400 \end{bmatrix},\quad G = \begin{bmatrix} -0.2520 & -0.0323 \end{bmatrix},$$

and

$$\begin{aligned} K_{10} &= \begin{bmatrix} 0.0108 & -0.1047 \end{bmatrix}, & K_{20} &= \begin{bmatrix} 0.0010 & -0.1028 \end{bmatrix}, \\ K_{11} &= \begin{bmatrix} 1.0194 & -0.0534 \end{bmatrix}, & K_{21} &= \begin{bmatrix} 0.0596 & -0.0551 \end{bmatrix}, \\ K_{12} &= \begin{bmatrix} -0.0016 & -3.3723 \end{bmatrix}, & K_{22} &= \begin{bmatrix} -0.0042 & -3.2842 \end{bmatrix}, \\ K_{13} &= \begin{bmatrix} 0.0099 & -0.1177 \end{bmatrix}, & K_{23} &= \begin{bmatrix} -0.0006 & -0.1164 \end{bmatrix}. \end{aligned}$$

in which the relevant parameters are given as the population size $N = 80$, the maximum of generations $T_{\max} = 50$, the single-point crossover probability $p_c = 0.8$, the single-bit mutation probability $p_m = 0.1$, the ranges of the parameters as

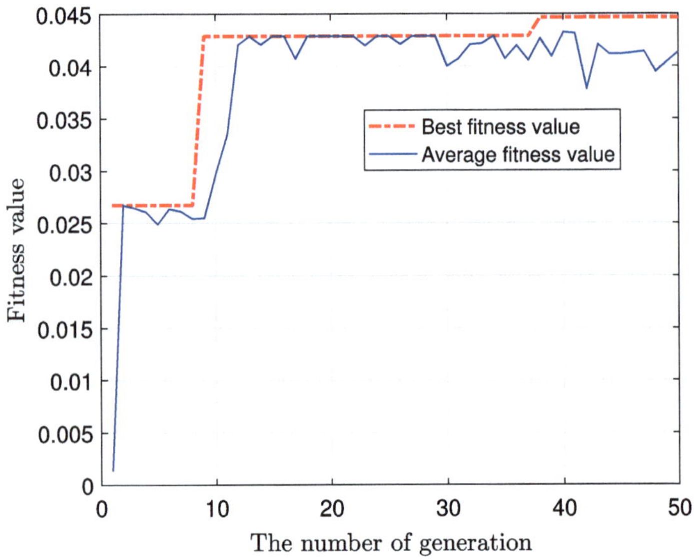

Fig. 10.2 The evolution of fitness value

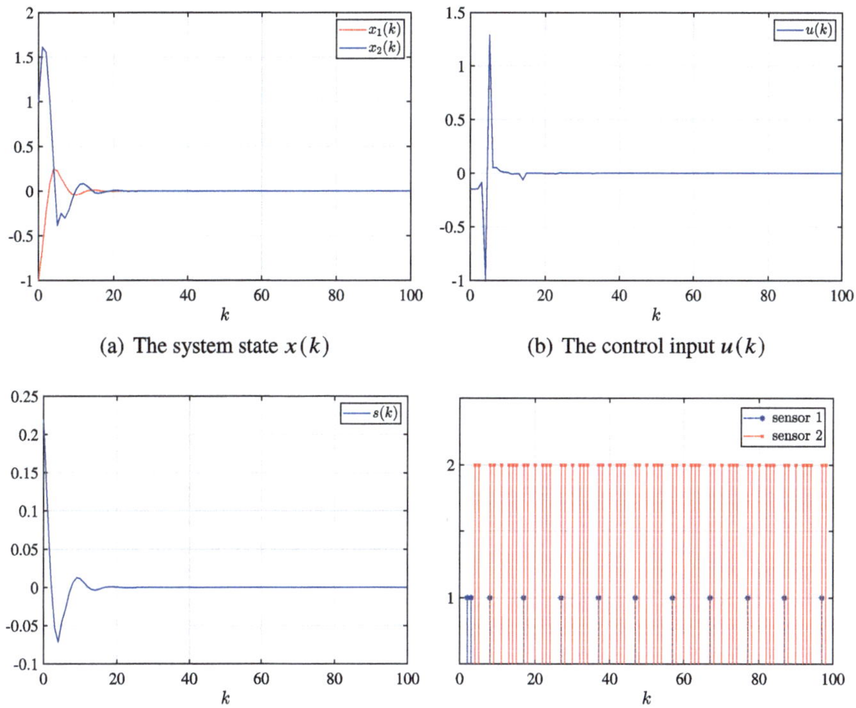

(a) The system state $x(k)$ (b) The control input $u(k)$

(c) The sliding variable $s(k)$ (d) The selected sensor node

Fig. 10.3 The simulation results under the ET-WTOD protocol

$\mathfrak{p}_{11}, \mathfrak{p}_{22} \in [0.001, 0.5], \mathfrak{p}_{12}, g_{11}, g_{12} \in [-0.5, 0.5]$, and the length of the binary string encoding each element as $L_1 = L_2 = 20$. Figure 10.2 shows the evolution of the fitness function.

Under the initial condition $x(0) = [-1 \ \ 1]^T$, the simulation results are shown in Fig. 10.3a–d. The selected sensor nodes are shown in Fig. 10.3(d), where the transmission number for the states $x_1(k)$ and $x_2(k)$ is 12 and 58, respectively. Moreover, it can be observed that at some instants, no signal is transmitted, at other instants, multiple sensor nodes are granted simultaneously to transmit signals.

10.7 Conclusion

This chapter investigates the SMC problem for IT2 T-S fuzzy systems under a novel ET-WTOD protocol for alleviating communication burdens while enhancing network resource utilization efficiency. A scheduling-signal-dependent IT2 fuzzy SMC law has been developed by reconstructing MFs to ensure practical implementability. By analyzing three transmission cases generated by two thresholds and using MFD analysis techniques, some relaxed conditions on the stability and the reachability have been derived. Meanwhile, the GA-based solving algorithm has been provided to solve the control design problem subject to the nonlinear constraints and optimize the convergence domain around the sliding surface.

References

1. Zou, L., Wang, Z., Gao, H., Liu, X.: State estimation for discrete-time dynamical networks with time-varying delays and stochastic disturbances under the round-robin protocol. IEEE Trans. Neural Netw. Learn. Syst. **28**(5), 1139–1151 (2017)
2. Zhang, Z., Niu, Y., Karimi, H.R.: Sliding mode control of interval type-2 fuzzy systems under round-robin scheduling protocol. IEEE Trans. Syst., Man Cybern.: Syst. **51**(12), 7602–7612 (2021)
3. Zhang, Z., Niu, Y., Lam, H.K.: Sliding-mode control of T-S fuzzy systems under weighted try-once-discard protocol. IEEE Trans. Cybern. **50**(12), 4972–4982 (2020)
4. Zou, L., Wang, Z., Han, Q., Zhou, D.: Ultimate boundedness control for networked systems with try-once-discard protocol and uniform quantization effects. IEEE Trans. Autom. Control **62**(12), 6582–6588 (2017)
5. Peng, C., Ma, S., Xie, X.: Observer-based non-PDC control for networked T-S fuzzy systems with an event-triggered communication. IEEE Trans. Cybern. **47**(8), 2279–2287 (2017)
6. Su, X., Wen, Y., Shi, P., Wang, S., Assawinchaichote, W.: Event-triggered fuzzy control for nonlinear systems via sliding mode approach. IEEE Trans. Fuzzy Syst. **29**(2), 336–344 (2021)
7. Song, J., Wang, Z., Niu, Y., Dong, H.: Genetic-algorithm-assisted sliding-mode control for networked state-saturated systems over hidden markov fading channels. IEEE Trans. Cybern. **51**(7), 3664–3675 (2021)

Chapter 11
Conclusion and Further Research

This chapter draws conclusions on the book and points out some possible research directions related to the work done in this book.

11.1 Conclusion

This book mainly focuses on analysis and synthesis for IT2 T-S fuzzy systems under various communication protocols. The following control design and analysis problems have been discussed in detail from the perspectives of event-triggered schemes and channel scheduling protocols.

- Chapter 2 has addressed the dynamic event-triggered SMC problem for continuous-time IT2 T-S fuzzy systems with unmeasurable states and fading channels. A dynamic ETM has been implemented in S/C channels to regulate the transmission of measurement signals, and fading phenomenon has been considered for the triggered measurement output signal. Based on the faded triggered output, a fuzzy state observer has been designed to estimate unavailable states. An integral sliding function and observer-based sliding mode controller have been designed to ensure the input-to-state practical stability in the mean of the closed-loop systems. Finally, Zeno behavior has been analyzed for ensuring the practical implementability of the proposed event-triggered control scheme of continuous-time IT2 fuzzy systems.
- Chapter 3 has investigated the component-based dynamic event-triggered SMC problem for IT2 fuzzy singularly perturbed systems. Individual ETMs have been designed for each sensor node so that the triggering sequence of each sensor node can be determined independently. Based on the triggering instant of each sensor and the mapping approach, a triggered-state-based fuzzy sliding mode controller has been designed such that sufficient conditions on the stability of closed-loop systems and the reachability of sliding domain can be derived. Finally, a dynamic PSO algorithm has been utilized to minimize the sliding domains and enhances the control performance.

Y. Yang et al., *Analysis and Synthesis of Interval Type-2 Fuzzy Systems Under Communication Protocols*, Studies in Systems, Decision and Control 636,
https://doi.org/10.1007/978-981-95-2355-9_11

- Chapter 4 has developed an event-triggered SMC scheme with dynamic quantization for IT2 fuzzy systems. A novel state-based and membership-function-based triggering condition has been derived. Under the MFD event-triggered scheme, a fuzzy sliding model controller and a dynamic quantizer with online adjustment parameters have been developed so that the desired control performance can be guaranteed. Meanwhile, MFs' local information has been utilized in stability analysis, and an optimization framework has been provided to balance communication rates and control performance.
- Chapter 5 has proposed a probability-dependent adaptive event-triggered SMC scheme for IT2 fuzzy systems suffering from DoS attacks. An adaptive threshold updating mechanism has been developed to counter random DoS attacks. A secure event-triggered fuzzy sliding mode controller has been designed to ensure the stochastic stability of the closed-loop systems and the reachability of sliding domain. By utilizing the global boundaries of MFs, co-design conditions of triggering parameters and fuzzy control gains have been derived.
- Chapter 6 has established a self-triggered control framework for IT2 T-S fuzzy systems. A self-triggered scheme has been proposed for determine the transmission of system state, meanwhile, the triggering instant can be calculated in advance by using the available triggering information and MFs. A fuzzy controller has been designed to ensure uniform ultimate boundedness of the closed-loop systems, and MFD stability conditions have been developed by utilizing the local boundaries and relationship of controller's and system's MFs.
- Chapter 7 has considered SMC design for IT2 fuzzy systems under multi-node scheduling protocols. In both S/C and C/A channels, the SS protocol governed by discrete-time Markov chain has been utilized to select multiple transmission nodes. By means of the mapping technique of Markov chain, a scheduling-dependent fuzzy sliding mode controller has been designed to ensure the stochastic stability of closed-loop systems. By integrating MFs' relations, sufficient conditions have been derived, and the SA optimization algorithm has been employed to optimize controller parameters.
- Chapter 8 has proposed the co-design scheme on SS protocol and sliding mode controller for IT2 T-S fuzzy systems. Based on the scheduled system state, a fuzzy sliding mode controller has been designed. The co-design scheme on transition probability matrix of SS protocol and control gains has been developed to ensure the desired control performance. By means of the relation between the MFs in fuzzy system and the ones in controller, sufficient conditions have been derived, and a PSO-based iterative optimization algorithm has been given for joint protocol-controller synthesis.
- Chapter 9 has developed the output-feedback SMC scheme for IT2 T-S fuzzy systems under multi-node SS protocol. The co-design framework integrating output-feedback controllers with TPM of SS protocol has been developed. By leveraging the global bounds of mismatched MFs between the system and controller, sufficient conditions have been established to ensure stochastic stability of the closed-loop system and the reachability of sliding surface. Additionally, an optimization-based control solving algorithm has been developed to enhance control performance.

- Chapter 10 has developed the SMC scheme for IT2 T-S fuzzy systems under an ET-WTOD protocol. By introducing two thresholds, transmission nodes can be dynamically adjusted to permit a subset of nodes to access communication channels. A scheduling-signal-dependent fuzzy sliding mode controller has been designed, and stability analysis has conducted for three transmission scenarios. Utilizing MFD methods and introducing slack matrices, sufficient conditions are derived. A GA-based optimization scheme has been developed to minimize the sliding domain and enhance control performance.

11.2 Further Research

Related topics for the future research work are listed below:

- The probability-dependent adaptive event-triggered strategy has been proposed to resist randomly occurring DoS attacks. In practice, there may exist some other types of attacks (such as deception and replay attacks, etc.), and the occurrence of attacks may not meet a known certain statistical property. Hence, future research can focus on designing effective event-triggering strategies under different network attacks, and proposing attack identification and communication recovery mechanisms.
- By combining a static ETS and a signal-node WTOD protocol, the ET-WTOD protocol has been developed to dynamically adjust transmission process and improve the control performance. Actually, the signal transmission might be implemented over multiple network conditions. To optimize both bandwidth allocation efficiency and control system performance, a promising research direction is to develop hybrid communication protocols through the integration of dynamic/adaptive ETS with existing WTOD/SS/RR protocols.
- This book has employed various MFD techniques (such as global/local boundary information, and inequality relationships between the controller's and system's MFs) to address mismatched MFs and reduce the conservatism of stability conditions. However, these methods may increase computational complexity and lack specific selection principles for the controller's fuzzy rules and MFs. An interesting research direction is to balance computational efficiency and control performance by establishing specific selection principles for controller's fuzzy rule and MFs.
- This book has mainly focused on the theoretical research concerning IT2 fuzzy control scheme under communication protocols. Future investigations could extend these theoretical frameworks to practical applications in modeling and controlling real-world uncertain nonlinear systems (such as smart grids and multi-agent robots).